RWTHedition

RWTH Aachen

Walter Eversheim

Innovation Management for Technical Products

Systematic and Integrated Product Development and Production Planning

Springer

Editor

Prof. Dr.-Ing. Walter Eversheim
RWTH Aachen
Laboratorium für
Werkzeugmaschinen
und Betriebslehre
Steinbachstr. 19
52074 Aachen
Germany
w.eversheim@wzl.rwth-aachen.de

ISBN: 978-3-540-85726-6 e-ISBN: 978-3-540-85727-3

DOI 10.1007/978-3-540-85727-3

RWTHedition ISSN: 1865-0899 e-ISSN: 1865-0902

Library of Congress Control Number: 2008935546

Cover design: deblik, Berlin, Germany

Printed on acid-free paper

9 8 7 6 5 4 3 2 1

springer.com

Preface

RWTH Aachen University is recognised as one of Germanys leading Universities of technology with a long tradition of close collaboration with national and international Industry on the one hand and globally relevant research and education organizations on the other. The national top position in third party revenues underlines the commitment and position of the University with respect to social impact.

Within the German high-tech initiative RWTH Aachen University has been granted the status of a University of excellence. In this context a strategy has been developed witch strives for the integration of not only technology and natural science but also to systematically include disciplines like economics, humanities and languages. Thus the University accepts the challenge of ultimately reaching the level of a fully integrated technical University.

In light of this RWTH Aachen University and Springer-Verlag Publishing have agreed on a publication series titled "RWTH Edition". The series shall demonstrate the quality of individual researchers and results on a global platform. The present work covers a topic of special relevance for RWTH Aachen University: The process of organising innovation. The contend on the one hand generalizes systematically acquired knowledge up to the "Aachen Innovation Model (AIM)" and one the other hand documents real world application case studies.

As Vice-Rector of Aachen University and Pro-Rector for research I express my sincere gratitude to the authors for this top level analysis as well as to Springer Verlag for its commitment and ongoing cooperation. I wish the book the impact it deserves in the scientific community and in modern education of our students and graduates.

Sincerely

Prof. Dr. rer. nat. R. Poprawe M.A.

Foreword

On the basis of several research and application projects in cooperation with industry we brought together our findings and know how in a German version of this book, edited by Springer Berlin, Heidelberg, New York in the year 2002. Due to the fact that the book became a bestseller as German edition we decided to prepare an English version which additionally has been updated, reworked and combined with an electronic software tool, called EDEN™. This tool is supporting our "Aachen Innovation Model (AIM)" by a multi user network system.

The origin of our innovation road map-method has been built by Dr. Frank Brandenburg, who developed in his doctoral thesis systematically a complete model for the whole network of processes, beginning with strategic planning and finally ending with successful products on competitive markets. All relevant methods available have been analysed by Dr. Anne Gerhards in her thesis and she allocated these to the relevant steps in the roadmap.

The "AIM" and the roadmap-method have been used and tested in cooperation with several industrial companies. In chapter 4 the representatives of five companies, Dr. Uwe Böhlke, Schott Glas; Prof. Dr. Winfried Huppmann and Dr. Stefan Nöken, Hilti AG; Rudolf-Henning Lohse, Dräger Medical; Dr. Daniel E. Spielberg, Suspa Holding; and Dr. Christian Voigtländer present their case studies and describe the results they achieved by using the W-Model.

Many thanks to them and the other authors and contributers who all did a wonderful job and are named in the list of authors and editors at the end of the book.

Finally we have to say that this English version of the book would not be available without the help and assistance of Prof. Dr. Dimiter Dimitrov and Konrad von Leipzig, both from University of Stellenbosch (South Africa), Dr. Gerhard Gudergan, Ralf Frombach and Dr. Volker Stich (Research Institute for Operations Management at RWTH Aachen University), as well as Mrs. Christine Kirschfink, who all took care of the translation into the English language and the rework of the new edition.

We appreciate especially the contribution of Prof. Niek du Preez, University of Stellenbosch (South Africa) and Prof. Dr. Eva-Maria Jakobs, Institute for Technical Communication, RWTH Aachen University, who both with their teams prepared the software-tool EDEN™ from INDUTECH Stellenbosch for an interactive support of the W-Model on an ontology basis.

Many thanks to all team members for their individual contributions and enthusiasm.

Aachen, August 2008 Walter Eversheim

Table of Contents

1 Introduction

WALTER EVERSHEIM, ELKE BAESSLER, THOMAS BREUER

Dynamics of markets result in an increasing stress of competition for the manufacturing industry. The aggravated competition conditions manifest in shortened product lifecycles that lead faster to obsolescent products.

The pioneer and growth phase of new product technologies is shortened and the maturity phase starts earlier. An indication for the increasing maturity of a product is the appearance of the "dominant design" (Utterback 1994): A product technology is established on the mass market and is offered on the market by almost every supplier.

(Product) technological innovation only emerges in niche applications, whereas the majority of innovations are reduced to efficiency increase. The few remaining suppliers are in a price competition with similar product concepts (Gassmann 1996). At the same time, the innovation leaps get shorter and just enable incremental improvements. These incremental improvements go beyond the optimization of the physical product. To an increasing degree, these improvements are realized in the service sector of the product.

Many companies embark on a differentiating strategy to avoid price wars, but this is becoming increasingly difficult to achieve. Little improvements are essential to existing products and offer short- term advantages on the market. However, regarding medium and long terms, no sustainable additional customers benefit can be realized. A lasting success will be achieved by the development of radical new products that lead to opening ups of new market potentials (cf. Sommerlatte 1997).

In reality, those product innovations originate mainly by chance (cf. Cooper 1993; Agamus 1998; Fraunhofer 1998; Management Partner 1999). Ideas for new products rather seem to be a qualitative then a quantitative matter: reality shows that only few of the gathered ideas are success-promising with regards to quality (Droege 1999). In many companies the idea selection is based on subjective preferences and coincidental knowledge of individuals. The choice is seldom systematic and objectively comprehensible.

The approach to product innovations should not only be efficient but also effective to ensure future market shares. The "right" product ideas must be generated systematically and transformed to a successful product. The aim is to educe the innovation needs with a strategic focus. Most of the time, the classic market pull approach just offers short- term product innovations. Furthermore, a derivation of middle- and long-term statements for future innovation needs is necessary. Regarding just the market, this can be done by observing not only the direct market but also the "market before the own market": an automotive supplier surveys the end customer; a producer of drill hammers orients towards material trends when planning the next but one product generation. The resulting product ideas must be systematically pushed forward, followed by an integration of a strategy-conformal implementation plan.

A guidepost for the systematic product innovation planning is the InnovationRoadMap methodology, a major content of this book. This methodology consists of an approach, the so-called W-model, which describes the whole process from goal setting to successful implementation planning (Chap. 3) (cf. Brandenburg 2002). Each phase is linked to existing and new methods of innovation management in a "method tool kit" to support practical application. That is how this model shows *what* has to be done and *how* it has to be done by using the assigned methods. Some of the significant and comprehensive methods are described in detail in chapter 4. Additionally, all methods are explained with method data sheets in the appendix. Chapter 5 is a collection of case studies and experiences that show the possibilities of preparing the ground for successful product innovations: How does a raw material supplier develop to become a system supplier? How to import existing competencies to new markets? How to investigate a growth market for innovation potentials? How to implement innovation processes in a globally operating enterprise?

Chapter 2 deals with the Integrated Innovation Management as a part of the St. Gallen Management Concept. Based on this concept, a framework for innovation management will be created. The model particularly enables an enterprise to realize a holistic alignment of its innovation management.

In this way, the present book offers the planner in the industry a comprehensive guide for the innovation management of technical products. The modular design allows the reader to start at any interesting point.

2 Integrated Innovation Management

WALTER EVERSHEIM, ELKE BAESSLER,
THOMAS BREUER

Innovation Management deals with a complex and multilayered field of activities whose successful processing requires a holistic and integrated approach.

- The Aachen Innovation Management Model AIM represents the reference framework for the innovation management. It enables identifying integration gaps and educing needs for action (Chap. 2.1).
- The Innovation Portfolio is used to adjust a "healthy" innovation project mix (Chap. 2.2).
- The Aachen Strategy Model for product innovation integrates the market pull and the technology push approach (Chap. 2.3).

Content complexity

The system-theoretical approach of the St. Gallen Management Concept SGMK allows the adoption of its general management approach to subsystems, i.e. enterprise units and enterprise tasks[1] (cf. Schuh and Schwenk 2001). The Aachen Innovation Model AIM was designed as a part concept of the SGMK and represents the reference framework for innovation management. Thus, logically delimitable scopes of duties can be accentuated, which are dealt with by the management (cf. Bleicher 1999). With the AIM, enterprises can lean against existing methods and models, but can also develop own forms and procedures for their innovation management. A detailed description of the AIM can be found in chapter 2.1.

Goal: enhanced innovation ability

The goal of the Integrated Innovation Management is to guarantee the innovation ability of an enterprise. It is defined as the ability of an enterprise to generate new ideas by using new knowledge or market understanding, and to successfully

---

1 Amongst others, the SGMK was translated into the quality management by Seghezzi (Seghezzi 1996) and into the complexity management by Schuh (Schuh et al. 1998; AWK 2002).

put them on the market (AWK 1999). Thus, the innovation ability of a producing enterprise is measured as the share of new products of the business activities based on the average product lifetime (Brockhoff 1985). This relative evaluation is much more meaningful than the consideration of the pure share of sales of new products, which generally misjudges enterprises of branches with short product lifetimes as capable of innovation (Eggers 1993). Whether future product innovations primarily refer to incremental improvements – e.g. perfective maintenance – or radical innovations – e.g. usage of novel technologies – is strongly influenced by the orientation of the innovation management.

Grade of Novelty of products: technical evaluation

In this context, the issue is to evaluate the novelty of products. On the one hand, the focus of the evaluation can be purely technical. Thereby, the grade of product innovation is evaluated according to the state of the art[2]. But this perception is lacking particularly in in-house perspectives.

Holistic evaluation

A holistic view concerning the novelty of a product requires a concurrent consideration of external and internal factors. The innovation portfolio includes both, internal, enterprise-specific and external factors given by the market (Brandenburg and Spielberg 1998). This is explained in chapter 2.2.

---

2 Altschuller for instance distinguishes between five invention levels, from "partial solution" to "discovery" (Altschuller 1984).

2.1 The Aachen Innovation Management Model AIM

The Aachen Innovation Management Model AIM was developed at the Fraunhofer IPT with reference to the St. Gallen Concept of Integrated Management. It represents a reference framework for the issues of innovation management and enables identifying integration gaps and main focuses for adjustments.

Starting from the vision, which is based on the enterprise philosophy, the sum of horizontal and vertical perspectives forms the integrative and holistic reference framework of the AIM. The goal of the AIM is to achieve stable innovation ability (Fig. 2.1). Within the decisive field of activities and its detailing by means of influencing factors, enterprises are able to position themselves. Another step of the model enables the statement of the nominal condition, which is oriented towards the corporate development. Due to the discrepancy between the nominal and the actual condition, strategic needs for action and the corresponding goals can be educed. The result is a holistic innovation management oriented towards the corporate development.

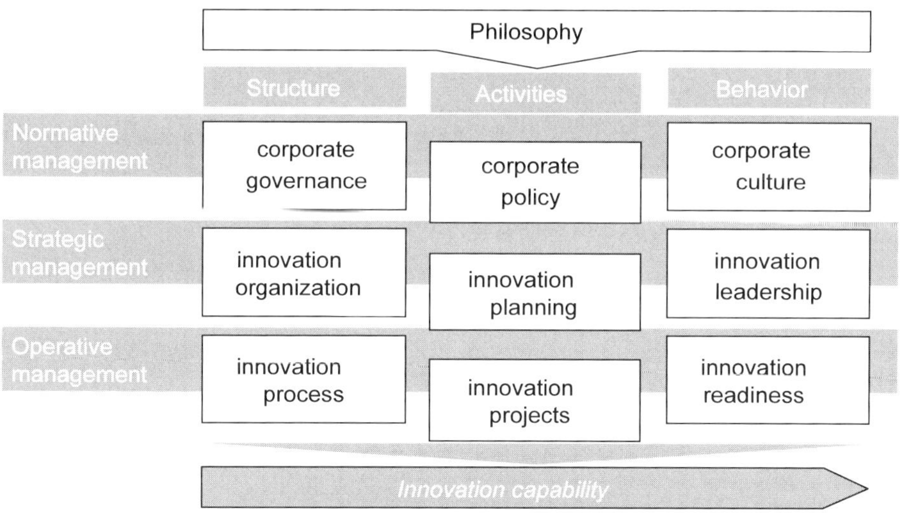

Fig. 2.1 The Aachen Innovation Management Model AIM (cf. Bleicher 1999)

Management Philosophy

The "lodestar" for the entrepreneurial business is the management philosophy. It comprises the basic settings, convictions and values that affect the way in which the managers of an enterprise plan and act (Ulrich and Fluri 1992). The horizontal level consists of the normative, strategic and operative management. Normative and strategic management form the conceptual framework, where the situational direction of the operative management takes place.

Structures, behavior, activities

Additionally to the horizontal approach, management levels can be differentiated vertically as well. Structures, behavior and activities are the three pillars that pervade the normative, strategic and operative levels. They discuss the integration of conceptual and creative intentions as well as its cooperative implementation. Content-related aspects are concretized within the pillars across the management levels (cf. Bleicher 1999).

Activities

In respect of the concretion of the activities, company guidelines for the innovation management will be developed out of the normative level and will then be concretized on the strategic level by the *innovation planning*. These guidelines, which are relevant for medium and long terms, will be defined on the operative level in forms of *innovation projects*.

Structures

Within the pillar "structures", the innovation management will be legalized on the normative level and will be concretized on the strategic level by the design of the *innovation organization*. On the operative level, the structural aspect is expressed by the resource-connected procedure of *innovation processes*.

Behavior

Finally, both aspects, activities and structures, have an impact on the human behavior. On the normative level, past enterprise cultures characterize future strategic and operative actions of enterprise staff members. Thus, the normative level states the reason for behavior. The enterprise culture is concretized by the strategic level in view of the management tasks. The operative "innovation readiness" is focused on the behavior of the staff during work flow.

The following view is concentrated on the strategic level. The strategic management is targeting a sustaining advantage in the market by an intensive use of resources. This is done by the consciously created preconditions that enable the enterprise to gain prosperities in long terms (cf. Pümpin 1968). The innovation management fulfills tasks in three areas: concerning structures, it is innovation organization, concerning activities, it is innovation planning, concerning behavior, it is innovation leadership.

2.1.1 Innovation planning

The innovation planning sets the direction of impact for future innovations. It is characterized in four dimensions: the temporal alignment, the competence orientation, the external orientation and the planning framework. Each dimension is defined by two axes, which, in extreme degrees, embody typical patterns.

Temporal alignment

The *temporal alignment* of innovation planning describes the methods of planning as present-related or future-related (Fig. 2.2)

The temporal alignment follows from the time pattern and the information profile of planning activities. The time pattern is differentiated between short term and long term. The character of the available information is divided into clear cut, detailed and fuzzy, not so detailed information. In this context, information is internal knowledge concerning internal or external issues like technologies, capacities, sales figures, market situation, etc. A short-term planning horizon and detailed information lead to *present-orientated innovation planning*. *Future-orientation* follows from long-term planning with less detailed information.

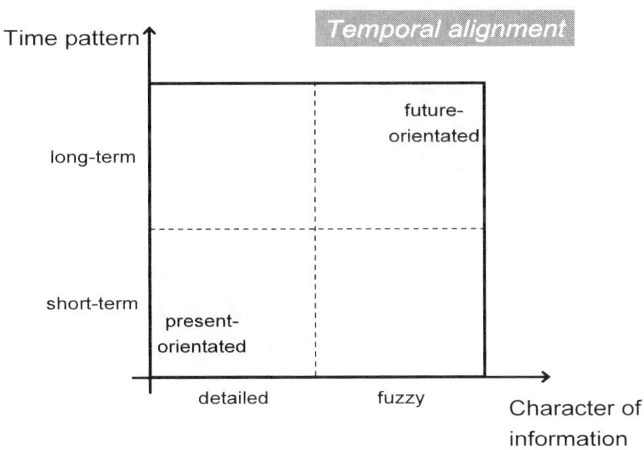

Fig. 2.2 Temporal alignment of innovation planning

Competence orientation

The competence orientation of an enterprise is another facet of innovation planning control (Fig. 2.3). The competencies of an enterprise are considered by technology and market view. A technological *competence establishment* is realized when knowledge and abilities are acquired within new technologies – i.e. new to the enterprise. This allows competitive advantage on the long run, if the technology is at the beginning of its lifecycle[3]. Capturing unknown markets is associated with the establishment of knowledge about branch, customers, competitors, etc. Additional *market competence* is established.

The contrast to *competence establishment* is *synergy utilization*: Common technologies are used in common markets. The production of goods and services can be executed with high efficiency, which is for example verified by cost reduction, an increase in quality etc. The strategy of using synergies is suited for the establishment of competition edges and entry barriers.

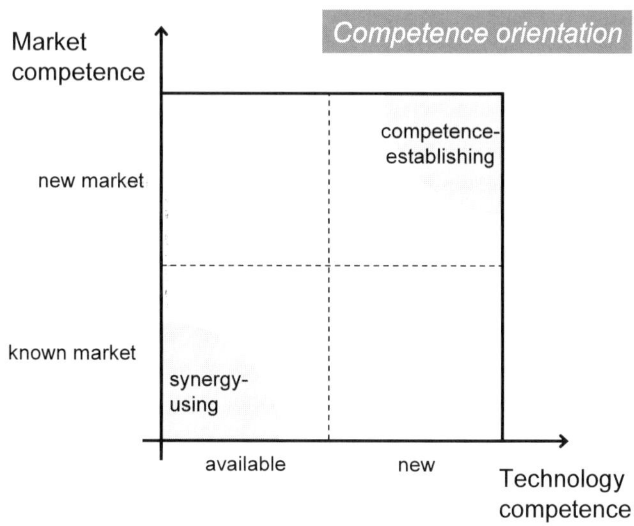

Fig. 2.3 Competence orientation of innovation planning

External orientation

The *external orientation* of innovation planning manages the general dealings of development competences related to supplier and customer (Fig. 2.4). The *supplier orientation* describes the way and intensity of collaboration with suppliers. The *customer orientation* is focused on the collaboration with customers during the product development. Supplier orientation as well as customer orientation may be of external-cooperative nature. Concerning supplier orientation, this means for example to go into development cooperation with

[3] For lifecycle models see chapter 4.5, figure 4.19.

suppliers. An example of external-cooperative customer orientation is the intensive contact with customers of a lead user cooperation[4] in an early stage of product development.

It is called an *autarkic product development*, when the cooperation with suppliers is limited to the placement of orders, e.g. production as per manufacturing drawing, and the customer is involved not before market launch, e.g. through a customer survey. In contrast to the lead user concept, the product is tailored to the "average customer" (main-stream).

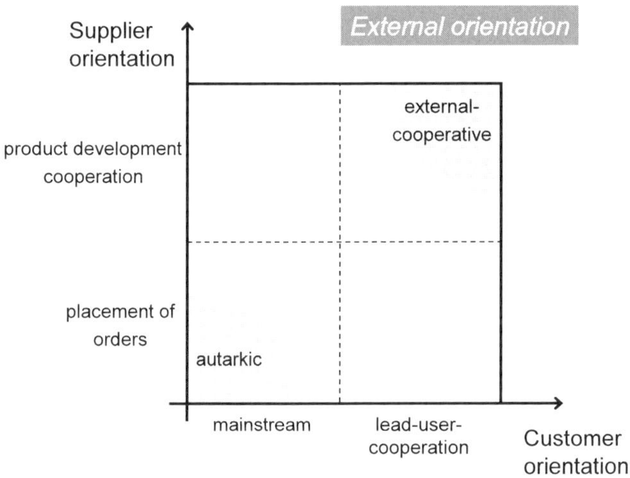

Fig. 2.4 External orientation of innovation planning

Planning framework

The forth dimension of innovation planning is the *planning framework* (Fig. 2.5). The focus here lies on the complexity of development planning tasks and on the required flexibility of development processes. A heuristic planning framework with few regulations and a problem-oriented systematic allows coping with the high complexity of the task by keeping a maximum of flexibility in the development process. The heuristic planning is suited for the solution of systematic tasks[5] that are typical for medium- and long-term innovation projects.

---

4 A description of the lead user concept is part of the method data sheet (appendix).

5 Systematic tasks are characterized by low structuring (that is few knowledge about essential input, targeted output, cause-and-effect-chains etc.) and less separability (high interdependency of partial tasks, interdisciplinary activities). (cf. Gassmann 1997).

Algorithmic and schematic approaches are suited just for the development planning with low complexity and low flexibility.

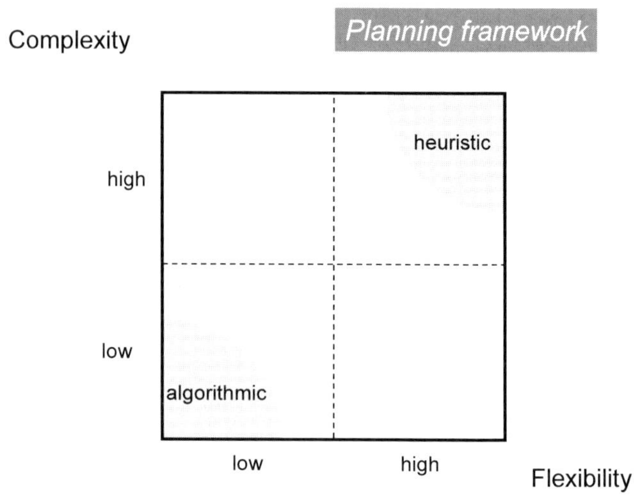

Fig. 2.5 Planning framework of innovation planning

Dimensioning of innovation planning

For the purpose of an Integrated Innovation Management, the partial strategies of the different dimensions of innovation planning have to be harmonized. This is done by illustrating all four dimensions and their values in a total view (Fig. 2.6).

Consolidating vs. changing profile

The extreme values of the dimensions of innovation planning result in two oppositional profiles: the consolidating and the changing profile. The consolidating profile (Fig. 2.6, inner circle) follows from a present-orientated temporal alignment, a synergy-utilizing competence orientation, an autarkic external orientation and an algorithmic planning framework. Enterprises with such a profile are called *Continuous Improvers.*

The consolidating profile: continuous improver

Enterprises that in the past have diversified or placed multiple new products may loose the overview, which results in insufficient attention to their product groups (Voegele 1999). In such a situation, a strategic orientation towards a consolidating profile would be a good idea.

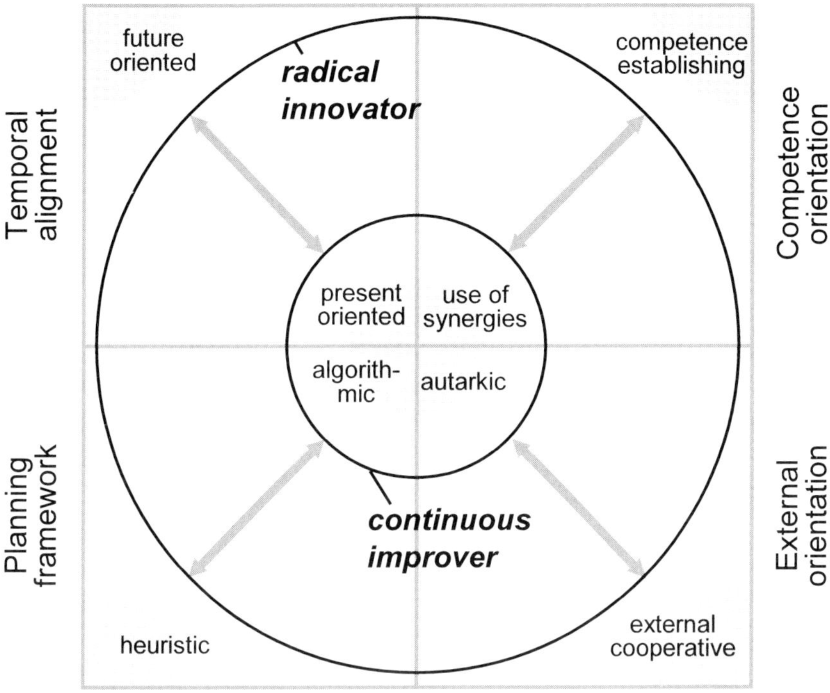

Fig. 2.6 Dimensioning of innovation planning

The major task of innovation planning is to utilize and expand the created market position in an optimal way. When utilizing existing competencies, the focus of activities lies on incremental innovations. Mainly, product management is done, and it is tried, through variant management for example, to reduce complexity and build up barriers for competitors to enter the market. Planning activities are usually present-related and should primarily address the increase of efficiency with the goal of maximizing the net-operating margin. Finally, the consolidating profile is characterized by innovation attempts with the goal of continuous improvements.

The changing profile: radical innovator

In contrast to the consolidating profile, the changing profile is characterized by a future-orientated temporal alignment, a competence orientation to establish abilities, an external-cooperative orientation and a heuristic planning framework (Fig. 2.6, outer circle).
Enterprises of such a type are called *Radical Innovators*.

If the improvement potentials of existing products are exhausted, the purpose should be to foster the next technology step or to enter new markets. At this point, it is important to shift the innovation planning to a changing profile. Cooperating with customers as well as with external research and development institutions is of avail when opening new benefit potentials. Furthermore, the changing profile is related to the goal of developing radical innovations. The appropriate planning framework is characterized by a problem-orientated and flexible approach.

Diversification

Enterprises pursuing the double strategy called *diversification*, i.e. the establishment of new competences in new markets, should be aware of the higher risks and therefore have a bankroll above average (Eversheim and Schuh 1996).

Fit of partial strategy

Both of the described oppositional types, radical innovator and continuous improver, should be seen as extremes of innovation planning. For the purpose of a holistic approach, the primary goal of an enterprise is to harmonize the partial strategies of innovation planning, i.e. to create a fit in the form of a concentric circle in the total view (Fig. 2.7).

Application in enterprises

1. Determining the as-is-profile

When applying the presented model, the enterprise is positioned in the dimensions of innovation planning. This requires the analysis of the enterprise with regard to the actual alignments of its innovation planning. For this, the partial strategies of the four dimensions (temporal alignment, competence orientation, planning framework and external orientation) are considered singularly and relatively to the extreme alignments. Because of the different distances of the analyzed dimensions to the point of origin, the arisen as-is-profile of innovation planning may have a shape different from a circle (Fig. 2.7).he as-is-profile visualized in figure 2.7 shows characteristics of the changing profile like external-cooperative orientation, future-orientated temporal alignment and competence establishing orientation. Just the planning framework with an algorithmic alignment is characterized by consolidating tendencies. Thus, a misfit towards the alignments in the other dimensions arises as visualized as a deformed circle. For the purpose of an intended, holistic and harmonic innovation planning, the misfit is not important.

2. Determining the to-be-profile...

If the partial strategies are harmonized, a concentric circle to the point of origin arises. The radius of the circle describes the overall strategy regarding the consolidating or changing profile.

To pursue a long-term strategy wisely, a circle-shaped to-be-profile has to be determined. Then, need for action and the aim of the partial strategies derive from the arising differences of the shapes.

3. Deriving need for action from to-be-as-is-comparison

Because of the essential differences of both strategies, the development of either incremental or radical innovation, the strategies should not be seen as static and permanent. Moreover, the strategy moves in an area between the poles of the extremes concerning the situational, temporal context.

Once the to-be-profile is achieved, the actual state should be kept as long as it is useful. Depending on the innovation ability and the intended level of innovation, the to-be-profile should be updated.

Thus, the innovation planning sets the direction of impact for future innovations. It updates and implements the corporate policy on the basis of the current situation of the innovation ability. This is followed by systematic developments of product and process innovations that are realized in innovation projects.

Direction of impact for future innovations

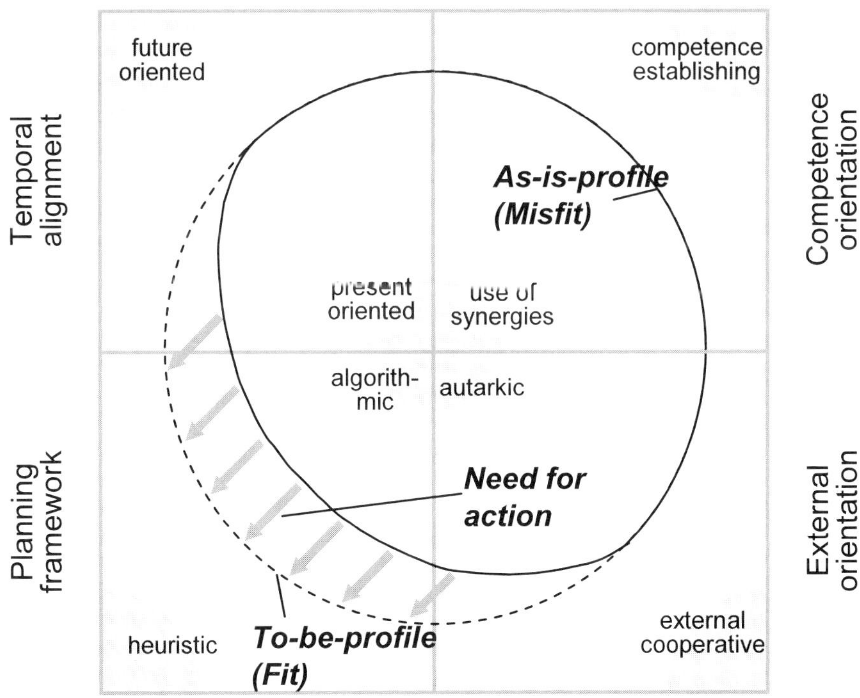

Fig. 2.7 Examples of an as-is- and to-be-profile of innovation planning

2.1.2 The innovation organization

Goal: establishing structures

The innovation organization as a structural and strategic component of the innovation management forms the framework for the innovation planning and the innovation control (cf. Bleicher 1999). The goal of the innovation organization is to establish structures that lead to an optimal innovation ability of the enterprise. Based on this goal, influence factors like *task positioning* and *information exchange* as well as *budgeting* and *resource assignment* are the dimensions of the innovation organization[6].

Task positioning

The aspect of *task positioning* distinguishes between focused and peripheral structures. Focused structures are given if a person-related orientation of the innovation task arises with the centered, organizational structuring. Radical innovations are fostered by concentrating the competencies in a central unit (Boutellier et al. 1999). The combination of task-oriented alignment with decentralized structuring is called peripheral task positioning.

Information exchange

The way *of information exchange* is often characterized as offensive or defensive. An open, internal communication and a bilateral dealing with information towards the environment – e.g. by a high presence in public – are attributes of an offensive information exchange. A defensive information exchange is characterized by unilateral and secretive dealings of information in both internal and external surroundings.

Budgeting

The dimensions of *budgeting* are aligned either autonomic or bound. Autonomic structures are characterized by a fixed annual budget for the innovation planning. A bound budgeting arises from a demand-oriented annual budget.

Resource assignment

The *resource assignment* concerning staff and time is either integrated – e.g. regarding the total innovation planning – or differentiated – e.g. regarding single innovation projects.

Dimensioning of innovation organization

Profile establishment

The consideration of the dimensions of the innovation organization in their overall context leads to a profile comparable to the innovation planning. A consolidating profile is characterized by peripheral task positioning, defensive information exchange, bound budgeting and integrated

[6] Additionally, see the dimensions of R&D-organizations of Saad (Saad 1991) and the dimensions of organizing R&D-teams of Gassmann (Gassmann 1997).

resource assignment. The features of a changing profile are focused task positioning, offensive information exchange, autonomic budgeting, and differentiated resource assignment.

2.1.3 Innovation Leadership

Staff behavior

Innovation leadership aims at developing a staff behavior which supports innovation.

The following dimensions are to be considered: *staff encouragement*, *decision making*, *performance evaluation*, and *communication behavior*.

Staff encouragement

Encouragement can be related to the specialist or the generalist type of staff. For generalists, leadership and cooperation skills are promoted in particular.

Decision making

Decisions are taken hierarchically from a designated decision maker with either little feedback or in a participative manner based on decision content.

Performance evaluation

Performance evaluation can be based on results or employee development. Result-based evaluation is characterized by a narrow scope and absolute measures. In contrast, development-based evaluation is characterized by a complex set of relative measures.

Communication behavior

Communication behavior can be distinguished into integrative an excluding patterns of behavior. Integrative communication is characterized by a holistic and explanatory attitude. Excluding behavior is characterized by *ex post* directing and task focus.

Dimensions of innovation leadership

Interrelation

Innovation leadership can have, similar to innovation planning, two characteristic patterns, which contain consolidating or respectively changing properties.

The consolidating profile is characterized by professional staff encouragement, result-orientated performance evaluation, and excluding communication behavior. In contrast, a changing profile is characterized by joint decision making, development-based performance evaluation and including communication behavior. This type of innovation leadership encourages innovative behavior which is best characterized by repeatedly challenging routine processes and their evaluation regarding performance contribution.

Innovation leadership is a powerful but challenging instrument to influence innovation capability. Innovation leadership is associated with high requirements for employees and can only be changed based on the broad acceptance of all persons involved.

2.1.4 The operational level

Operational implementation

Normative and strategic management are implemented at an operational level in innovation projects. There is also the innovation process for the structures (Fig. 2.1 left side) and the innovation willingness of the employees for the behavior side (Fig. 2.1 right side).

The task of the operational management is to transfer inputs from the normative and strategic level into practice (Bleicher 1999).

InnovationRoadMap method

The InnovationRoadMap method provides several practical tools for the operational tasks of the innovation management. In particular, it explains innovation processes and innovation projects in detail. (Chap. 3).

2.2 Innovation Portfolio

External and internal view

The innovation portfolio is based on a company-external and -internal view, which enables evaluating the innovation degree of innovation projects and products. The different views are illustrated in the innovation portfolio along the horizontal and vertical axis (Fig. 2.8). Within the internal view, there is made a distinction between the competence development and the use of synergies related to existing products and other innovation projects.

Competence development and internal use

In innovation projects, the criteria for competence development and for the use of synergies are fulfilled differently. Projects with high learning effects, i.e. with a high degree of competence development, usually do not allow making use of synergies. Products which are based on an intense application of existing knowledge, i.e. based on synergies to other products, usually do not contribute to competence development. An appropriate project selection is directed towards a balanced and situation-specific fulfillment of both internal evaluation criteria, because the success of the company is based on both, the use of synergies as well as an ongoing learning process related to the development of new competencies.

Innovation and improvement

From an external point of view, innovation projects are classified into improvements and innovations. Both, improvement and innovation are required for a company's success. Highly innovative projects always bear a risk of technical or market failure. The risk of market failure can be reduced through the simultaneous development of low-risk improvement products. On the other hand, the application of new technologies in innovative products is required because existing technologies are becoming obsolete in a specific phase of life. At this point, they only enable marginal and less profitable improvements.

Teachers, Stars, Basics, High-Risk

If company-internal and -external evaluation criteria are correlated to each other, four types of innovation projects or resulting products arise (Basics, Stars, Teachers, High-Risk). The characteristics of these types are explained in the following (Fig. 2.8).

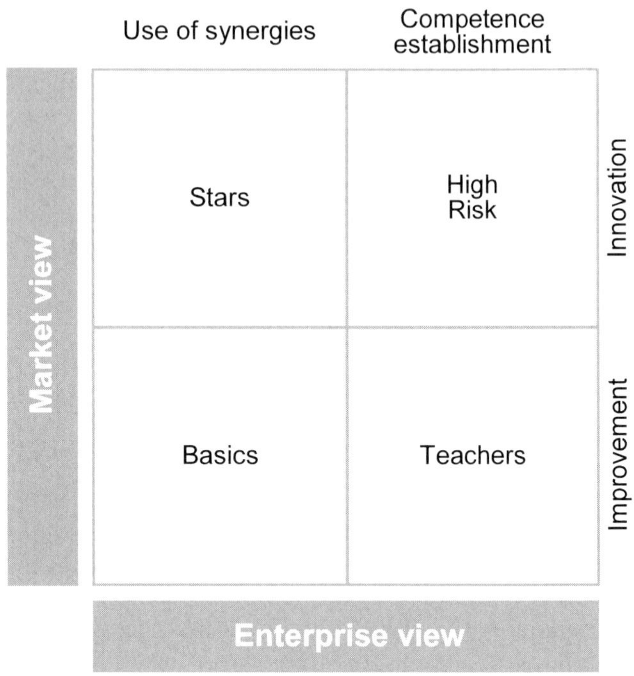

Fig. 2.8 Innovation portfolio (Brandenburg and Spielberg 1998)

Basics

Basics result internally from the use of synergies and represent the improvement of an existing product from an external point of view. They are related with a low risk, because technical feasibility and market potential can be estimated based on company experience. In the automotive industries, the facelift of existing types is an example for the type basic.

Stars

Stars are achieved internally through the use of synergies, however, they are perceived as an innovation from the external, market point of view. Due to their positive external evaluation, stars can considerably contribute to the company's success. However, they generate little internal learning. If a company generates a product for a foreign market based on its core competencies and takes over technology leadership, this product is likely to be evaluated as a star (Example: transfer of technologies from the automotive industries to two-wheeler production).

Teachers

Teachers are learning projects. With these projects, internal competencies are developed, which are evaluated as improvements from an external point of view. These projects allow entering into new markets or technologies with a high future potential. If for example a supplier aims at enhancing his competencies towards system integration and delivery, this could be initiated by realizing some teacher projects which are planned to integrate knowledge of new production technologies into the company.

High-Risk

High Risk Projects combine the risk potential from teachers and stars. Internally, they cause competency development and are perceived to be an innovation from external. The objective of a high-risk project could be to enter a new market by means of a new technology. High-risk projects can be very lucrative if they are successful. Because of the related risk, they are usually initiated sporadically[7]. (Example: the development of a four-wheel drives from a manufacturer who has produced sports-cars so far).

Balanced mix for company success

For company success, a balanced mix of all four project types is necessary. Successful high-risk projects ensure for long term company success, as they generate outstanding profitable radical innovations (Gassmann et al. 2001). Basics generate necessary cash-flow in a short term and thus balance the high risk related with the high-risk projects. Depending on the company situation, a focus can be put on the one or the other direction.

7 For the management of high-risk projects see GASSMANN (Gassmann et al. 2001).

2.3 Aachen Strategy Model for Product Innovation

The Aachen Strategy Model for Product Innovation has been developed by the Fraunhofer IPT in cooperation with the WZL of the RWTH Aachen, and is based on their experience from many innovation projects (Fig. 2.9). It is described here for the first time.

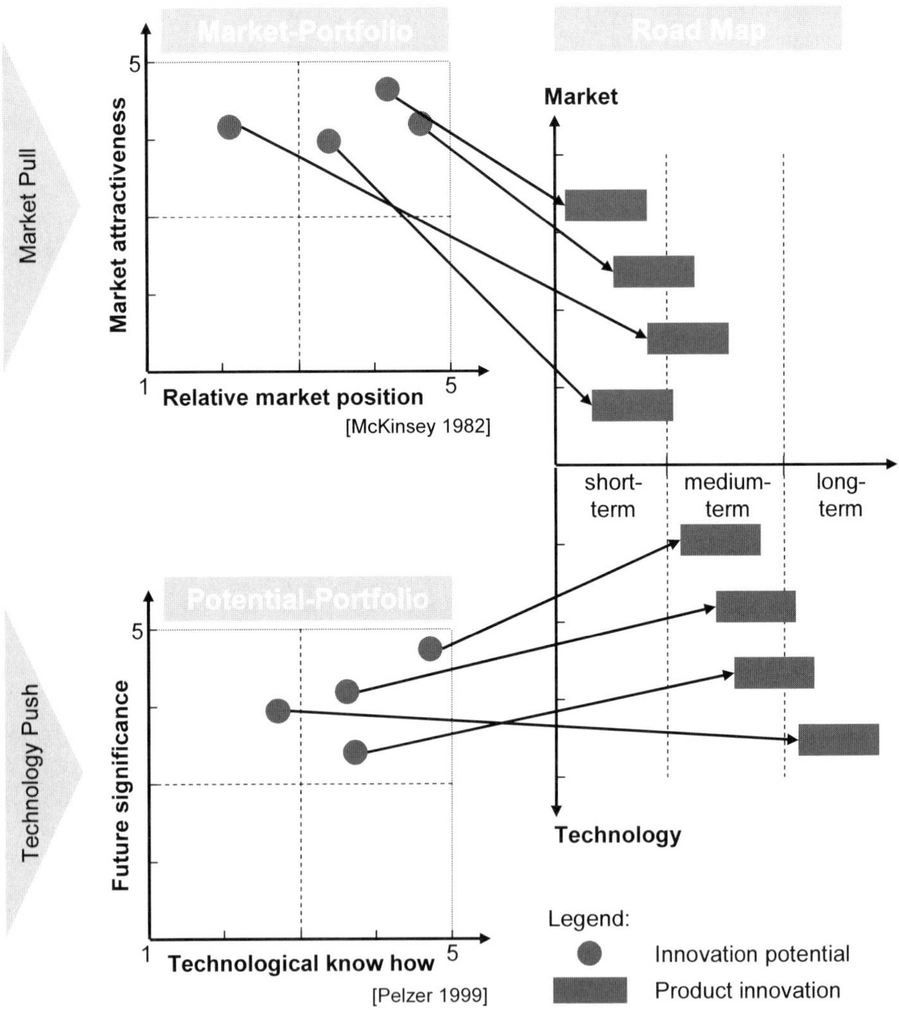

Fig 2.9 Aachen Strategy Model for Product Innovation (ASM-PI) according to EVERSHEIM

Market pull

If a company dominantly implements market-oriented strategies in order to search and select new product ideas, this generally creates short-term innovation potentials (Fig. 2.9, top). This temporal constraint can be removed partially by forecasting and auditing direct market opportunities, as well as by observing opportunities in indirect markets. A producer of camshaft grinders, for instance, has to observe precisely at what time engine developers accomplish to substitute camshafts by computer-chip-controlled-solutions (solution substitution). Engine developers have been searching for such concepts for about 30 years already.

Technology push

If a company has technological potentials, capabilities or know-how (e.g. patents/licenses), a technology push can be achieved if these potentials can be translated into product ideas. This strategy creates mostly medium- to long-term innovation potentials (Fig. 2.9, bottom).

Combined Strategy (ASM-PI)

Depending on the business targets, the branch, and the respective market position, both strategies (Market Pull and Technology Push) are used as a combined strategy. These combined strategies are summarized in the Aachen Strategy Model for Product Innovation.

3 The InnovationRoadMap Methodology

Walter Eversheim, Frank Brandenburg, Thomas Breuer, Michael Hilgers,
Christian Rosier

The *InnovationRoadMap methodology* allows the systematic planning of successful product innovation. Through this methodology, both the strategic as well as operational aspects of the integrated innovation management system of the St. Galler Management concept are addressed. A prime advantage of the methodology is the systematic approach, something that leads to complete continuity throughout the whole product development cycle, even though it is already a primary tool in the early phases of the cycle.

W-model as structure

The basis of the InnovationRoadMap (IRM) methodology is the *W-model* (Brandenburg 2002), an approach split into 7 distinct phases. This illustrates, starting from the enterprise strategy, how an action plan is developed that maps the medium- to long-term innovation activities of the enterprise. It is this guide into the future that is visualized in the *InnovationRoadMap*.

In order to arrive at a practical tool that facilitates the whole process of arriving at a successful product concept, the different process phases are coupled to existing and new methods of production planning. This not only describes the steps required (what should be done), but also shows how to do it practically with the aid of methodological/technical support.

Requirements for a methodology to plan technological innovation

In the following sections the most important requirements for a methodology to plan technological innovation will be described. On the one hand, these requirements have been deduced from the characteristics of the planning object "Product Innovation" and the planning process "Innovation Process". On the other hand, they have been deduced and/or changed based on practical experience as well as by analyzing available studies looking at innovation and the requirements of a planning methodology. As such, they are a reflection of the requirements of most manufacturing organizations. From the analysis, 9 requirements for a planning methodology have been found (cf. 3.1).

In chapter 3.2 a high-level description of the W-model structure, consisting of 7 main phases, is given. A detailed description of the individual phases, as well as the "how" (the methods), follows in chapters 3.3 to 3.9

3.1 Requirements for the methodology

The main requirements for a methodology used to plan technical product innovation can be summarized in nine points, based on BRANDENBURG (2002). These points serve as the basis for the IRM-methodology.

1. Set clear goals

Setting *clear goals* and following these systematically ensures that both the goals of the enterprise as well as those of the decision maker are known and are being followed. Only when innovation goals and the innovation strategy – based on the enterprise strategy – have been identified, defined, and communicated and the "fit" between strategic and operational planning has been done, can relevant solutions, i.e. product ideas that fit into the organization, be looked for or found (Saad 1991).

2. Quality of ideas more important than quantity

Innovation studies and project experience show that successful product innovation more often than not is the result of ideas from individual, motivated employees, and as such basically happens by chance. Enterprises have many of these randomly generated product ideas (Albers 1991), but the percentage of useful ideas from these is very low (Droege 1999). A methodology for innovation planning must therefore support the systematic generation of *qualitatively highly valuable ideas*.

3. Create for the future

(Only) the fulfillment of future and/or latent customer requirements through future-oriented or at least leading-edge technology creates uniqueness. An important prerequisite for not only the continued success of an enterprise but the improvement thereof through the continuous and precise fulfillment of customer wishes will be the ability to continually recognize or even create new or developing trends, and then to produce relevant products through a network of the most competent, creative and innovative enterprises (Warnecke 1997). Part of a strategy-driven product innovation planning therefore has to be *the identification or recognition of these latent and future customer- or market requirements* (Kleinschmidt et al. 1996) and from these the deduction of the innovation potential[8]. It has to be said, however, that these potentials are rarely identified using conventional market research or customer questionnaires. Therefore, methods that show up potential innovations have to be integrated so that these potentials can be translated into successful product innovations.

4. Use existing strengths

Good ideas lead to success if they are implemented. This can be achieved only if the company succeeds in leaning on or using its own particular strengths. In rating or choosing certain product ideas, besides the market potential, the "enterprise fit" of the ideas also needs to be evaluated.

5. Create transparent and standardized processes

Successful control of complex problems requires two distinct levels. On the one hand, the theoretical struggle with such a complex problem has to be simplified and done systematically, while on the other hand, the holistic overview needs to be kept in the detailed analysis. To be efficient, particular attention has to be given to rationalization potentials. These might typically be achieved by making sure that required tools and/or inputs, such as information, programs etc. are only developed once and are standardized as much as possible (cf. Schmidt 1996).

6. Objective, retraceable idea selection

The easier it is to physically imagine something, i.e. the more available specific information regarding a certain alternative is, the higher the chance that that idea will be regarded in a positive light (Tversky 1986). In practice, this can lead to the fact that existing, well-known solutions are rated higher than new, lesser known technologies (Lenk 1994). This results in ideas being further refined that seem to have a higher information availability, instead of exploring or searching for further information on possibly better alternatives (Dyckhoff 1998, Eisenführ 1999). It is therefore vital for an objective idea selection not to focus on the product- or company-specific history and even on well-known technologies, but to make sure that totally different but functionally equivalent opportunities are included in the analysis (Pfeiffer 1995).

In judging or rating different ideas, vague and qualitative factors need to be taken into account. Should the decision maker insist on using well-defined, quantified information, the implementation risk might be reduced, but in order to obtain this information, like for instance the introduction to the

8 Innovation potential refers to the relevant possibilities, requirements and technological potential within the product innovation planning of an enterprise.

market of new, similar products, leads to a high degree of time pressure in the product development phase, which in turn leads to the use of existing, successful technologies, but this ultimately results in very little scope for real differentiation (Brandenburg 2002).

7. Accepting uncertainties

In the early phase of the product innovation process one will always be confronted with the uncertainty of the eventual realization of envisaged developments. The assumptions can include for instance assumptions regarding the technical capability of implementing a specific alternative, or developments in the market and economy in general. In this environment of uncertainty and time pressure relevant methods are required (Staudt 1996).

8. Synchronize market- and technology requirements

In technology-intensive environments it is especially important to obtain a high technological competence. To be on the forefront of or even a market leader in new (product) technology requires the necessary resources for the development and refinement of these are made available. The ultimate purpose needs to be that the technology is already developed by the time the market requires these. In the development of these there is however an inherent risk. Obviously, technologies developed too late results in the non-satisfaction of a market demand. On the other hand, too early developments of technologies put undue pressure on financial and human resources that could have been used more effectively. It is therefore required that market and technology developments are synchronized through a methodical, systematic support system that makes the realization of necessary technologies on a just-in-time basis possible.

9. Keep an open mind and stimulate creativity

The planning process as well as the specific methods used during idea generation and the implementation of technological product innovations are unique for each specific planning object. There is no generic or "ideal" way; there is only a "best" approach for successful product innovations (AWK 1999, Schulz-Wild 1997, Zahn 1992, Sabisch 1991). This results in the requirement that one and the same methodology, if used systematically, will result in the satisfaction of all different or possible tasks required in the specific viewpoint. However, this methodology should be modularized, should allow the isolated use of specific building blocks, and should allow the user to start with any one of the phases. Specifically, it has to allow sufficient individuality and creativity, as successful innovations are normally not achieved through a purely systematic and logical approach (Schmitz 1996).

Model

The IRM methodology makes use of a largely *standardized approach*. This standardized approach has the advantage that, though providing for the specific requirements for creativity and intuition, the co-ordination of all participants is made easier, and the underlying structure of the approach does not need to be re-developed for each separate planning process (cf. Schmitz 1996).

It is important to emphasize that the IRM methodology does not exist for itself, but is a tool to be used in the development of successful solutions. It is not a supplement for achieved capabilities, situational knowledge or the theoretical exploration of creating something new: the methodology takes this as a pre-requisite, depends on it, and should stimulate this even further. It therefore provides a golden thread for the development of technical product innovations, and should be used creatively and intelligently. The advantage will then be reflected in the creative and thinking potential (cf. Haberfellner et al. 1999).

3.2 The W-model: Structure of the IRM-Methodology

Basic structure: seven phases

The IRM methodology is based on a structured foundation: the W-model. It allows the fulfillment of the requirements of a method to plan technical product innovations (Chap. 3.1). The W-model consists of seven distinct phases (Fig. 3.1). These different phases accentuate logically different planning units. In each of the planning phases various planning activities need to be performed. As the planning results are dependent on the dynamics of market-, technology-, and organizational developments, the main activities within phases 1, 2, and 7 need to be repeated periodically. Phase 3, 4, 5, and 6 are more or less continuous activities and need to be integrated into the (product-) planning cycle of the enterprise (Brandenburg 2002).

This differentiation according to the various phases within the model is a rather analytic exercise, as it is practically impossible to clearly distinguish the phases during the process of product innovation (Thom 1980). The sub-processes not only interlink with each other, but are not independent either, and therefore not only will but should be running concurrently. Within this context it is important to emphasize that only the basic model will be described, and this needs to be refined and modified for a particular company or situation.

Splitting the model into different phases therefore does not mean that either all the abstractly named phases will be of equal importance within the sequence described, or that all the activities will be present in a particular innovation process. More specifically, some of the phases can be worked on very intensively, while others might be totally absent within a given situation.

The integration of the different planning phases within the broader framework of the strategic and operational planning levels is shown in the high-level illustration in figure 3.1. Showing the distinct planning phases relative to the planning levels results in the optical presentation of a W – hence the name.

The planning phases are distinguished according to purpose, subject and informational relationship (cf. Brandenburg 2002).

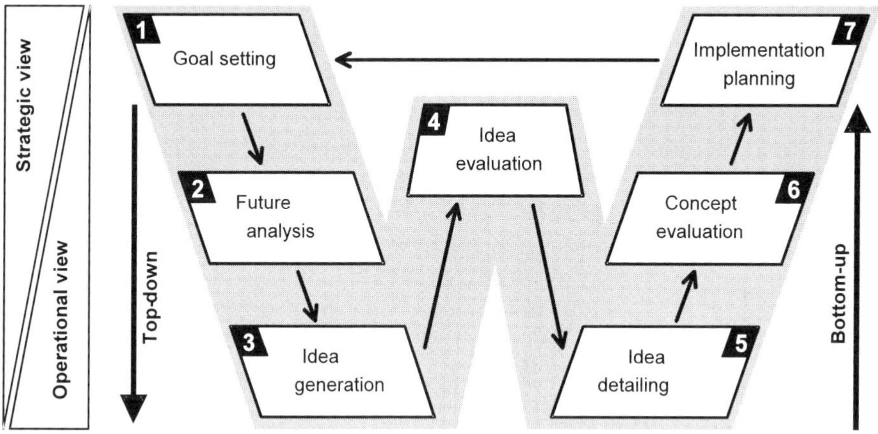

Fig. 3.1 The W-model as high-level sequencing concept (Brandenburg 2002)

Goal setting

During "goal setting", the strategic direction and innovation goals are defined, or are deduced from the overall company strategy. Besides company potentials, strategic fields or areas where the company will be operating in and which are relevant for innovation planning purposes are defined. Pieces of information required as input for this phase are company-specific external and internal data. On the output side is information regarding the innovation goals, the innovation strategy, the company potential as well as the chosen formation fields.

Future analysis

The purpose of the future analysis phase is the identification of innovation potentials and the formulation of specific innovation activities for the company. To start off with, general trends as well as more specific developments within the chosen formation fields are analyzed. Following this, the impact these will have on the formation fields and the company in general is projected. Based on this, and taking into account the company potentials, innovation potentials are deduced which will correspond with future market or technology developments. Output from this phase is therefore information regarding company innovation potential or more specific innovation tasks.

Idea generation

Based on the identified innovation potentials, in the "idea generation" phase, product ideas are developed. This starts with the identification of product ideas of the 1st order, i.e. problem and solution ideas in a creative-divergent phase. Within the framework of the deduced innovation potentials, the purpose is to generate a broad spectrum of future-oriented ideas. The result of the "idea generation" phase is the development of creative-convergent product ideas of the 2nd order, where every problem is combined with at least one possible solution. These ideas need to be documented for future planning phases.

Idea evaluation

The purpose of the "idea evaluation" or rating phase is the identification and evaluation of all product ideas that might be successful or look promising. The evaluation is based not only on market- and technology aspects, but also includes strategic conformity and the advantages for the company. The output of this phase is information that allows the allocation of all the ideas both in terms of time and content into the Innovation RoadMap (IRM).

Idea detailing

"Idea detailing" means that further market and technology information is collected for the chosen ideas. The purpose is to develop product concepts. To do this, product requirements need to be defined in detail, and specific detail solutions need to be found for the resulting technical tasks. Based on these, product concepts can be generated. The result of this phase are various concept variants for each product idea, and if possible these should already be validated through functional prototypes or "demonstrators".

Concept evaluation

The purpose of the concept evaluation or rating phase is the quantified evaluation of the developed product concepts. This starts with a basic repetition of idea evaluation, but including the additional and verified information from the previous phase. Over and above this, however, the basic economic feasibility needs to be calculated. This is based on either a cost-income or a cost-benefit analysis. The results of this detailed evaluation will again be used to further populate the InnovationRoadMap.

Implementation planning

In the "implementation planning" phase, the individual, company-specific activities for the developed product ideas and concepts are combined into one program, the InnovationRoadMap. This program is a list of wants that in the longer term leads to the development, use and maintenance of technological potentials. The previously detailed singular results are again aggregated in this phase, to illustrate the short-term, medium-term and longterm dynamics and interactions between the environmental and market requirements and the technological product solutions and their development.

Typical but idealized sequence of phases

The model as described with its various phases should be seen as an ideal but simplified sequence, in that a concurrent, inter-dependent and networked population of the different phases has to be strived for. Especially the phases "idea generation" and "idea detailing" will in practice either overlap, or will even be done in one and the same phase. A definite distinction between the two phases is not only practically impossible, but also not useful, as under certain conditions detail tasks will already be worked on and problems solved during the "Idea generation" phase. Similarly, the phases "idea evaluation" and "concept evaluation" will overlap to a large degree (Brandenburg 2002).

This integrated or even overlapping work being done in phases 3 and 5 or 4 and 6 happens especially in the following scenarios (Brandenburg 2002):

- there are only very few product ideas and all of them can or should be analyzed in detail;
- there are relatively simple or at least no complex products or product ideas;
- there are competing product ideas in a very clear and defined market segment.

Summary

The sequential structure of the W-model is the central building block for the IRM-methodology. The W-model consists of seven distinct planning phases: Goal setting (1), Future analysis (2), Idea generation (3), Idea evaluation (4), Idea detailing (5), Concept evaluation (6) and Implementation planning (7).

The IRM methodology aids in the description of the goals, what should be achieved, which alternative approaches (future analysis, idea generation, idea evaluation) can lead to the desired results, to describe these alternative routes over time in more and more detail (idea detailing, concept evaluation) and to make certain that the goals as originally formulated can in fact be achieved (implementation planning).

During the application of the methodology, the number of ideas - originally in the form of futuristic projections and innovation potentials and then in detailed product concepts - is constantly being reduced (Fig. 3.2). This reduction in the number of ideas through the so-called *idea funnel* is necessary as the required work content is increasing as the ideas are becoming more and more concrete – there is a reduction in flexibility and agility available per idea.

The methods used in the InnovationRoadMap methodology are mapped and modified according to this relationship between concreteness of an idea and number of ideas. This means that the more concrete the formulation of an idea – depending on the stage within the planning timeframe – the more detailed and specific the relevant methods used become (push for creativity, analysis, evaluation etc.). The result of this methodology is the InnovationRoadMap. A more detailed explanation and description of the different planning phases is given in the following chapters 3.3 to 3.9.

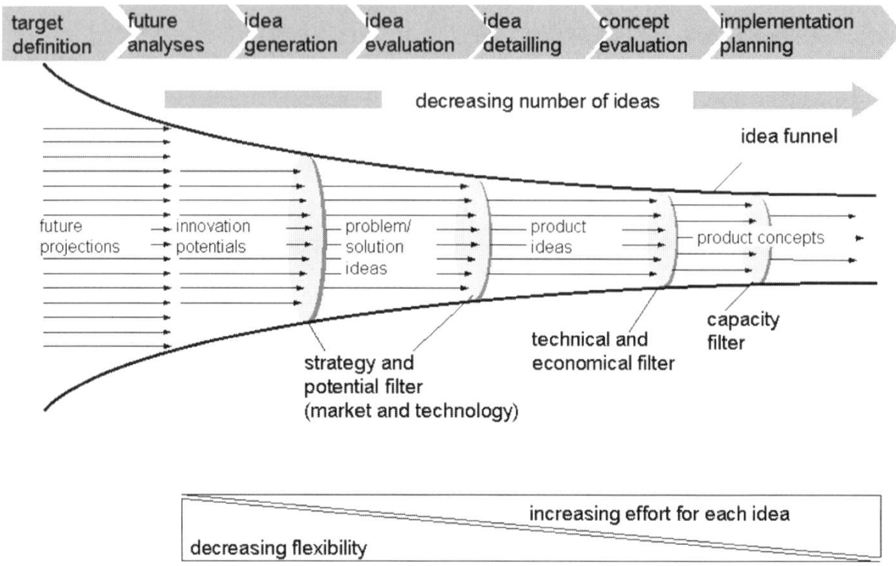

Fig. 3.2 Successive choice or elimination of ideas and refining / focussing them

The case study: Center Positioning Systems Ltd

The following, fictitious case study is intended to vividly describe all phases of the W-Model and to help in the understanding of the methodology.[9]

The case study: Center Positioning Systems Ltd

We will use the Center Positioning Systems Ltd as an example. It is a company in the German Black Forest and has about 750 employees. They produce positioning systems for the tooling industry, for robotics and automation. The company is led in the 2nd generation by Dr. Peter Palmer, who followed his father as the company's CEO after finishing his business degree. Since he took over the business he split the company's organization into three divisions, namely „Sliding carriage systems", "Precision positioning systems" and "Sensors and control". Every division is led by a *business unit (BU) manager*. Furthermore the staff position *New Business Development*, led by Mr. George Stone for three years, was

9 All sections in future chapters marked with the gray sidebar indicate a continuation of this case study.

created to gain access to new markets. Mr. Stone is directly responsible to the management board. He initiated several innovation projects within the company and works closely together with the different business unit managers. Since the creation of the staff position he is supported by 5 employees.

Initial situation

Center Positioning Systems Ltd has been a successful player at the market for several years but is now confronted with an increasingly difficult competition. Being a visionary businessman, Dr. Palmer identified this position as problematic and tried to confront it by gaining access to new markets using innovational products. Therefore Mr. Stone was commissioned to plan a new, image-enhancing product which enables Center Positioning Systems Ltd to expand into new and promising markets.

Wish for a systematic and directive approach

Mr. Stone was aware of the importance of his mission and tried to solve this task in the best possible way. In the past he and his staff often relied on their experience and intuition when looking for new products and markets. This time though he wanted a more systematic and directed approach to innovation planning. Looking for such an approach he had read a lot of literature lately without finding a satisfying solution though. Mr. Stone was looking for a method which was understandable, practical and delivered the necessary tools to support a consistent planning.

Decision for the InnovationRoadMap method

A few months ago Mr. Stone attended a seminar which informed about new methods and concepts in innovation management. There he took special interest in the InnovationRoadMap method and even had the chance to discuss its practical use with scientists from the Fraunhofer Institute for production technology. Based on this discussion, extensive information and working sheets on the IRM method he had the idea to use it for his own innovation plans. So he decided to present the IRM method at the next board meeting.

Dr. Palmer and the BU managers approved the method, a strong support by the management board therefore was ensured. Immediately an appointment was arranged to start with the first phase of the innovation planning process the next week…

The planning phases of the IRM Methodology

3.3
Goal setting

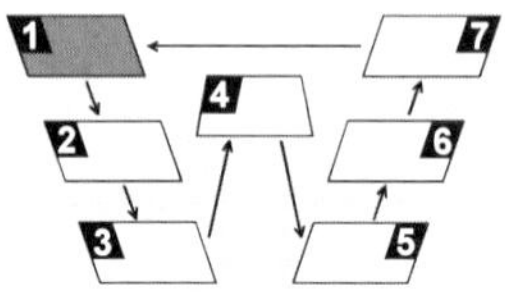

During goal setting, the strategic direction and innovation goals are deduced from the overall company strategy. Based on this,

- company potentials are defined and
- figuration areas are identified.

This specific or even singularly defined goal is a pre-requisite for the planning of successful product innovations.

Setting of generic boundaries

In setting goals it is vital to acknowledge the starting or present situation of the company, to deduce innovation goals and to define the strategic figuration areas. Proposals for solving these on a conceptional level are worked on during the following steps. The content of this first phase is therefore to set the generic boundaries for the product innovation methodology. Within these, the strategic innovation goals and innovation strategies need to be deduced.

As the planning is really a periodically revolving or recursive exercise, the innovation strategy has to be seen as input information, but at the same time also as the planning goal (Fig. 3.3). Coupled to this, the competencies and potentials within the company need to be acknowledged. These serve as the basis for product innovation on the one hand, but on the other hand show up any gaps or discrepancies between these competencies and the company goals and strategies. These discrepancies have to be rectified during the innovation planning phase (Brandenburg 2002). Both from an efficiency perspective as well as practically not all the possible figuration areas can be analyzed. One therefore needs to focus on those areas that, from a company-strategic perspective, are identified as the most important ones or the ones where the chances for success are deemed to be highest. Therefore this phase is split into three sub activities, as shown in the following methods tool kit (Fig. 3.3). The description or discussion of these activities is given in the relevant sections of this chapter.

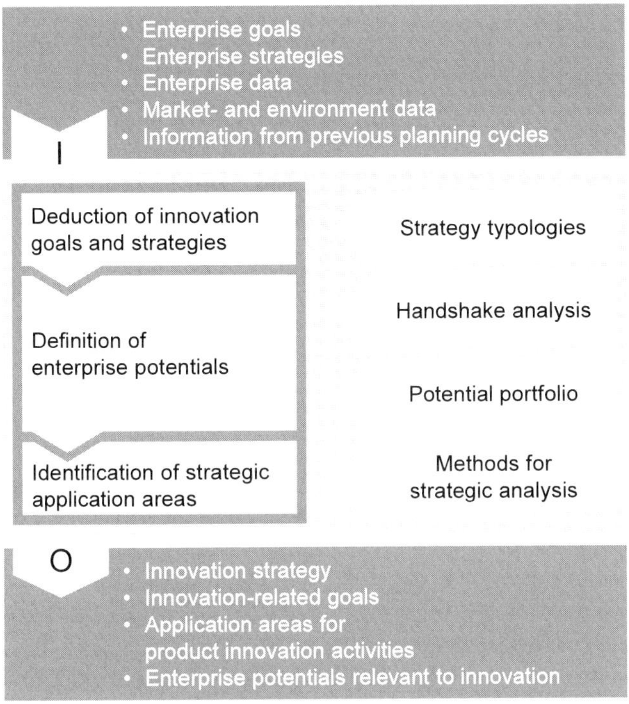

Fig. 3.3 Methods tool kit for "Goal setting" (cf. Brandenburg 2002)

3.3.1 The deduction of goals and strategies relevant to innovation

The goal and strategy formation have the purpose of guaranteeing the goal and strategy congruence of the planning activities that follow. The goals, which are pursued with technological product innovations, are different from one enterprise to the next. They result from e.g. the super-ordinate enterprise strategy, or the demand habits of the market to be served or the positioning in relation to the competition. Without going into the detail of fundamental proceedings for the listing of target systems, using the framework of BROSE, the following important goals of technological product innovations are differentiated (Brose 1982):

Profit increase

- Increasing the contribution amount through substitution of existing products and absorption of the relatively high contributions from product innovations.

Keeping or improving growth targets

- Developing new customer groups and/or new markets or market segments, and making up decreasing demand for older products through product substitution. At the same time the image of the enterprise is positively impacted on through product innovation.

Utilizing capacity

- Increasing the utilization of idle capacity, and balancing employment variations.

Distributing risk

- Becoming more insensitive against turbulence in the market (e.g. exchange rate fluctuations).

Use goals as evaluation criteria

Based on the goals specified above, evaluation criteria (e.g. for the idea evaluation) need to be developed. These make the comparison and evaluation of different solution alternatives as proposed in the planning phases to follow possible (Haberfellner et al. 1999). They therefore impact directly on the importance of the evaluation criteria, which are different from one enterprise to the next. The unambiguous definition and transparent representation of these strategic estimation or evaluation criteria is thus a central success factor for entrepreneurial innovation activity (cf. Pleschak 1996; Cooper 1993) and forms the starting point for the *top-down approach* of the W-model during the evaluation of technological product innovations.

Competitive strategies

To have successful product innovations, besides the orientation on the goal system one also needs a harmonizing *innovation strategy*. This needs to have a very close relation to the competition and timing strategy of the enterprise (AWK 1999). The expression competition strategy designates the choice of offensive and defensive measures required to be successful in this competition (Porter 1997). In the timing strategy, the moment of new product launch is the most important variable (cf. Meffert 1998; Buchholz 1996; Perillieux 1996; Pleschak 1996). The classic competitive strategies are (cf. Porter 1997):

Cost leadership

- Cost leadership: Making comparable products available at a lower cost in a specific market / market segment.

Differentation

- Differentiation: Striving for specific features (e.g. technology or quality leadership) at a competitive / similar price.

Focus

- Focusing on a very specific or narrow market segment.

Timing strategies

Timing strategies basically distinguish between "innovation leaders" and "innovation followers" (cf. Gassman 1996) and if necessary between early and late followers (Gassman 1996; Meffert 1998).

Innovation leader (Pioneer)

The innovation leader (First-to-Market) is the first to convert a product idea in the market by succeeding with the combination of the market and technology.

Early-follower strategy

The "early follower"-strategy (Second-to-Market) should decrease the innovation risk in that the experiences of the pioneer and the market prepared by him are analyzed and used. In addition, the early follower must enter into the market in as short as possible a time interval to the pioneer with at least the same product characteristics offer. Frequently the follower shows characteristics such as a refined and slick marketing concept, technological modifications, customer-specific application development or other add-on benefits compared to the pioneer (Gassmann 1996).

Late-follower strategy

Through a "late follower"-strategy (Later-to-Market) the innovation risk is supposed to be avoided completely; the market is only entered once standards are developed and the buyer behavior can be predicted with relative certainty. Here the differentiation from the competition takes place frequently via the product price (Eversheim and Schuh 1996).

3.3.2 Analysis and development of enterprise potentials

Deriving new products

Besides the strategic goals an analysis of the (technological) enterprise potentials is required for product innovation planning. Here the enterprise potentials refer to the totality of all company capabilities, in answering requests for problem solutions and reacting quickly to new market requirements as well as to develop and apply new products and commercial success (cf. VDI 1983).

Internal utility potentials

Regarding enterprise potentials, one can also speak of enterprise competencies or core competencies (cf. Prahalad 1991). PÜMPIN refers to the *internal utility potentials* (Pümpin 1991). These can manifest themselves in the environment, in the market or in the existing enterprise constellations, through the activities the enterprise can use for the advantage of its reference groups or relevant stakeholders (Pümpin 1991).

Methods to derive innovation potentials

Within the context of innovation planning, especially those potentials that have a strong influence on the innovation activities of the business are of particular interest. Methods for the derivation of such innovation potentials out of (manufacturing) technological competencies and the enterprise-specific potentials have for instance been developed by PELZER (Pelzer 1999), EßMANN (Eßmann 1995) and KEHRMANN (Kehrmann 1972).

Handshake-Analysis

A possible way of integrating the technological and market-oriented enterprise potentials has been developed by TSCHIRKY with the *Handshake-Analysis* (Tschirky et al. 1996). It serves to illustrate systematically and on a strategic level the connections between needs, technology use and product-market-combinations. Additionally, a system with several matrices is set up, which systematically structures and defines the interactions between the connections. Each single matrix shows the relationships between two distinct views or indicators. Over defined junctions, relations on a higher level can also be represented. A pre-determined reading structure makes for easy reading and understanding of this matrix system, and guarantees the correct understanding of the documented connections.

3.3.3 Identification of strategic formation fields

Chances and necessity prescribe the choice of application area (formation fields)

Building on the results of the goal- and strategy development as well as on the derivation or ascertainment of enterprise potentials, in the next step the formation fields have to be developed. These serve to show the relevant view as well as the boundaries within which the specific planning activities will take place and serve as the general reference point. Within this context, AEBERHARD speaks of the task-specific environment, which includes the direct partners with which the enterprise interacts (Aeberhard 1996): customers, suppliers, competitors, investors, research partners or collaborators etc. The purpose is to select those formation fields where either the chance or the necessity exists that with product innovation, the required business goals can be achieved. They can be determined by means of integrative strategic analyses that combine different aspects of complex circumstances and show their relationships with each other. Frequently, methods such as a Gap or SWOT analysis, which have been practically proven successful, are used (cf. Webster 1989).

Gap analysis

In the Gap analysis, revenue over the time is applied or calculated, while 3 different areas are distinguished (Fig. 3.4).

1. Secured or guaranteed revenue with available products,
2. Planned revenue with existing and new products ("operational gap"),
3. Discrepancy between the revenue specified (goal value) and the revenue plan value with old products and new products ("strategic gap").

According to the size and occurrence in time of the strategic gap, corresponding measures such as product innovation activities need to be initiated.

SWOT analysis

The SWOT analysis aims at analyzing strengths and weaknesses of the enterprise as well as opportunities and risks of the business environment. From that, key problems and/or key tasks (Strategic Key-Issues) for the future are deduced.

Further methods that can be used here, such as "portfolio analysis", "life cycle analysis" and "scenario techniques" are described in chapter 4.

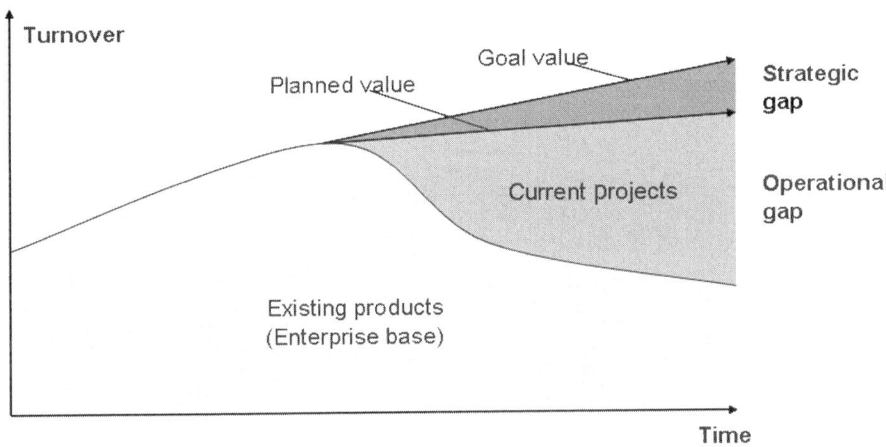

Fig. 3.4 Closing the strategic gap through product innovation

Defining objectives at Center Positioning Systems Ltd

… "What do you need to know Mr. Stone?", Dr. Palmer asked his chief of staff. Mr. Stone was well prepared for this discussion as the results of this first phase would influence all upcoming steps of the innovation planning process. As a basis for discussion Mr. Stone brought a file with relevant business data along which represented the current position of the company in numbers. He had brought the sales data, the current cash flow and financial statements as well as own analyses on the *market situation, data on the business' surrounding*, e.g. on the *competitive environment* and *assessments of experts* regarding the general situation of the branch.

Derivation of goals and strategies related to innovation

"Using the InnovationRoadMap method we can plan innovations which are genuinely new and fit our company as well" Mr. Stone started the discussion about strategy. "Therefore we have to adjust our innovations to our business' philosophy and strategy in the best way possible. If they are defined in an obvious way we can compare our own potential with the demands of the market. Doing so, it is possible to gain access to new markets without violating our business strategy. Therefore, I would like to derive the innovation strategy and objectives related to innovation today."

Taking up the innovation strategy

Dr. Palmer leaned forward: "Mr. Stone! We have to broaden our product range, we must identify new markets ... and then strike! That's how it works! We have always been the technology leader ... and we have to remain on top!"

Mr. Stone knew these strategy plans, his boss expressed them in preceding discussions. Therefore he could easily classify the visions of his chief. Under the headword *innovation strategy* he noted that they wanted to become technology leaders. Furthermore, a diversification of the product range had to be achieved to expand to new promising markets.

Identifying objectives related to innovation

"We have identified the strategy we are going to use to expand to new promising markets", Mr. Stone led to the next topic – *objectives related to innovation*. "Our financial statements indicate", he continued, "that preceding innovation projects only aimed at raising our income. How are we going to prioritize in this case?" This question led to an intensive discussion about general objectives of the company like ensuring growth, diversification of risks or the enhancement of the company's image. All of them were related to the new innovation planning. "Mr. Stone, I don't have to explain to you how severe our current situation is", Dr. Palmer said, "the tooling industry is not what it used to be! When did we introduce something truly genuine the last time? I want expert groups to speak about Center Positioning Systems Ltd again! We have to get back to the top!" Next to "traditional" goals like increasing the company's income, Mr. Stone noted further goals like image gain, access to new customers and securing the company's growth. Those guidelines were supposed to help him in setting up the new planning phase, the identification of business potentials.

Mr. Stone set up a meeting with all BU managers for the next Monday. According to the agenda it was about identifying *business potential related to innovation*. Excited about the next steps all BU managers arrived to the meeting on time. After a short introduction by Mr. Stone the discussion started. "Gentleman, think about the strength of your business units", Mr. Stone tried to motivate the managers. "You, Mr. Khan, are responsible for our high precision positioning systems, how good are you at that?" "So far we have been quite successful with our systems...", the BU manager replied slowly. "Exactly! Now that is a potential we can build upon!", Mr. Stone said. "Our sliding carriages are really good as well though!", Mr. Plumb, head of the "Sliding carriage systems" responded. Suddenly a vivid discussion started which resulted into many comments on business potentials. After the meeting Mr. Stone wrote down a list with important potentials relevant to innovation:

Discussions are used to regularly acquire information

- Center Positioning Systems Ltd has got a very good technology know-how in the fields of "precision positioning systems", "sliding carriage systems" and "sensory and control".
- The products "positioning tools" and "sliding carriage systems" produce bestsellers in various fields.
- The current product range can be used to realize the functions *lift*, *transport* and *drop*.
- An area-wide distribution network exists.

"Doesn't this strongly correlate with the "Pick and Place"-functions I have heard about lately, but wasn't that in the tooling industry ...", Mr. Stone was thinking while reviewing the company potentials, "... but somewhere in the medical sector ", he remembered. "Well, that's not really our branch", and thought he might go on to the next planning step.

Expert thinking creates mental barriers

"Ah, this is about finding new figuration areas for the identified business potentials", Mr. Stone summarized as he was reading the next step of the IRM method. "So, where do you need the functions lift, transport and drop?", he was thinking while looking at his computer. Suddenly he had the idea to just search for those ideas on the internet.

Identification of figuration areas

Use of the internet to acquire information

Mr. Stone was quite surprised when he realized how late it already was. The last few hours had been extremely interesting. He had so many ideas regarding lift, transport and drop functions. "But I have to write them down!", he thought, "before I don't know what it was all about tomorrow." "Pretty interesting", he thought finishing his internet research, "those

Documentation of ideas

pick and place functions are going to be needed in so many branches."

The next morning Mr. Stone called Dr. Kent, a fellow student who had studied at the RWTH Aachen as well and now lectures "Automation in the tooling industry". "Arthur!", he greeted his friend, "Tell me, don't you research those pick and place functions, they seem to have become quite popular lately! Could you tell me where those functions are currently used and where they might be used?"

The answer surprised Mr. Stone. His friend just told him that medicine technology is one of the most promising markets for pick and place functions. His internet research had come to the same results, but he didn't really believe in it before. His friend however was an expert, which convinced him. Now he could also estimate the potential of other promising markets which he could identify in his internet research.

Figuration areas

The next day Mr. Stone introduced the figuration areas for new product innovations to his staff: Medicine technology, micro systems and bio technology.

After the meeting with his innovation staff Mr. Stone started to prepare the next phases of the IRM method. He already had an idea how he could analyze the future prospects of the figuration areas. He remembered a student who had talked to him about a master thesis about trend research…

3.4 Future analysis

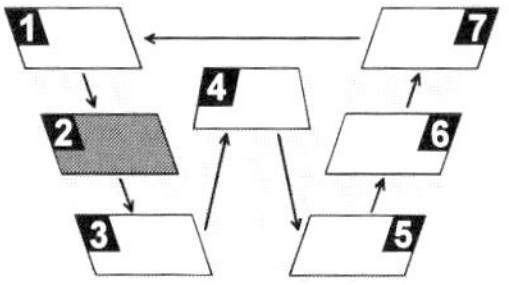

During the future analysis innovation potentials are deduced in a systematic way from trends and business potentials. The procedure enables the pro-active and timely impetus of product innovation activities. The formation area of the business is analyzed within a matrices system, and seminal innovation potentials that look promising insofar as success is concerned are identified. The derivation of the innovation potentials is done in three steps:

- Finding or defining future requirements,
- Analyzing chances for success,
- Defining the tasks required.

The results of this phase are selected innovation tasks that, with regard to their future and business coherence, their timely incorporation as well as the focus of the further (idea-finding) activities are described comprehensively.

The goal of the "future analysis" is to deduce innovation potentials within the defined formation field. Innovation potentials are problem areas where product innovation could lead to a definite market potential, and where relations to existing enterprise competencies either exist already or where they could lead to the development of new competencies. The future analysis encompasses three distinct steps (Fig. 3.5) (cf. Brandenburg 2002):

- In the *future requirements finding*, the formation field is analyzed regarding its future development. To do this, firstly the trends within the environment are looked at, starting very broadly and becoming more and more focused or detailed. Then the effects of the trend on the formation field are documented as "future projection" and are valued or prioritized according to their value or possible impact. (Finding future requirements)
- In the second step, the *analysis of chances*, based on the future projections, the innovation potentials within the formation field have to be identified. (Analyzing chances)
- During the *task definition* the identified innovation potentials are analyzed further in both an internal- and external-oriented manner, to quantify the market potential of the innovation idea as well as to form an idea on the impact on or contribution to company- (Task definition)

specific strengths and the strategic intend of the enterprise to transform the idea into a product.

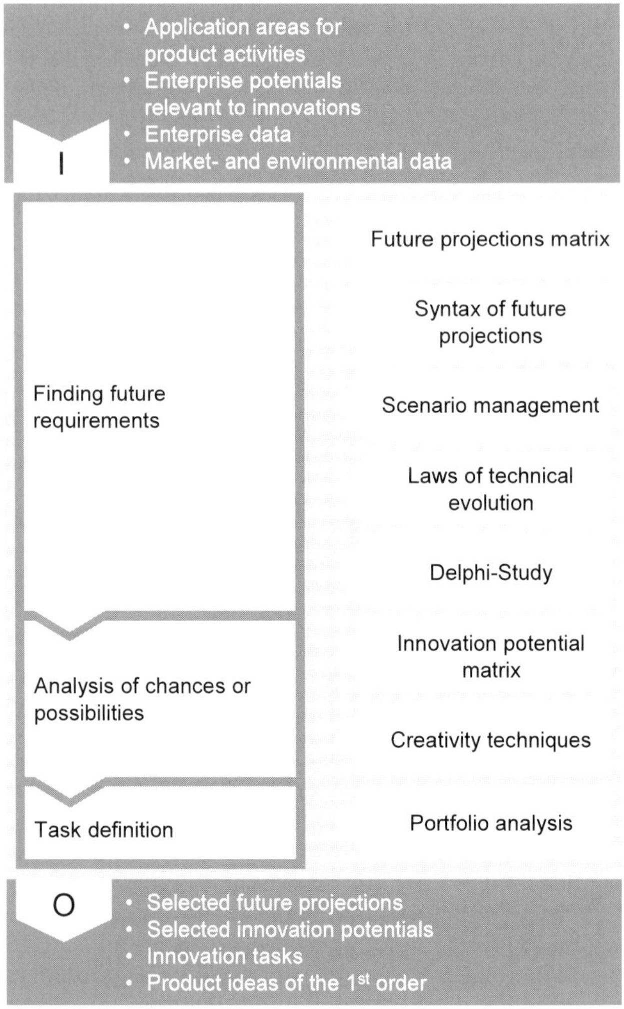

Fig. 3.5 Methods tool kit for "Future analysis" (cf. Brandenburg 2002)

For the execution of the future analysis, special techniques and tools were developed which, together with various specific methods, enable the efficient preparation or development of the planning task (Fig. 3.5). The itemized entrance information is required in the three planning steps within the respective methods used in order to generate the represented exit information.

Future projection-matrix

In order to be certain of a systematic approach and complete information processing, a specific recipe, the future projection matrix (ZP-matrix) (Fig. 3.6) was developed for the future analysis In the first matrix, the results of the future requirements findings are documented, while the second matrix processes the results of the chance analysis. The portfolio representation (below to the right) helps in the strategic value estimation. In the following section, the action steps of the future analysis are described.

3.4.1 Finding future requirements

Deriving future projections

Goal of the planning step "Finding future requirements" is the derivation of future projections. These projections are statements regarding future developments within the view area, which company-internally are regarded as true or highly likely. These are generated in a creative process out of the trends documented earlier. The goal is the population of the future projection matrix.

Trend scanning

The analysis and collection of trends is a continual process used in the early clarification. A trend can be described as the basic direction of either a development or a development bias. The "trend scanning" takes place in different observation areas, which together form the observation field. It represents the global environment of the formation field. In the ideal case, detailed information on developments in single observation areas already exists in the business so that these can be analyzed formation field-specifically. Basic trends as well as possible developments in various subject areas can be developed through for instance a Delphi-study (Delphi 1998).

Observation field of future analysis

Various models exist to subdivide the observation field. An overview of such a structural model is given by AEBERHARD (Aeberhard 1996). In the synthesis of the models, AEBERHARD structures the global environment in five observation areas (Fig. 3.7) (Aeberhard 1996).

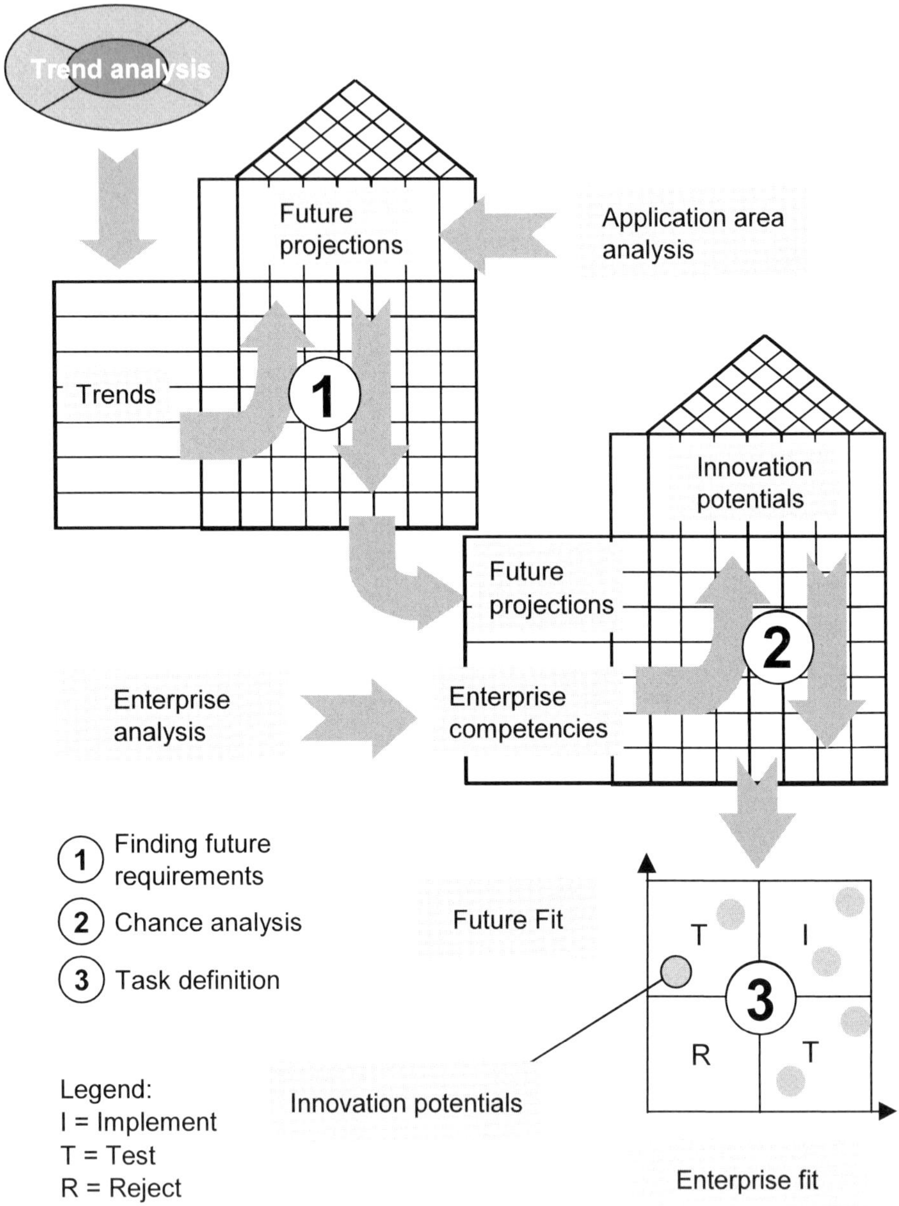

Fig. 3.6 The matrix system of the future analysis

1. Economic environment

The economic environment comprises those trends which relate to the national economic development and impact on the relevant markets of a business. Macroeconomic factors that need to be examined include for instance (Aeberhard 1996):

- Development and changes of the gross national product,
- Normal variations or cycles within the broader economy,
- Growth rates and productivity trends within specific industries,
- Income developments.

2. Socio-cultural environment

The analysis of the socio-cultural environment concerns an indirect approach to the detection of future customer wishes (Brandenburg 2002): Consumers orient or base their need on the existing social value system, which is constantly subject to change. Indicators for this change are the social trends that are typically identified and named by trend researchers (Horx 1996). The occupation of so-called trend researchers emerged in the eighties. The best-known representatives, for instance NAISBITT, POPCORN and HORX regularly publish their observations (e.g. Naisbitt 1991; Horx 1996; Popcorn 1995). The trends as identified by these researchers normally last for 10 years (Popcorn 1995). The trends resulting from changes in social values need to be distinguished from so-called "hypes"[10] und "fashions"[11], which last substantially shorter and therefore play a subordinated role in the derivation of product ideas.

---

[10] A hype is a short-lived trend phenomenon with a high effect width or broad base. Because global information accessibility is becoming simpler and easier, such a short-lived phenomena can reach and influence an enormous amount of persons. Examples of a hype are the Tamagotchi or seasonal colour trends (cf. Buck 1998).

[11] A fashion vis-à-vis a hype is a short-lived trend phenomenon with a narrow effect width or base. Fashions are society-specific and are competing with other fashions. Characterisic for fashions is that they are normally following a wave pattern and are recurring. Examples: flares, platform shoes or high heels (cf. Buck 1998).

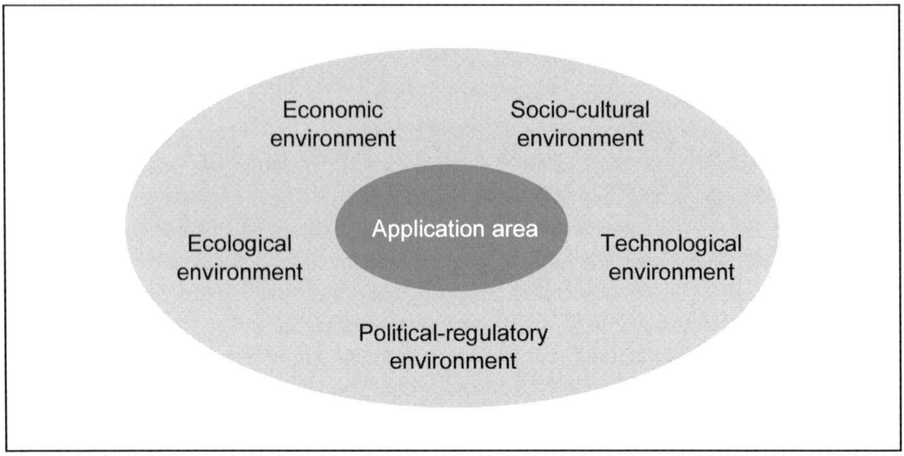

Fig. 3.7 Observation areas for the future analysis

The identified trends are explained through social phenomena and manners. In order to be able to use these trends in the derivation of future projections, it is vital to understand the explanatory patterns lying behind a trend.

3. Technological environment

The main task of the technology analysis is to identify early on any transition from an existing to a new technology so as to provide the company concerned with sufficient time to react (Aeberhard 1996). Techniques and/or tools facilitating the recognition of these transitions are the technology life cycle and the S-curve concept (one of the eight laws of technical evolution – chapter 4).

Because of the increasing sensitivity of the population for the indirect consequences of new technologies, especially with regard to ecological and social impact, the estimation or quantification of possible technology consequences becomes more and more important (cf. Servatius 1992; Wolfrum 1994). The goal of the technology impact estimation is the systematic development and analysis of all secondary and tertiary effects that could be evoked by development, use and distribution of new technical problem solutions. The impact estimations should make it possible to anticipate, evaluate and assess potential external effects, which result from the use of new technologies for general or specific social groups or areas, to avoid potentially negative effects on the firm's image or sales numbers (Wolfrum 1994). Atomic energy, asbestos-based building substances or gene technology are examples of a belated technology impact estimation.

4. Political-legal environment

The political-legal environment includes developments that are purported by the activities of the country as well as of entities with legislation authority. Such activities are primarily the national legislations for specific countries. In the course of the globalization of the markets, international legislation, e.g. EU-guidelines, are becoming more important (Aeberhard 1996; Hill 1989). Coupled to this are international and/or country-specific product test requirements, e.g. prescribed crash-tests in the automobile environment. Should new legislation or requirements be developed in these areas, they have to be recognized early and, if required or appropriate, incorporated and integrated into the product planning.

5. Ecological environment

The increasing discussion of ecological demands, like for instance the reduction of the consumption of non-renewable resources and the manufacture of environmentally friendly products, makes the consideration of ecological concerns more and more a prerequisite for strategic success. This also includes or overlaps with aspects of the technology impact estimation that was clarified already in the description of the technological environment above.

Trend scanning

If the trend-scanning in a business is established, a list of trends from the various observation areas exists, that is reviewed, updated permanently and/or analyzed regarding its relevance. In addition to the available trends, the observation areas can be examined formation field-specific. Obviously, a complete analysis of the collection of trends is required.

Examining the resemblance of trends

In the trend test, the examination of the resemblance of the trend(s) identified in the individual observation areas is important. During the resemblance analysis, the goal has to be to objectively rate the weighting of the trend while avoiding a latent overrating and/or undervaluation of a trend direction by the consideration of several similar trends. If similar trends are taken up in the ZP-matrix, this could lead to an overrating of future projections. Moreover, an unnecessary matrix enlargement is avoided with the grouping of similar trends (cf. Gausemeier 1996). The resemblance analysis furthermore promotes the possible description of the trend as unambiguously as possible.

Should a trend have been described rather inexact in the first description, a resemblance to other trends will quickly show up. As an example, the socio-cultural trends *fitness* and *wellness* can be taken. Both trends describe the increasing health consciousness of society. As such they are extremely similar and need to be summarized. In a more exact investigation, differences in the contents of the trend arise or become apparent: related to the type of the health consciousness and hygiene, the type of physical activity and the specific age groups in which the respective trend appears to a higher or lesser degree. In a concrete trend description, once these differences are grasped and understood, they are separately included as trends in the matrix.

Scenario management

In the scenario management according to GAUSEMEIER, the resemblance analysis is carried out based on exact calculations (cf. Gausemeier 1996). Here, however, an object-related examination appears to be sufficient (Brandenburg 2002).

After the trends are described sufficiently and the resemblance analysis has been performed, the trends are finally selected for the ZP-matrix.

Assessment and rating of trends

It is possible to weight the trends according to their meaning for the formation field. With the weighting factors, the relative impact or importance that each trend should have in the estimation of the future projection is decided on. Frequently these weighting factors are determined intuitively, however systematic methods can be used to support the decisions (e.g. pairwise comparison) (cf. annex).

During the trend investigation in the single observation areas, ideas arise frequently for new formation fields that were not considered within the situation/business analysis previously. These ideas are to be kept and brought into future planning cycles.

With this, all the trends are taken up and weighted in the ZP-matrix. In the next step, future projections are generated.

Formulation of future projections

Future projections describe the effects of the trends from the different observation areas on the formation field. For the formulation of the future projection, knowledge of the trend and its background as well as the connections to the formation field is necessary. In contrast to the trends, the future projections are related specifically on the formation room and therefore form the base for a purposeful and systematic ideas finding. The future projections are to be derived in interdisciplinary teams and need to be based on consensus as they represent or describe how the business estimates future developments.

Syntax for the formulation of future projections

For an unambiguous and exact description of future projections, the establishment of a syntax in which the future projections are formulated is required (Fig. 3.8). Next to a slogan or a headline, the future projection is specified through additional explanations and references.

The formulation of usable future projections

In the year 2010 voting for the Bundestag will be done via the internet

Trend in a specific environment	Indicator	Time horizon	Development	Certainty	Intensity
Which area is looked at? Which trend has led to the assumption?	Which indicator is impacted?	Which specific date or time period is involved?	How will the indicator develop?	How certain is this development?	How intensive is the development?
• Technology • Ecology • …	• Supplier reliability • Bit transfer rate • Data security	• 2005 • 2036 • In ten years • Next year	• Rise • Stay • Fall	• Definitely • Probably • Could • Can't be excluded • Must	• Slowly • Revolution-ary • Becomes more important • Explosive • Huge

Fig. 3.8 Syntax for the formulation of usable future projections (cf. Micic 2000)

Creativity techniques

Intuitive creativity techniques can support the process of the formulation of future visions - here brainstorming is probably the best known method. Independent from the technique(s) employed, a systematic approach along the observation fields should be done.

Method of „networked thinking"

To facilitate the generation of the future projection, the method of "networked thinking" can be used. This method aids in the detection of complex connections (Probst 1991) and thus supports the comprehensive investigation of the trend effects on the formation field. To each trend, at least one future projection is to be formulated that describes the effect of the trend on the formation field. Here no contradictory or mutually excluding statements are allowed, unlike for instance in the scenario-technique where it is not only allowable but intentional (Chap. 4.1). This difference to the scenario-technique results from the different objective. While in scenario-projects several scenarios are supposed to be prepared by means of the future projection, here the goal is to derive *innovation potentials* by means of the future projections, and this should lead to concrete product ideas. In addition unambiguous statements are necessary. Nevertheless, in investigating contradictions, it has to be considered that especially in the socio-cultural area contradictory trends can exist next to one another. This can result in future projections that are contrary, but not necessarily contradictory as they are developed for instance for different market segments or different age groups.

Formation field analysis

Parallel to the collection of the trends out of the observation areas, a formation field analysis is to be carried out in order to document or deduce possible further future projections that result directly from the formation field. These are for instance niche trends that (up to now) influenced only the target market. These are included directly in the future projection. The step "formulation of future projections" is terminated with the resemblance analysis of these future projections (see above)

Weighting and evaluation of future projections

After the formulation of the future projections, these need to be weighted in order to determine their importance within the formation field. The weighting of the future projection takes place in two separate areas. In the connection matrix (Fig. 3.9), the connection between the future projections and the trend is valued, and results in the so-called "trend-fit". The trend-fit shows how strongly or to what degree a trend from the future projection from the observation areas is carried forward into the formation field. In the correlation matrix the relationship between the future projections is analyzed, and a "formation field fit" calculated. This gives an indication of the conformity a specific future projection has in relation to all the other future projections. A high value indicates that the specific projection is in harmony with the others, and might even give an indication that a basic trend exists.

Trend fit

Formation field fit

Calculating strengths of relationships

The strength of the relationships can be one of three possibilities. A strong relationship is given a weight of 9, an average relationship gets 4, and a weak relationship counts 1. Should there be no relation between a trend and a future projection, that field in the relationship matrix is left empty. The trend-fit can then be calculated as the sum of the individual relationship points between the future projection and the individual trends. The relationship values are obtained from multiplication of the trend meaning and the relationship between trend and future projection. In the correlation matrix – the roof of the ZB matrix – the correlation can be either positive, negative, or neutral. Should two future projections support each other, they are positively correlated and get a value of 1. Should two projections contradict or even exclude each other they get a value of -1, i.e. they are negatively correlated. If there is no relation between them, the field stays empty. The sum of all these values gives an indication of the formation field fit.

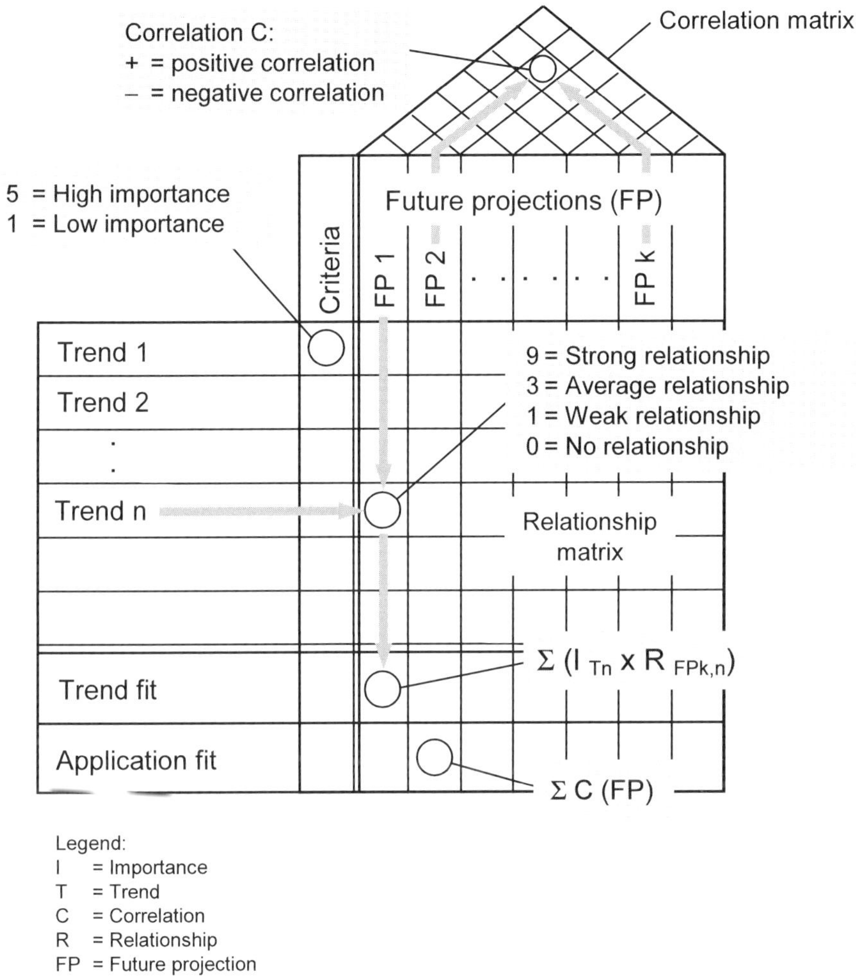

Fig. 3.9 The deduction of future projections during the future analysis

Formation-specific future vision

With the formulation of the future projections the user has drawn up a formation field-specific future vision. Based on this, the innovation potentials can be deduced systematically. From the company competencies further innovation potentials can be deduced. The steps required in the chance analysis are described in the following sections.

3.4.2 Chance analysis

Deducing Innovation potentials or making them visible

The purpose of the chance analysis step is to show or focus attention on the innovation potential (Fig. 3.10). The information on which this is based is given by the future projections from the planning phase "finding future requirements" and the definition of the company specific competencies in the "goal setting" step.

An innovation potential is brought to the fore through either a problem idea or a solution idea. A problem idea defines a future problem or a customer requirement, which comes about as a result of a future projection and for which there is not yet a complete or satisfactory solution. An idea solution on the other hand describes the principle of a new technological solution for which new or additional usage areas are sought. From this identification of problem or solution ideas one obtains a product idea which consists of a problem idea and at least one accompanying solution idea. The combination of problem and solution ideas is in principle already part of the solution finding process. The solution idea is actually not the most important aspect in the chance analysis, but rather the search for innovation potentials.

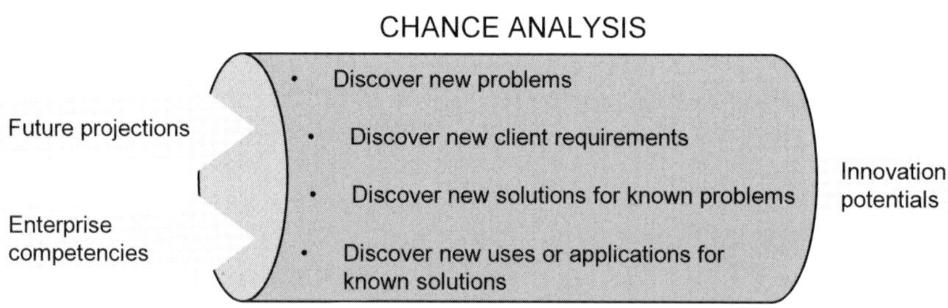

Fig. 3.10 Making innovation potentials visible during the chance analysis

Methods for analyzing future projections

In the process of analyzing future projections and company competencies, besides intuitive creativity techniques one can or should use methods which look at boundaries or constraints coupled to the future projections:

- *Thinking inside* or seeing it with the "eyes" of the target group or targeted user[12] is a method to show important unsolved problems of the projected user. In this method, the use of workshops is recommended, as these support a goal-oriented work environment. In these workshops questions such as "What does a typical day of our targeted user look like in future?" are answered in a structured and moderated way. It has to be taken into account that the targeted future user might look completely different from a present day user, even if he/she is of similar age.

Inside thinking

- *Customer Process Monitoring (CPM):* During CPM the process and environment where a customer uses the product is analyzed, so that improvements in for instance the detailed use of the product can be found (cf. Schröder 1998).
  Within the context of the chance analysis the planner can analyze the behavior of a specific client given the influence or impact of the defined future projections, so as to identify possible problems the client might have within the future situation to be created.

Customer Process Monitoring

- *Reframing*: Reframing means: "to put something into a new reference framework". This thinking process has its origin in the neuro-linguistic programming (NLP) and changes the frame of reference of a specific statement so as to convey a totally different meaning (O'Connor 2000). Within the context of its use here, reframing is uncoupling the future projections from the present framework and to put it into a future context. Given a different meaning, it becomes possible to deduce different or additional innovation potentials from these future projections.

Reframing

As an example for the deduction of a future projection and the generating of an innovation task one can look at the area of "living" (Fig. 3.11).

[12] The expression targeted user in this context is seen as synonimous for a representative potential target group or even a specific market segment.

Trend	Future projections	Innovation potential	Innovation task
The mobility of the population increases. An end of this trend is not in sight.	The mobile population will be adapting their lifestyle and buy equipment and furniture suited for frequent changes of homes.	Mobile kitchens and bathrooms Built-in kitchens and bathrooms present problems when moving. They are too expensive for short-term use. Present solutions are too inflexible to take along, both because of their fixed layout and the amount of labor required to re-install them.	Mobile kitchen systems Planning kitchen systems that take cognizance of the flexible and changing requirements

Fig. 3.11 From a trend to an innovation task

Before one can rank or grade the innovation potentials, their similarities to each other have to be examined. Because the innovation potentials are deduced from two different sources, (future projections, enterprise competencies), overlaps can exist which will only be shown or can only be seen once a detailed analysis of possible similarities is done.

3.4.3 Task definition

Rating of correlation between innovation potential and future projections through future and enterprise fit

The rating or comparison of innovation potentials is done in the innovation potential matrix, much the same as the system or technique used for the future projections. The weighting of future projections has to take cognizance of the rating given to the future requirements. The enterprise-specific competencies can be weighted either intuitively or systematically. Rating the interrelationships between innovation potentials and enterprise competencies leads to the values of the "future fit" and the "enterprise fit" as shown in Fig. 3.12.

Future and enterprise fit

The future fit gives an indication of the potential seen from the external environment. The enterprise fit on the other hand shows the relationship between innovation potential and enterprise competency. Here the ability of the enterprise to actually implement the innovation potentials is judged. In a so-called correlation matrix the dynamics and inter-relationships between the various innovation potentials is evaluated. From these, one can derive the synergies that exist between these potentials. The future fit and enterprise fit can be visually represented in a portfolio figure (cf. p. 54, Fig. 3.6, below right). In the figure, the future fit is shown along the vertical axis, with the horizontal axis representing the enterprise fit. From this, the strategic importance of the innovation potentials is rated or deduced.

Deduction of action proposals

Depending on the placement of the innovation potentials in the portfolio, the following actions can be taken or can be proposed:

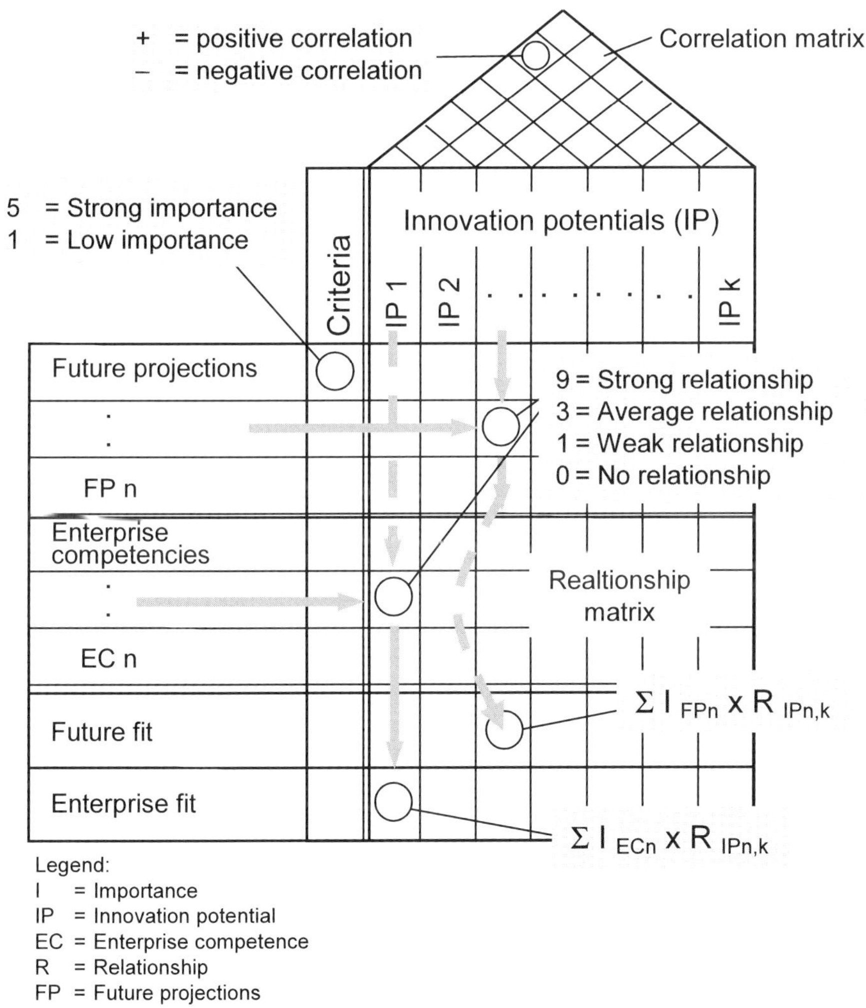

Fig. 3.12 Innovation potential matrix in the chance analysis

Put on ice/Discard

- The deduced innovation potential has a very low future fit, and very few, if any, enterprise competencies can be used. Further development or refinement of this innovation potential is therefore not recommended.

Evaluation

- The innovation potential has a high enterprise fit, but a very low future fit. Here it is important to see how the innovation potentials that have a high correlation to the specific potential have been evaluated or judged: is it possible to implement the idea with very little additional effort by making use of existing synergies? In general the innovation idea should be looked at again within the context of the list of criteria resulting from the earlier goal setting.

  The innovation potential has a high future fit, but existing enterprise competencies can only be used to a very limited extend. Therefore the following questions need to be answered: Can competitors use there internal competencies to a larger degree than we? Does implementation of the innovation potential lead to achieving strategic intend(s), for instance the opening up of new customer groups or market segments or even entering into a new business fields? Could the innovation potential be classified as a so-called "learning project" through which new enterprise competencies could be developed? Here too the alignment with the company strategy needs to be evaluated.

Implement

- This proposal for action is aimed at those innovation potentials that have a high future fit and where use can be made of existing enterprise competencies. If there is a high alignment between the innovation potential and the company strategy or goals, then they have a high potential to contribute towards the success of the enterprise and should therefore be implemented. If, however, there is no or very little alignment between the innovation potential and the company strategy or goals, the contradictions have to be critically reviewed, for instance one should look at the possibility that during the definition of strategic goals or directions future developments were ignored or not taken into sufficient consideration.

Formulation of innovation tasks

As a last step in the future analysis the innovation potentials selected through the strategic rating have to be translated into specific innovation tasks. This task defines the requirement(s) for the idea generation step. Depending on whether the innovation potential includes a so-called product idea of the 1st order (i.e. a product- or solution idea), it becomes the task of the planning phase "idea generation" to find solution ideas for a problem idea, or to couple problem ideas to specific solution ideas. Formulating this as a question, the task becomes (cf. Trux et al. 1985):

- Here is the problem, where is a solution? or:
- Here is the solution, where is the problem?

Summary

In the "finding future requirements" future projections are formulated for the various formation fields. These describe the envisaged developments of a specific formation or application fields from the point of view of the enterprise. Based on this, innovation potentials that could lead to future or new products are deduced. A matrix system was introduced that can facilitate the systematic deduction of these potentials by using and integrating relevant methods or techniques and visualizing the process. This systematic selection of innovation potentials is further supported through the rating of the future fit, the enterprise fit, and the strategy alignment. For the selected innovation potentials, detailed tasks are formulated, which have to be completed or solved in the idea generation phase. Through the early deduction and definition of innovation potential, a longer time period to translate these potentials into marketable new products is made available.

Analyzing the future at Center Positioning Systems Ltd.

... A couple of months ago Jennifer Smith had written an application to write her master thesis at Center Positioning Systems Ltd and was invited shortly afterwards by Mr. Stone. Talking to Mrs. Smith he was able to convince her to write her master thesis at his company.

Future projections

Global trends

Trend scanning

The student of economic engineering was heavily involved in searching future trends and tendencies afterwards. First she had started to analyze global trends to afterwards focus on a few promising areas. She divided her search into five areas – economic, socio-cultural, technological, political-legal and ecological environment. She clearly could identify trends and potential changes in the different categories. In a first trend scanning for example she analyzed potential currency fluctuations and the development of income taxes in the economical environment. The analysis took the specific demands of each category into account. The socio-cultural environment for example deals with developments in education, whereas education politics are part of the political-legal environment. Mr. Stone was particularly interested in data about the technological, economical and ecological environment as those included trends within product and production technologies (e.g. new CAD technologies, new production systems etc.), tax development and politics and ecological aspects (e.g. the use of resources, restrictions by laws etc.).

Market trends

Apart from analyzing global trends, the economy student also analyzed market trends. In her research she identified potential customer demands. The focus of the market trend research was the figuration areas medicine technology, micro systems and bio technology, which were proposed by Mr. Stone. The identified trends were presented as future scenarios by Mrs. Smith.

Today, four months after her first talk with Mr. Stone, she gave him her completed thesis which was titled "Technology and market trends for positioning systems". She wanted to thank Mr. Stone for the good cooperation and the successful completion of her thesis. Therefore they sat together drinking coffee and eating cake which she had baked just for this event.

Mr. Stone nearly knew the content of the thesis by heart as he had followed the progress of the work intensely over the last few months. Personally he was especially interested in the results of chapter five in which future trends for Center Positioning Systems Ltd were described as a case study. Once more he read the selected future projections:

Micro systems

- The sales of micro systems will grow approximately 30% every year during the next 10 years. Micro systems will not only affect "classic" branches related to engineering.

Medical technology

- Within medical technology (especially in minimal invasive surgery) most surgery methods will switch to remote controlled surgery. The surgeon will take the role of a guard who is supported by intelligent systems which automatically perform some surgeries and assist in others.

Biotechnology

- Biotechnology will revolutionize many areas. Especially in the DNA analysis promising results are expected. The fast growing number of analyses will be enabled by (automatic) functions, which are designed for a specific task. The technology which supports existing (analysis) functions already exists so that new systems will set standards on a short- or medium-term basis.

Scaling technologies

- Within scaling technologies the scaling of increasingly small tools will be enabled by using measuring apparatus which themselves will become smaller as well. In-process checks will become state-of-the-art. The

- Analysis of the innovation task,
- Collection of and generating product ideas of the 1st order,
- Ideas structuring und compressing
- Formulation of secondary product ideas.

Create space for initiative and creativity

The purpose of idea generation is the collection of product ideas of the 2nd order, which are the result of creative thought processes. As the ideas should ideally be developed without any restrictions, it is advisable to allow the employees especially in this phase almost unlimited freedom in as far as initiative and creativity are concerned (cf. Albers and Eggers 1991).

Methods for supporting creative thinking processes

The creative thought processes can be further supported through multi-disciplinary teams and the use of intuitive-creative or analytic-systematic methods[13].

A minimum of structure and concretization is required

It is required to have at least some structure and concrete or practical suggestions so that the ideas can be evaluated in the planning phase. To facilitate this, a number of tools and methods supporting the various planning steps are shown in Fig. 3.14.

3.5.1 Analysis of the innovation task

Preparation and analysis of required information

For the analysis of the innovation task all available information is made available, prepared and structured for the planning steps to follow. The task is further structured in such a way that a goal-oriented generation or search for ideas is not only possible, but that the ideas also complement each other and are condensed (cf. Brandenburg 2002).

Structuring the innovation tasks through a functional analysis

In structuring the innovation task a functional analysis has to be performed first. Here the task is looked at from a functional perspective, and is split or broken down into the different purpose-tools relationships. Only then the various potential solutions need to be analyzed with regard to their relation to the innovation task – this relation can be either a technological or a functional one. This would guarantee that existing solutions are not either imitated or "re-developed" for a second time (Brandenburg 2002). Testing or analyzing the

13 An overview of further creativity enhancing techniques which supports the choice of a particular and effective technique given the problem-specific situation is given in the appendix to this book (also cf. Friese 1975, Hauschildt 1996).

solution can be supported through the analysis of patents (Wagner and Thieler 1994) or a value-analysis (VDI 1995) of competitor products.

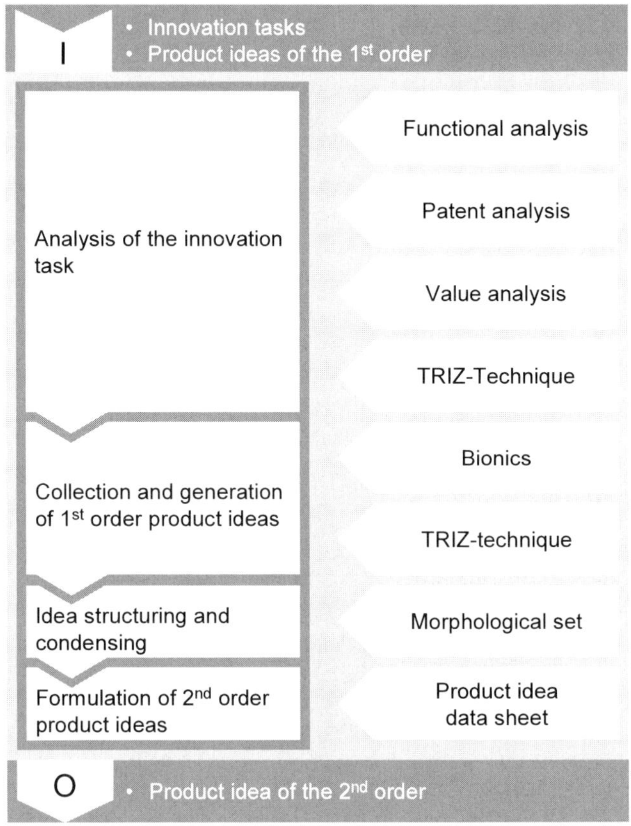

Fig. 3.14 Methods tool kit for "Idea generation" (cf. Brandenburg 2002)

Based on the analysis of existing competitive solutions, relevant expectations, requirements and goals need to be concretized, so that the innovation task can be classified.

Formulating the ideal product

Ultimately, the key functions are developed and the "ideal" product is formulated. This ideal product is a completely theoretical construct. Its specific characteristics are that it can fulfill the required key functions without using any resources or having negative consequences or by-effects. The definition of this ideal product leads to not only the scope and boundaries of the search area, but also provides some orientation for the deduction of higher order solutions. By doing this, existing mental barriers can be overcome (cf. Linde and Hill 1993).

3.5.2 Collection and generation of primary product ideas

The central activity in the planning phase "idea generation" is the collection and generation of product ideas of the 1st order (Brandenburg 2002). On the one hand, existing information is collected and further analyzed, while on the other hand new information is collected or made available.

Methods and techniques

A whole collection of tools and techniques regarding creativity enhancement exists in literature. Most of them are based on the principles of association, abstraction, combination, and variation. In using these principles, the area of bionics, where the principles of solutions used in nature are transferred to the given problem definition, is especially relevant. In the last number of years, methods allowing contradictory ideas and solutions are finding more and more support. Here one should mention for instance the *TRIZ-method* (the theory of the explorer problem solving, chapter 4.4), which combines a methodical approach of problem solving with various methods and supporting techniques (Altschuller 1998). These building blocks can be used later again in either the idea generation or the idea detailing steps (Brandenburg 2002).

Example: TRIZ-method

A complete description of the TRIZ-method as well as other approaches towards idea generation will be given later in chapter 4.

3.5.3 Structuring and condensing of ideas

Structuring and condensing of ideas

The problem and solution ideas presented as results at this stage are different insofar as their level of detail and concretization, their principal direction or basis (problem or solution idea), as well as their relationship with each other (competing or synergistic) is concerned. In order to be able to refine these into useful and applicable combinatorial solutions in the form of product ideas of the 2nd order, the various individual product and solution ideas have to be structured and condensed. To do this, the ideas are analyzed and modified according to the original innovation task (Brandenburg 2002).

Complete idea: morphological combination of problem- and solution idea

The actual combination of problem and solution ideas results from the morphological interaction of the functional structures and solution ideas. After breaking up the whole system into functional units or characteristics, possible solutions for all these are shown in a so-called morphological set. Every combination of a solution idea for a specific characteristic with the solution idea for a different characteristic provides an at least theoretically possible solution. If different required characteristics are now combined, various solution concepts are generated. The purpose of this methodical, formal and systematic approach is to generate out-of-the-box solution concepts that provide a wider solution field from which the most exiting can be chosen (Haberfellner et al. 1999).

Pool of ideas

In the planning step "idea generation" it is possible to develop ideas that do not have a direct relation with the specific innovation task. Even if they are only part-solutions (solving only part of the problem) these should not be discarded but rather be taken up and documented in a company-specific pool of ideas (Brandenburg 2002).

3.5.4 Formulation of secondary product ideas

Company-specific information model: a uniform level of detail for the ideas

The planning phase "idea generation" ends with the development and formulation of 2nd order product ideas. So as not to have to compare and evaluate ideas of different detail levels in the following phase (idea evaluation), the different parts of functions of acceptable ideas have to be detailed to a comparable level (Breiing 1997). To achieve this, a company specific and tailored information system showing the level of detail required should be developed and used (Brandenburg 2002). In the following paragraphs the development of such a system will be explained briefly.

Development of an information system

In order to generate transparent planning activities and fulfill the requirements for the following planning steps, a multitude of information must be collected and contextualized. In practice, however, it is difficult to obtain and aggregate all the relevant data and information, as more often than not the data cannot readily be quantified or concretized through for instance patents or detailed studies. The largest part of the information exists informally, in the way of expected values or employee know-how. Furthermore, the information carriers are both internal and external to the company, and the information content is often not what is required, either because relevant data is omitted or does not exist, or that instead irrelevant information is provided. The aspects discussed result in the fact that huge amounts of resources have to be spent in the research effort, or, even worse, that the foundation for important decisions is incomplete.

Information model

The information model proposed has to ascertain that the relevant product and planning information required is detailed. In using this information model, the user or planner will find various positive spin-off effects regarding the methodology. These include for instance (cf. Brandenburg 2002):

- A guarantee for an efficient and effective planning based on the available purpose and requirements of the planning information,
- All employees or people involved in the planning have access to the most actual and uniform information,
- The acquisition of data over functional borders as well as the sharing and evaluation of information is supported and encouraged,
- The idea provider has a "golden thread" aiding the identification of which information is still required[14],
- A knowledge base is developed and populated which provides systematic support for the following planning activities and phases.

Product idea data sheets

The purpose of this planning-oriented information model is to have all relevant product- and planning-related information within a single information platform or carrier. As typical carrier that has been used successfully mention should be made of so-called product idea data sheets (Fig. 3.15) (Brandenburg 2002).

Structure of the product idea data sheets

The data for the product idea data sheets are divided into four areas *organization*, *idea description*, *technology* and *market*. These areas obviously interact with the relevant planning phases in that they both require and provide information (Brandenburg 2002). The example below shows possible information content for the four areas (Fig. 3.15). Neither the structure nor the information content can necessarily be absolutely complete; they are shown to initiate a company-specific design of product idea data sheets. A further example of a completed product idea data sheet for the fictitious case example used in text is shown in the appendix.

[14] An overview of techniques used in practice is given by, amongst others, HABERFELLNER (Haberfellner et al. 1999).

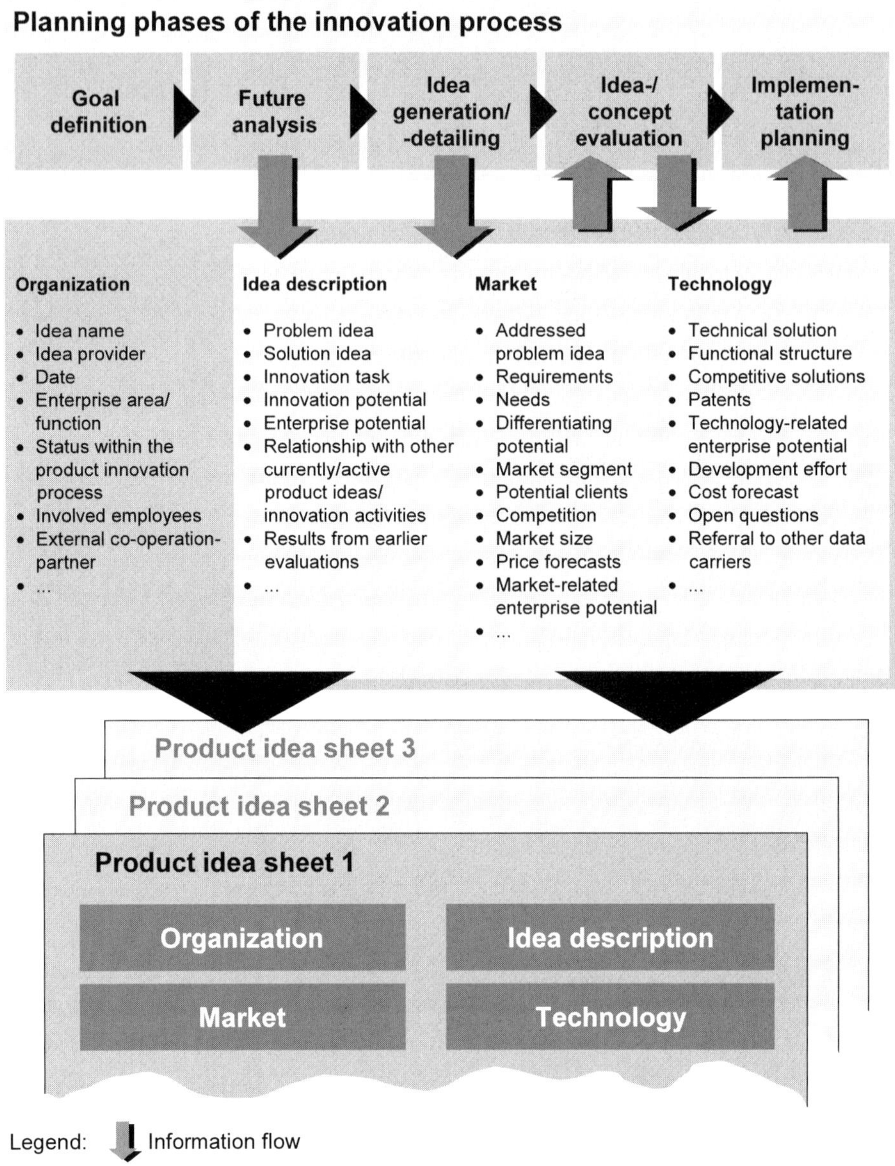

Fig. 3.15 Information structure of the product idea data sheet

Result of "idea generation": selected 2nd order product ideas

The results of the planning phase "idea generation" are structured product ideas of the 2nd order. They have a uniform level of detail and concretization with regard to the planning and evaluation requirements (Brandenburg 2002). Next to the positive effects described above, an additional advantage lies in the fact that at the later selection of promising ideas these have a similar formal level. The final decision about which idea should be developed further is however independent of the way and manner in which the idea is described.

Generating ideas at Center Positioning Systems Ltd

... "Good bye, Professor Lambert!", Mr. Stone bid farewell to one of the experts on endoscope diagnostics. During an interesting discussion he had just now learned about many details on the usage and possibilities of endoscopes to draw a comprehensive picture about methods for diagnosis and treatment. He had heard for example that there is a huge research potential for minimal invasive medicine, i.e. for tools which only harms human tissues to a minimal degree. The manifold usage of endoscopes for diagnostics and surgeries results in a promising future for those tools.

Primary market research

The main problem for today's endoscope methods is founded in the restricted focusing area of cameras which can only be broadened by extensive and pain-causing rotations. A reasonable innovation in that field would be a system which could rotate a camera at the head of an endoscope without moving the endoscope in whole. The discussion also pointed out alternatives to a camera itself; a gripper used in minor surgery for example. A suitable positioning system for such a gripper had to be developed though. Mr. Stone and Mr. Lambert had already talked about fully automated surgeries. Thinking about those ideas Mr. Stone had to smile. He yet had to get used to the thought of an automated endoscope which could support "operate by wire"-surgery as Mr. Stone called it. After all, he worked in the field of positioning systems for the tooling industry until now. "We'll see if automated endoscopes will soon be part of the Center Positioning product range", he thought while walking back to his car to drive home to the Black Forest.

Collecting 1st order product ideas

Back in his office Mr. Stone examined some reports of his staff which were taken from interviews with specialists in various fields. He read about simple improvements but also about truly genuine innovation ideas. Each interview started with a rough product idea which was concretized during the discussion.

TRIZ-method: "the perfect product"

To perform the interviews Mr. Stone and his staff had thought about different methods to define problems. The method to define the perfect product, one of the TRIZ-tools, was outstanding in these discussions and Mr. Stone was especially informed about that method. This method to define problems proved itself useful in the discussion with Professor Lambert. The "automated endoscope" was one of the most promising ideas among many others.

Patent analysis

To ensure the potential implementation of ideas at an early stage, Mr. Stone commissioned a patent analysis. Therefore he passed the 1st order product ideas on to one of his employees and asked him to check those ideas for pending patents.

Structuring and condensing ideas

The next day Mr. Stone had another meeting, this time with experts from his own company. Aim of the meeting was to find solution ideas to the 1st order product ideas. Therefore Mr. Stone came up with a specific method.

Creativity techniques: Brain writing

He wrote one product idea on a transparency and presented it. After that every expert was asked to write down a solution idea and pass the sheet on to his neighbor. This way everyone could refer to the ideas of his predecessors and concretize or enhance them if necessary. This procedure was repeated for every product idea resulting in at least one solution for every problem. Using a morphologic combination the problem ideas were connected to alternative solution ideas resulting in 2nd order product ideas.

2nd order product ideas

After reviewing the meeting's protocol afterwards he noted the following 2nd order product ideas:

- Positioning system for prosthesis: A positioning system controls a mechanic which enables the movement of a prosthesis (component), e.g. a finger.
- Positioning systems for endoscope diagnostics: A positioning system within the head of an endoscope positions a camera.
- Positioning systems for DNA-analyses: Several pipettes are put onto a crab. The absorbing or delivering of probes is realized by facilitating the capillary action of the pipettes. The transport of pipettes is done by using a precise positioning system.
- Dynamic positioning in the espionage industry: A mocking bird is equipped with an artificial wing beat. A camera transmits pictures to a ground station.
- Positioning systems for the micro perforation of coffee filters: The perforation tool is moved step by step by using a positioning system.
- Positioning systems for micro assembly robots: A positioning system should be facilitated for a fast and automated assembly of semiconductor elements.
- Positioning systems for micro systems: Production and assembly systems have to be supplied to produce small

component structures. Positioning systems supply the necessary realization of micro traverse paths.

- ...

"Some of the ideas are quite smartish", Mr. Stone thought while reviewing the product ideas. "I'm curious what comes out of the valuing of ideas...", he thought and checked the "Generating ideas" headword on his project plan which was titled with IRM method. Just now he had finished the generating ideas step...

3.6
Idea evaluation

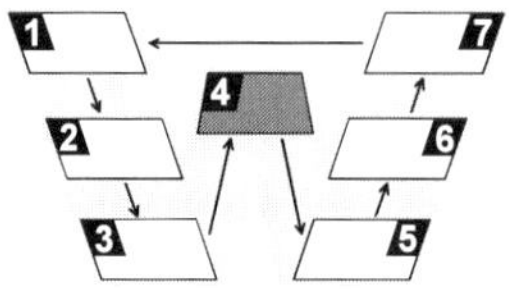

Idea evaluation is used to identify promising product ideas. Innovational principles of products have to be assessed according to their company, market and technology potential. Uncertain information and unspecified schedules characterize the assessment in this phase. The rating takes place in two steps:

- Set-up of a rating and evaluation system
- Evaluation of product ideas and their according solution principles

Information collected in this phase is used for the classification of product ideas within the InnovationRoadMap.

Needed information input

The basis for evaluating new product ideas is laid in the preceding phases of the W-Model. The innovational strategy, as formulated within the goal setting and adjusted in the future analysis, is the input for evaluating the product ideas that were found in the idea generation chapter. Each idea already has at least one possible solution.

Evaluation criteria

The yet only roughly formulated product ideas have to be evaluated in terms of a systematic innovational planning with regard to their *company profit*, *technology potential* and *future prospects*. Goal is a ranking of product ideas to purposeful spread limited resources to the further work on ideas.

Classification within the InnovationRoadMap

The result of this phase is a classification of product ideas within the InnovationRoadMap (IRM) in terms of content and time. The IRM itself will present all necessary actions to implement a product idea on a time axis. The idea evaluation takes place in two steps, the set-up of a rating and evaluation system and the evaluation of product ideas and solution principles itself. (Brandenburg 2002).

Methods and Instruments

There are various methods and assets that can be used to evaluate ideas. Their allocation to the chapters can be seen in the chart below (Fig. 3.16).

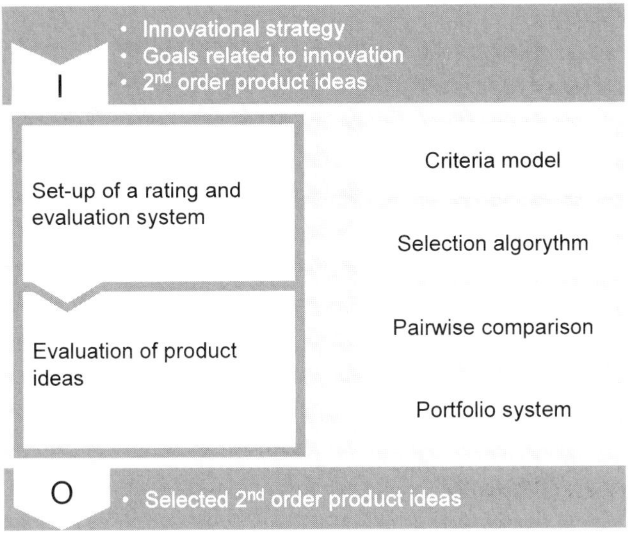

Fig. 3.16 Methods tool kit "Idea evaluation" (Brandenburg 2002)

3.6.1 Set-up of a rating and evaluation system

Rating and evaluation system

The rating and evaluation system consists of several company-specific and -unspecific criteria, which are used to rate and evaluate product ideas[15]. A selection algorithm is used to divide more important from less important criteria, which are rated in a pairwise comparison with regard to the objectives and the overall business strategy. By using checklists, quantitative criteria can be cardinally acquired and used for evaluation. The evaluation results are visualized within two portfolio representations, from which actions for the further improvement of product ideas can be derived.

The uncertain information which is accessible during the early phase of product innovation planning is problematic for evaluation though. To nonetheless utilize this information for the evaluation of product ideas, one has to use evaluation methods beyond traditional economic efficiency calculations (Staudt et al. 1991).

[15] Here the connection of a criteria's characteristics to attributes is understood as *rating*. Judging if instruments can be used to achieve a certain goal is described as evaluating. (Brandenburg 2002; cf. Lenk 1994; Brose 1982).

The evaluation situation, which confines or even prevents a purposeful and boundless usage of known methods, can be characterized as follows (Brandenburg 2002):

- The information security related to product ideas is not high enough.
- The gained insights for the further planning process are not satisfying.
- The practicability is not given because of the high count and complexity of evaluation objects.

Information can be characterized by four essential types of uncertainty (Brandenburg 2002; cf. Kern 1977):

General uncertainty of results

- The *general uncertainty of results* deals with the question if required information can be found without paying attention to cost and time aspects.

Time uncertainty

- The *time uncertainty* describes if information is available or can be obtained at a certain time.

Effort uncertainty

- The *effort uncertainty* describes under which effort information can be achieved at a certain time.

Utilization uncertainty

- The *utilization uncertainty* deals with the uncertainty of the economic utility of innovational management results.

Problems through uncertain information

Nonetheless, to satisfy the holistic and strategic approach to innovation management, especially the different degrees of uncertain information and the time horizon have to be taken into account. This proves to be an enormous challenge to product innovation in early development stages: the time horizon is wide, the uncertainty about future developments is high and the information about possible advantages of product ideas is low (Brandenburg 2002). Furthermore the information to judge product innovations, with regards to cost, effectiveness and further criteria, can only be estimated. Therefore evaluating different strategies by comparing alternatives using quantitative rules is insufficient. Consequently the degree of uncertain information has to be taken into account when evaluating ideas, as well as using appropriate methods to deal with the tensions between uncertainty and time horizon. To deal with these problems, an evaluation systematics was developed at the Fraunhofer Institute for Production Technology (Fraunhofer-Institut für Produktionstechnolgie - IPT), which copes with the demands of this situation (Brandenburg 2002). This evaluation method will be described later on in this chapter.

The Innovation-RoadMap is the result of the innovation planning process

According to the W-Model the methodological goal of the innovation planning process is the InnovationRoadMap, which represents its results. Therefore the results of evaluating the ideas must lead to that roadmap, taking the already explained insecurity issues into account. It should represent the essential product ideas and the innovational activities connected to them. Resulting from the process of evaluation there are several description parameters (Brandenburg 2002):

IRM-Description parameters

- *Maturity:* The maturity of defined activities with the characteristics "short-, medium-, or long-term" has to be brand- or company-specific. A reasonable allocation would be for example < 1 year/ 1-3 year(s)/ 3-5 years.
- *Priority:* The priority of a product innovation activity can be defined, according to the portfolio strategy, as "implement", "check", "consider" and "reject".
- *Kind of activity:* The kind of activity refers to the next planning phase; within a phase you furthermore can differentiate between different activities like market or technological activities.
- *R&D-application:* The application of R&D for a product idea can differ between self-dependent R&D, R&D cooperation and no own R&D (e.g. acquisition or waiting)
- *Key argument(s):* Key arguments for the textual and time-based classification could be a law coming into effect, the expiry of a patent or the rollout of a competitive product.

Portfolio evaluation

The multitude of possible target figures and factors that influence the product innovation planning, respectively the evaluation of product ideas in a company, speaks in favor of a systematic approach in evaluation using an appropriate criteria model. The problem in choosing criteria is to find an optimum between the extremes of "pragmatic regulation for a single case" and "too high generalization using a systematic approach" (Pfeiffer 1995). In principle this optimum can be achieved by using a combination of a textual and methodological definition. The latter arises from the portfolio concept that demands an influenceable and a not-influenceable main parameter, whereas every parameter can be further defined by appropriate criteria (Pfeiffer 1995).

3.6.2 Evaluation of product ideas

Information input

The information necessary for evaluation comes from the preceding phases of goal setting, future analysis and idea generation, i.e., they are the product ideas that come at least with one solution idea that fits best to the objectives set within the "Defining objectives" chapter. The chosen information differs greatly with regard to expense and return impact. The product idea for example can be the substitution of an already existing product. In this case the expenses and the expected return can be estimated quite easily. But the product idea might also be about accessing completely new markets or market segments. In these cases the expenses and the return can only be roughly estimated (Brandenburg 2002).

Evaluation criteria

The criteria for product evaluation basically come from the areas of company benefit, future prospects and technology potential (Brandenburg 2002).

Portfolio representation

By evaluating ideas you create a criteria model which enables the idea evaluation. The portfolio technique enables the appropriate representation of product ideas as values (Fig. 3.17).

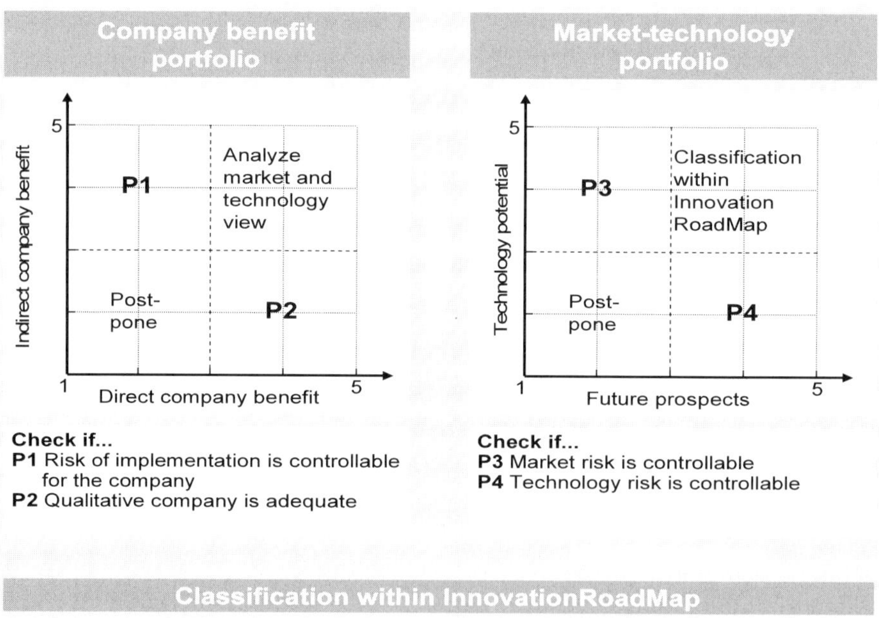

Fig. 3.17 Portfolio system to derive recommendations for actions (Brandenburg 2002)

Those can be represented numerically as well as linguistically differentiated, for example with terms like "light", "medium" or "high". From the portfolio representation there can be norm strategies derived, i.e. recommendations for actions. The procedure mentioned above enables recommendations for actions to be put into the InnovationRoadMap. Thereto the following procedure should be utilized.

Company portfolio

First a *company portfolio has to be set up* in which the direct and indirect benefit for the company is opposed and evaluated. The classification of a product idea follows a criteria system. The system's criteria refine the respective axis of the portfolio (Fig. 3.18).

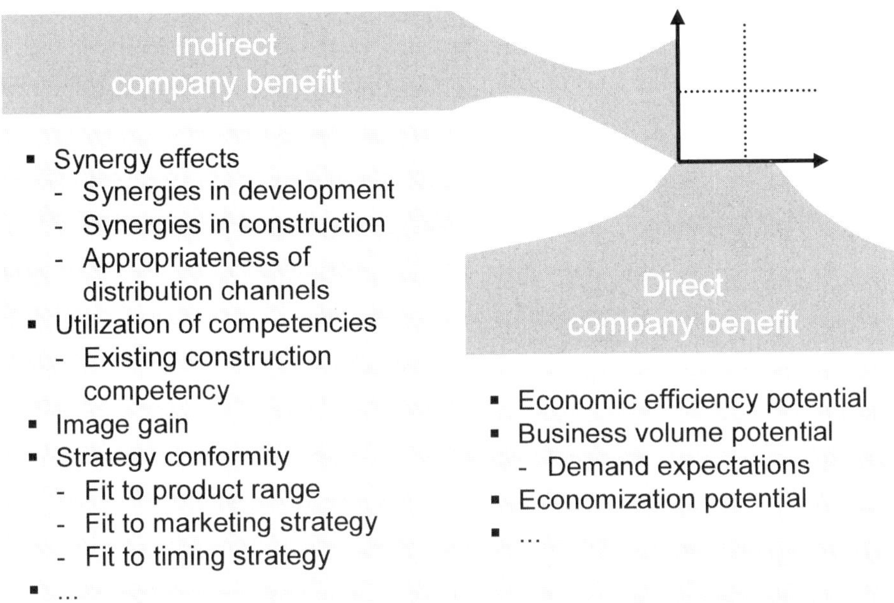

Fig. 3.18 Criteria of the company benefit portfolio

Evaluation criteria: Direct company benefit

The company benefit enfolds the *direct advantages* or *company benefit* understood as the achievement of objectives regarding business volume, benefit, economization, etc. – basically the monetary goals as defined in the "Defining objectives". This implies a basic concept of the desired sales to estimate for example the business volume. Although the impact of economical factors on the product innovation can only be roughly estimated in this early phase of the planning process, it nonetheless has to be taken into account.

Evaluation criteria: Indirect Company benefit

The other company benefit criterion to be evaluated is the *indirect company benefit* which is made up by (Brandenburg 2002):

- Synergy effects in all company branches (purchasing, development, production, sales),
- Utilization and extension of competencies,
- Positive image gain for the company and
- Strategy conformity.

The indirect company benefit is also influenced by uncertainties during the early phases of product innovation planning.

Important criteria and classification within the portfolio

A meaningful evaluation and representation of product ideas within the portfolio requires the analysis of evaluation criteria with regard to their conformity towards company goals and strategies. The method of pairwise comparison is used for the analysis. Thereby the criteria are pairwise opposed and relatively weighted against each other. If one criterion is more important for the company's goals or strategies it gets a higher weighting than the other. This results in the relative meaning of criteria with regard to the company's goals and strategies.

After the meaning of criteria is set, a value benefit analysis is required which determines the utility of a product idea. Within this analysis the product ideas are opposed to the criteria and each criterion is provided with a degree of implementation. By multiplying the weighting of the criteria and with each criterion's degree of implementation one gets the utility for each criterion. By then summing up all single utilities one gets the total value – the utility of a product idea. The pairwise comparison and the value benefit analysis are made for the direct and indirect company benefit. The described procedure results in numbers that allow the classification of a product idea within the portfolio.

Market-technology portfolio

Analogue to the company benefit portfolio a market-technology portfolio is set up which opposes and evaluates the main criteria technology potential and future prospects. The axes of the portfolio are specified by criteria so that you can classify product ideas within the portfolio (Fig. 3.19).

Evaluation criterion: Future prospects

The main criterion *future prospects* corresponds directly with the results gained in the future analysis chapter. Essential aspects in this case are the customer benefit, i.e. the extent to which product innovations satisfy future demands of a customer, the potential of diversification, which describes the level of diversification to existing or future alternative solutions, and the security to substitutions, the "resistance" against copies or substitutions which results from market impediments like patents, the attachment to specific systems and the like (Brandenburg 2002).

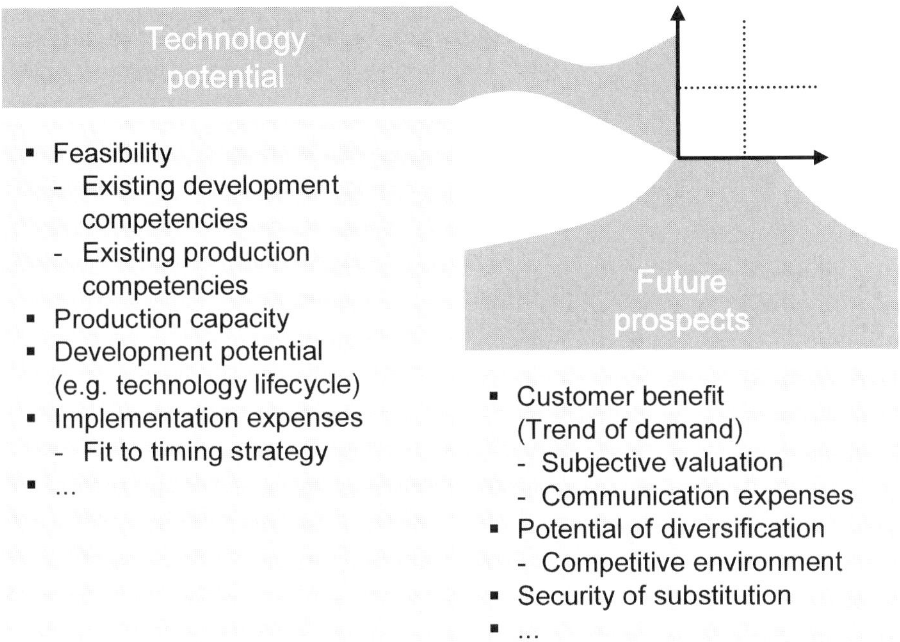

Fig. 3.19 Criteria of the market-technology portfolio

Evaluation criterion: Technology potential

The main criterion *technology potential* includes basically all feasibility studies made within the chapters "idea generation" and "idea detailing", which are determined by criteria like technology control (on both development and production side), reproducibility of solutions, availability of resources (raw materials), environmental impact and others. Furthermore the development potential of a product idea has to be evaluated with regard to technological aspects of the future analysis (e.g. concerning its position in the technology lifecycle and its development dynamic). Finally, analogue to analyzing the utility, a rough prediction of the implementation expenses has to be done. For this the necessary R&D expenses, investments and other initial expenses have to be evaluated based on the company potential analysis (Brandenburg 2002).

The criteria lists presented in figure 3.19 should be seen as a basis for a company-specific criteria system. They are meant to be open lists to which individual criteria can be added.

Classification within the portfolio and derivation of action strategies

After all criteria are weighed and evaluated, every product alternative can be classified within the portfolio. Then a strategy can be derived for every position in the portfolio, which leads to different classifications of the product idea within the InnovationRoadMap (Brandenburg 2002).

Idea pool: Make all ideas available

As the industrial environment is constantly changing, ideas which are not promising today might become valuable innovations within a couple of years. By saving all ideas within an idea pool they can easily be taken into account again. The InnovationRoadMap enables the saving of ideas by presenting possible product ideas in correspondence to future markets and time (Chap. 3.9). An instrument that supports the systematic and concise collection of technology-specific ideas is the technology calendar (Chap. 4.9).

Summary

By evaluating ideas, promising product ideas are rated and chosen. Therefore first a rating and evaluation model has to be constructed which deals with the overall uncertainty of information. The evaluation of problem and solution ideas itself is done by facilitating a company-specific criteria system. The evaluated ideas will then be presented within a portfolio. By classifying ideas within portfolios, actions can be derived. During the next phase of product innovation planning the evaluated ideas need to be refined to product concepts.

Evaluating ideas at Center Positioning Systems GmbH

... "Good thing that I wrote down all information in detail", Mr. Stone thought while reading in the IRM-method that he would need information from both the goal setting and the future analysis chapter for the evaluation of ideas. "After all that's crucial to evaluate product ideas", he explained to himself the reason why he would need information from the first two planning phases. "Interesting idea", Mr. Stone smirked, "when I think about that we talked about the importance of goals before we even had an innovational strategy which defines goals related to innovation." Remembering earlier innovational projects Mr. Stone knew that decisions are often made on interpretations of personal experiences and subjective knowledge. Using the IRM-method decisions had become much more objective. The guidelines defined during goal setting would simplify the work of the evaluation team and lead to a more objective evaluation of ideas.

Create awareness for the benefits of a systematic innovation planning

Set up an evaluation and rating system

As preparative to evaluate product ideas Mr. Stone began to set up a company-specific evaluation and rating system. On a large sheet of paper he noted the four main criteria *indirect company benefit, direct company benefit, technology potential* and *future* prospects.

Pairwise comparison

After that he used the pairwise comparison to analyze the sub-criteria of each main criterion. "We want to be the leader in technology and develop new fast growing markets", Mr. Stone thought, "so the criteria *strategy conformity* should be more important than *synergy effects in production*", he continued the train of thoughts and marked a "1" on the according place in the pairwise comparison. After a while he had identified and weighted the criteria "strategy conformity", "use of competencies", "image gain" and "synergy effects" in the "indirect company benefit" category with descending importance.

Weighting criteria

After that he did the same thing for the other three main criteria. For the category "direct company benefit" he identified "company potential" as most important. Within the category "technology potential" he made a ranking for the criteria "existing development and production competencies" and "development potential". Finally he identified the descending order "diversification potentials", "substitution security" and "customer benefit" in the "future prospects" area.

Check the strategy conformity of criteria weighting

Relying on this exploratory work Mr. Stone went over to Dr. Palmer. Together they wanted to discuss the weighting of the criteria and change them if necessary. Mr. Stone had done a good job though, so only minor adjustments were necessary. In fact Dr. Palmer only wanted to check whether the weighting of the criteria was according to the company goals, strategy and philosophy or not. Being the leader of the company he felt especially obliged to the strategic elements of the IRM-method.

Evaluate product ideas

For the next step Mr. Stone was planning to start with the evaluation of ideas itself. Therefore he invited all BU managers again to discuss and to weight the criteria with them.

Create common understanding

The weighting was quite easy for some product ideas; long discussions were necessary though sometimes. Mr. Stone realized that some managers interpreted the importance of some criteria differently. In that case he stepped into the discussion and explained the background and the importance of the criteria again. He also noted for himself that next time he would explain the criteria first before inviting the managers to a discussion.

Portfolio system

Already during the discussion Mr. Stone had thought about the results of each evaluation with regard to the four main criteria and sketched the portfolios as suggested by the IRM-method. (Fig. 3.20).

Chosen 2nd order product ideas

Looking at his sketches Mr. Stone could see that some product ideas that seemed to have an extraordinary potential were already highlighted whereas others did not seem to make sense. "Probably there won't be many changes", he thought while looking at his sketches and notes. He could derive the following 2nd order product ideas from the evaluation:

- Positioning systems for DNA-sequencing,
- Positioning systems for prostheses applications,
- Positioning systems for endoscopic diagnosis,
- Positioning systems for the assembling of micro components,
- Positioning systems for slide systems of micro tool machines.

The classification of ideas within the portfolios proposed different plans for action.

Activity option: Classification within the InnovationRoadMap

The product ideas "positioning systems for endoscopic diagnosis" and "positioning systems for DNA-sequencing" could be classified within the InnovationRoadMap. Therefore Mr. Stone consulted his notes about the future analysis to see when the market demand in medical and biological technology would be sufficiently high. From today's discussion he could derive a rough timetable for the realization of according positioning systems and write it into the InnovationRoadMap. "The technical feasibility will definitely get more concrete during the concept evaluation", Mr. Stone thought and put the first sketch of the InnovationRoadMap back to his files.

"Concerning the positioning systems to be used in prostheses we first have to check if the implementation risk wouldn't be too high for our company", Mr. Stone summarized the recommendation for actions regarding this product idea. "I could have a talk with Professor Lambert tomorrow and see about his opinion", he thought and put a yellow note-it next to his screen.

He swiftly browsed the measures that resulted from the classification of ideas into the portfolio to then archive that information within his IRM files as well.

Anticipatory action and planning

In the meanwhile it had become evening and Mr. Stone took the IRM guide once again from his shelves and read it. "Now we have to refine the product ideas to concepts and evaluate again", Mr. Stone visualized the further actions for himself. "We'll see what comes out of the 'idea detailing' chapter… ", he thought and turned the lights out ...

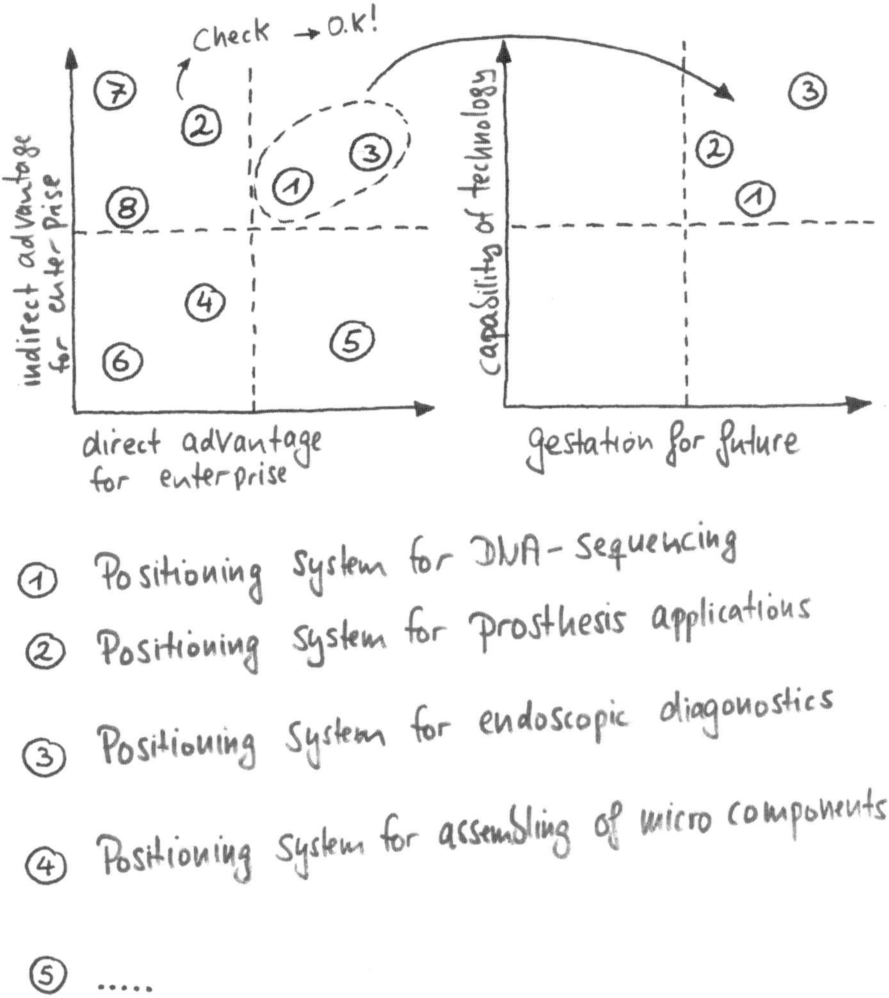

Fig. 3.20 Portfolio system for idea evaluation (case study: Center Positioning Systems Ltd)

3.7 Idea detailing

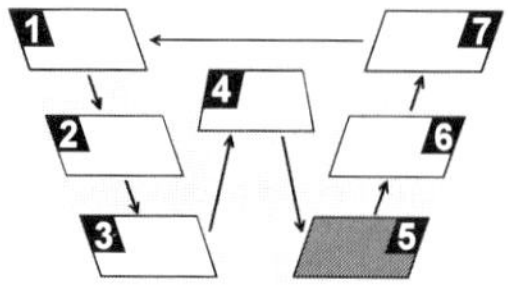

In the planning phase "Idea detailing" further market- and technology-related information is acquired for the selected product ideas, with the aim of developing product concepts. The following planning steps are required:

- Acquiring product idea relevant information,
- Concretizing product requirements,
- Deducing product characteristics,
- Collecting and generation of detail solutions,
- Generating product concepts.

The results of this planning phase are various concept variants of the product ideas. Some of these can be validated through demonstratable products.

Goals of idea detailing

The risks associated with innovation can be significantly reduced by acquiring relevant market information at a very early state and checking the technological feasibility (Geschka 1999). Given this background, during the planning phase idea detailing one has to further develop those ideas of the second order that have earlier on been identified as having the highest future potential (Brandenburg 2002).

Methods tool kit Idea detailing

To support this in a systematic manner, various methods and supporting techniques are available (Fig. 3.21).

Product idea relevant information gathering

For the detailing of the second order product ideas, information relevant to the planning is collected. The search is based on two perspectives, namely market and technology.

Primary and secondary market research

In order to understand client requirements, during the idea generation step, secondary market research has been done already. Based on these, primary market research is done during the detailing of ideas to not only verify the results, but also to generate new, or not yet available, data.

The aim is to fill in the column with client requirements in the first house of quality (HoQ) of the QFD-method (for QFD refer to 4.2).

Kano-Model

With the methods from the primary market research those client requirements referred to by Kano as capability characteristics are acquired. (Chap. 4.2). Kano differentiates in his model between basic, performance, and excitement client requirements. To refine and complete the client requirements, methods need to be used that focus on the collection of these basic and "order winning" characteristics.

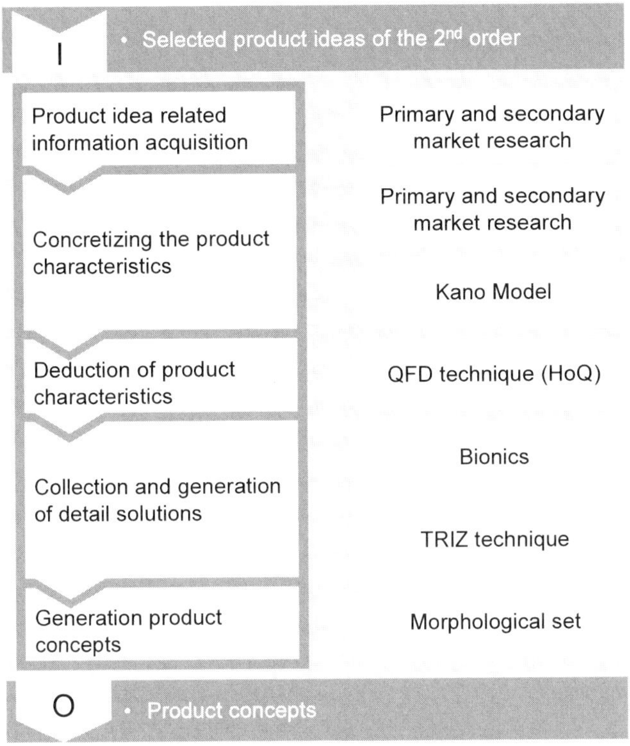

Fig. 3.21 Methods tool kit for "idea detailing" (cf. Brandenburg 2002)

TRIZ technique

The inclusion of the order-winning characteristics is supported by the TRIZ tools "ideal product" and "laws of technical evolution" (Chap. 4.4). The "ideal product" has already been used in earlier phases, so that those results can be re-used here. By approaching the "ideal product" or by creating a further step in the evolution process according to the laws of the technical evolution or the bionical evolution laws, order-winning characteristics previously unknown to clients can be generated (Chap. 4.5 and Chap. 4.6). To collect primary or basic requirements, earlier product models can be used. Detailed descriptions of the use of these methods are given in chapter 4.

Structure and weighting of client requirements

Structuring client requirements is done according to Kano's model. Weighing the various requirements demands a conjoint analysis (Chap. 4.8). Through this, the question of how to best or optimally structure the new idea according to the client requirements is answered (Meffert 1998). To do so, a real purchase situation is simulated where clients have to weigh up or compare different product characteristics with each other.

Generating product characteristics through creativity techniques

After the weighing of the client requirements, the technical product characteristics are generated. Here the methods or creativity techniques based on intuition are used mainly. At least one product characteristic for each identified client requirement has to be generated. The ratio between characteristics and requirements should however not be higher than 1.5 to 1.8 (Teufelsdorfer 1998).

So, as to be able to prioritize the development of product concepts, the product characteristics need to be weighed as well.

Defining technical problems

Based on this weighting, the definition of the technical problem(s) is done. Here it is useful to look at the interaction and/or correlation between the different characteristics so that possible synergies or contradictions can be identified early on. For this, the roof of the HoQ is used.

Solve technical problems with the TRIZ technique

The TRIZ method is helpful again in solving the technical problem definition. Especially contradictions in the roof of the HoQ can be solved with the help of the "contradiction matrix". Other tools and techniques for solving technical problem definitions do exist too (cf. e.g. Walter 1997).

Setting up product concepts

In the last step of the idea detailing, the solutions for the technical problem definitions are combined and summarized (Brandenburg 2002). Combinational methods, and especially the "morphological analysis" (Schlicksupp 1995) are used in this context. Here the complete system is divided into various characteristic parameters, such as characteristics, functional units, subsystems etc.. For each one of these parameters, the possible solutions are put into a scheme – the so-called "morphological analysis". Each combination of a single solution for a parameter with that of another parameter gives a theoretically possible solution combination. The combination of these part-solutions gives rise to solution concepts. The method should facilitate the development of unusual solution combinations, so that a wider selection becomes possible.

Result

The product concepts developed are the result of the idea detailing.

Idea detailing at Center Positioning Systems Ltd

... Curiously Mr. Stone opened the envelope he had just received. One and a half weeks ago he had gained some product idea related information at different producers. This time a producer of medical components had sent detailed information on endoscopes and their production. By now Mr. Stone had very detailed data on the production of endoscopes.

Gathering product-related information

Using primary and secondary market research he had commissioned many product idea related information researches and done quite a few himself. Do detail the ideas on "positioning systems for endoscope diagnostics" he had made many phone calls with Professor Lambert for example.

Quality Function Deployment

To implement the detailing of information Mr. Stone wanted to use the QFD method of which he had heard many successful implementations already (Chap. 4.2). As it was too difficult to gather all chosen doctors for a meeting he had set up a video conference to let the experts built a "House of Quality". Therefore he had already sent an agenda, blank style sheets for the "House of Quality" method and an explanation about the video conference to everyone.

Concretizing demands

At 10 am it finally started. The participating experts appeared on the flat screen of the video conference system. After a short greeting and an introduction he asked the experts to tell him all relevant requirements for an automated positioning system within the head of an endoscope. Mr. Stone already had prepared a list with several demands which included most of the necessary details. He read the demands on after another and asked everyone for their opinion. During the discussion a few more requirements arose whereas others were canceled. In the end he had a specified list of all demands.

Weighting of customer demands

The second part of the conference was about weighing demands which Mr. Stone later on decided to call customer demands. Once again a vivid discussion took place. Professor Lambert, also taking part in the conference, stressed the criteria of *pain free appliance* whereas some of his colleagues considered the *simple usage* as more important. Nonetheless a compromise was found in the end.

Mr. Stone had written down all the demands during the conference and was now reviewing the results.

Identify technical characteristics

"Now the engineers have to do something", he thought while watching the technical characteristics part of his "House of Quality" sketch. He sent a copy of his sketch to the technology-orientated BU managers and asked them to think about technical characteristics which could be relevant if producing a positioning system for endoscopes. After that he had to compare the customer demands to the technical characteristics and discuss the outcomes.

Identify technical contradictions by using the "House of Quality"

"Hey, that was fast", Mr. Stone was quite happy when looking at the completed "House of Quality" form on his desk on Wednesday (Fig. 3.22). His colleagues had done a great job. The connections between customer demands and technical characteristics had been weighed according to their intensity. Technical contradictions could easily be seen on the roof of the "house of Quality". Even a competitive comparison had been made by his colleagues.

Generate product concepts

With the help of this information it was possible to condense the product ideas to concepts:

- Product concept 1: "Intelligent endoscope – IntEnd": A micro positioning system is build into the head of an endoscope to move optics consisting of a camera and lighting. It has to be remotely controlled to keep the shaft of the endoscope as thin as possible. Energy supply is realized by using a flexible conductor within the shaft of the endoscope. Once in position the endoscope should not be moved anymore, only its optics. Therefore 320° pivoting horizontally and 120° pivoting vertically has to be implemented. It must be possible to position the optics continuously and without bucking or start-up acceleration to not change the position of the endoscope. Furthermore all used components must be according to hygienic regulations for medical tools. The positioning system should be manufactured as a system which can be implemented into a standard endoscope.
- ...

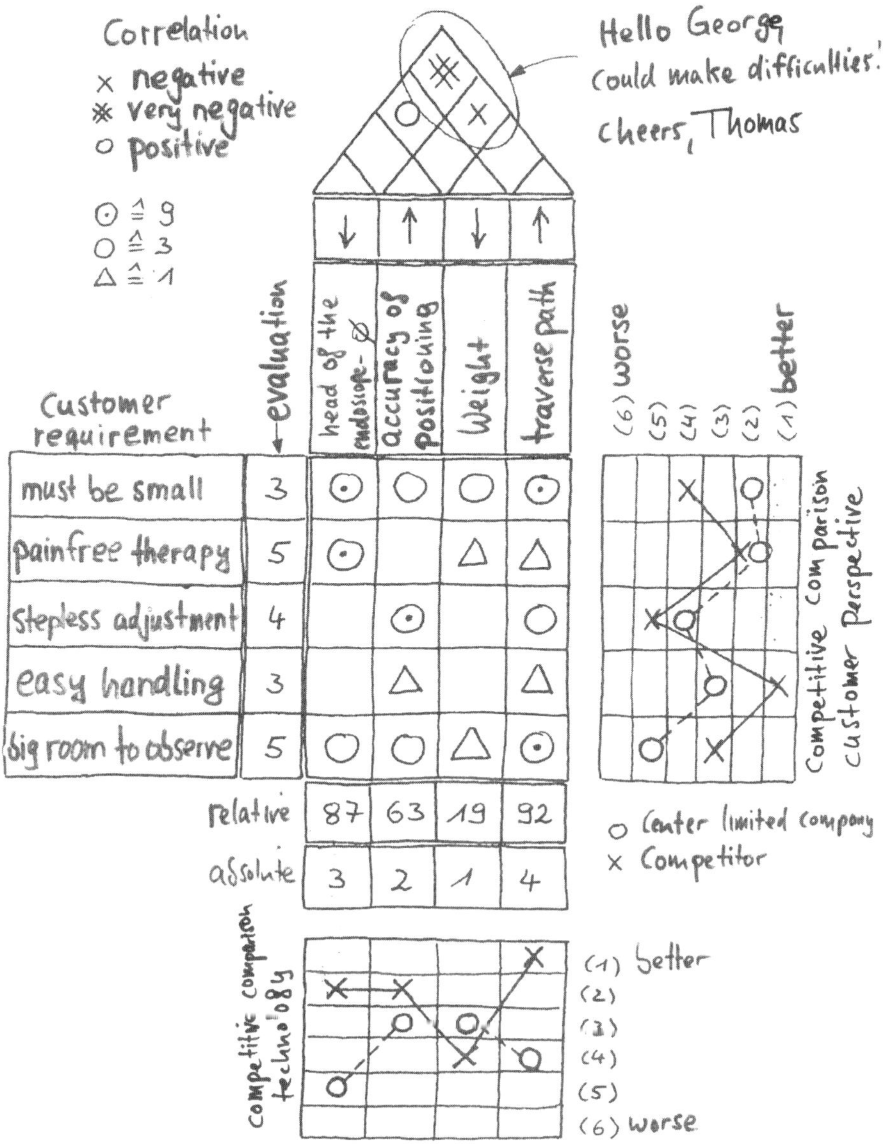

Fig. 3.22 House of Quality (Case study: Center Positioning Systems Ltd)

"That sounds quite promising", Mr. Stone thought while reading the product concepts for the intelligent endoscope. Suddenly his phone rang.

It was Thomas Plumb, BU manager of "sliding carriage systems" who wanted to have feedback on his work on the

"House of quality". He was troubled by the technical contradictions which meant an end to the intelligent endoscope in his point of view. Mr. Stone had a different view though.

TRIZ method

Mr. Stone had read that the TRIZ method could be used to resolve technical contradictions. "Thomas, I have read about TRIZ workshops which are held at the Fraunhofer Institute for production technology. What do you think, shouldn't we take part in one of those?" "If we can find a solution to the identified technical contradictions", he continued, "we could finish the product concept and move on to the next phase of the planning!"

Mr. Plumb liked that idea, so he called the TRIZ experts at the Fraunhofer IPT. He wanted to take part in one of those workshops as soon as possible to finally finish the idea detailing...

3.8 Concept Evalution

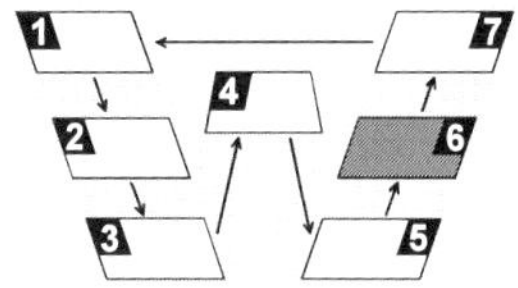

In this phase the product concepts developed are evaluated quantitatively. For this, the criteria used for the idea evaluation are not only taken again to rate the more mature product ideas, but are also further refined and expanded based on the verified information from the idea detailing. The fully developed and detailed product innovation ideas are also rated according to their technical feasibility and economic profitability. The evaluation of the concepts is done in three steps:

- Evaluation of the fulfillment of requirements,
- Description/rating of the technical feasibility,
- Evaluation of profitability.

After this step, selected product concepts that are both technically feasible and economically reasonable are available.

Contents of the concept evaluation

In the concept evaluation the additional information obtained from the idea detailing is used to more concretely rate or evaluate product innovation ideas. Besides the results of the previous idea detailing step, especially the innovation strategy from the goal finding and the goals developed in the future analysis are used as the base or input information for this phase. The result of this planning phase are selected concepts that are both technically and economically geared towards the enterprise strategy and that can contribute towards organizational goals. Because of the refinement of the information, more detailed and accurate evaluation methods can be used than previously in the idea evaluation phase for instance (cf. Brandenburg 2002).

How to approach the concept evaluation

The concept evaluation is done in three steps, during which it is made certain that not only the client requirements are met, but also that the technical and economical feasibility is given. More particularly, one distinguishes between the following three planning activities (Brandenburg 2002).

- Evaluating the fulfillment of requirements,
- Rating the technical feasibility, and
- Economic considerations.

The different steps are supported by various methods (Fig. 3.23).

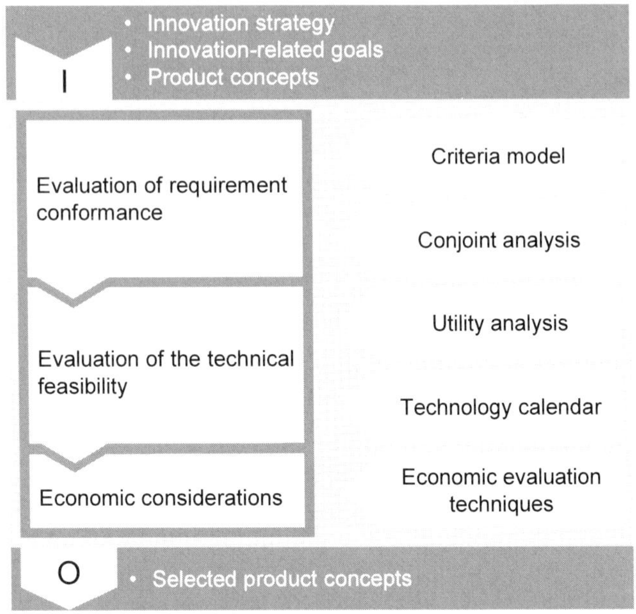

Fig. 3.23 Methods tool kit for "Concept evaluation" (cf. Brandenburg 2002)

The important difference to "valuing ideas" is the profitability calculations

As was done in the phase "idea evaluation", the innovation ideas are evaluated according to a specific evaluation and rating model. The difference, however, lies in the level of detail, which is far greater than in the earlier phases. Therefore a more concrete evaluation of the product concept alternatives becomes possible. More specifically, this is not only a more complete evaluation, but the economic evaluation in this instance allows the use of detailed economic and/or financial models and techniques.

Technology calendar

With the aid of the technology calendar method the informational capturing of the necessary data is supported, and thus aids in the validation of the technical and economical feasibility of the product concept alternatives (Chap. 4.9).

Utility analysis

This pairwise comparison allows the relative weighting of the various criteria, and results in a ranking according to these. The weighting is done according to the enterprise-specific goals and strategies. Here for example the strategy will give an indication on which criteria, cost reduction or production time reduction, is more vital or important within the enterprise environment. Based on this pairwise comparison, in this utility analysis the ranking between each alternative is done by calculating the total utility of each product concept.

Reference figures of the evaluation

The ranking and the specific values are not the same any more like during earlier evaluation phases. Where previously only vague concepts and undetermined timeframes were available, here the availability of concrete data requires more and different evaluation criteria to be used. These can be characterized as follows (Pleschak and Sabisch 1996):

- The evaluation must result in well-motivated statements which have to be valid over the long term.
- The evaluation methodology must take into account the high complexity of the product ideas and concepts.

Multi-dimensional evaluation techniques

For these reasons, more dimensional evaluation methods are required, which need to take cognizance of a number of different goals. The higher the number of these different goals, however, the more difficult the evaluation becomes, as for each alternative a number of advantages and disadvantages exist. One should therefore strive to reduce the number of goals to only the most important ones, making it possible to get to a definite synthesis which is still multi-dimensional. This reduction can be achieved by translating and/or transforming the various technical, organizational and social characteristics required for the evaluation of the product concepts into economic variables like profitability or amortization.

3.8.1 Evaluation of the requirements fulfillment

Client orientation

The fulfillment of the specific and concrete client requirements through the product concept needs to be evaluated for each one (Brandenburg 2002). In this step the correlation between the original client requirements as captured earlier on and the product concept is calculated. By doing this, only those concepts that achieve a high requirement fulfillment are pursued further. The ideas are also compared to the requirements and technical characteristics as formulated according to the IRM method in the House of Quality in phase 5 (idea detailing) of the innovation planning process. By having more detailed information, the client requirements can be evaluated against detailed technical criteria. Also, outstanding or further data required for a complete analysis can be obtained from detailed client questionnaires or for instance a conjoint analysis (Chap. 4.8).

Detailed technical and economical evaluation is possible

These more extensive concepts and calculations can be incorporated in the criteria catalogue from the idea evaluation, which leads to a detailed technical and economic evaluation.

3.8.2 Evaluation of the technical feasibility

Prototype

To demonstrate the technical feasibility, in most cases, a so-called demonstrator (a physical, or nowadays more often than not a digital prototype) should be built or modeled (cf. Brandenburg 2002). In the last couple of years, rapid product development techniques are used to hasten this demonstration. Also, commercially available simulation software can be used successfully. This allows that prototypes of the product can be generated virtually on a computer screen, and the evaluation of these according to their functional characteristics. (cf. AWK 2002). This practical testing of the product concept serves as proof for the technical feasibility of transforming the innovation idea physically.

Further aspects to be regarded for the technical feasibility

Next to the purely technical or physical feasibility of being able to make a product, other characteristics such as estimated sales, batch sizes, variants, geometry, quality characteristics etc. have a role to play. Choosing a specific manufacturing or production technology therefore needs to take these into account. One method combining technical manufacturing aspects as well as market-oriented strategic technology planning is the technology calendar, explained in detail in chapter 4.9.

Selecting the evaluation criteria

The first step in evaluating the technical feasibility is the choice of criteria. If there are specific requirements, such as environmental laws, social considerations etc. the evaluation criteria have to be expanded to take the social and environmental aspects into account. To keep the evaluation as specific and uncluttered as possible, only the most important criteria and those having specific relevance to the organizational strategies and goals should be incorporated.

Relative weighting of criteria

After the identification of the evaluation criteria, they are compared pairwise so that the relative weighting of each one can be established. This allows the calculation of a dimensionless evaluation number, giving an indication as to the degree the original requirement has been or will be achieved. The sum of the product of the relative weighting of a criterion and the evaluation number gives the *utility of the product concept*. This is required only if there are at least two

Utility of a product concept

different and competing product ideas which need to be compared or evaluated.

Advantage of the evaluation technique

The advantage of using the multi-dimensional evaluation method lies in the fact that complex evaluations can be done, which lead to a definite rating of the product concept. Furthermore, this method is easy to apply, and can be customized to take cognizance of the specific uncertainties inherent in the concept.

3.8.3 Economic considerations

Methods for profitability analysis

The phase concept evaluation is concluded with the economic feasibility study. For this, a number of methods have been developed and used successfully[16] (Brandenburg 2002)˙ The decision for or against a product innovation can be equated to an "investment". Therefore the economic evaluation is based on well-known concepts from investment, profitability, and amortization calculations. Typical methods, techniques and evaluation criteria will therefore only be mentioned below: for a more complete description the reader is referred to the relevant subject literature.

Static and dynamic methods

Observations regarding the economic feasibility can be done either statically or dynamically (Schierenbeck 1993) (Fig. 3.24).

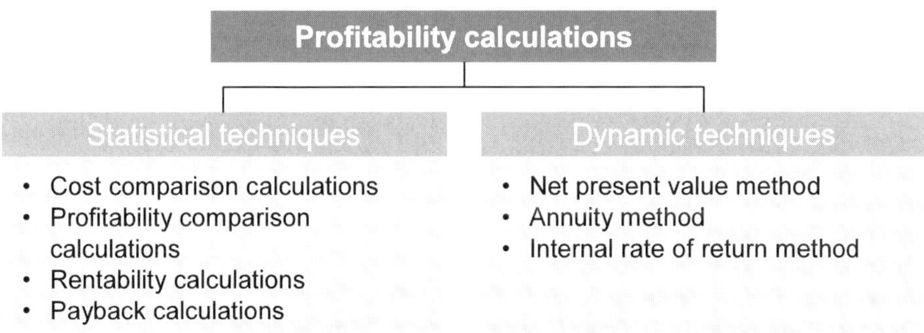

Fig. 3.24 Overview of static and dynamic profitability calculations

[16] For methods regarding economic considerations of product innovation refer to STUMMER (Stummer 1998), MARTINO (Martino 1995) or BROSE (Brose 1982).

Cost comparison calculation

- When using the *cost comparison calculations*, the product innovation leading to the lowest cost is chosen. Here, the direct or operating costs as well as the capital costs required for the product realization need to be taken into account. Because this method does not look at expected profits, it is often used in cases of replacement or "rationalization" investments, where similar profit expectations exist for the various alternatives (Kruschwitz 1993). A similar lifespan and similar capital investments for the different products is however required for this method to be useful or fair.

Profit comparison calculation

- An expansion of the cost comparison method is the *profit comparison calculation*. Here the decision criterion is the expected investment profit, defined as the balance of cost and income per time period. This method is especially useful when the qualitative outputs of the various concepts are different. Here too, the comparison needs to be based on innovation objects with similar if not identical life spans and required capital investments.

Return calculations

- *Return calculations* need to be done when innovation objects, or in this case innovation projects, require different amounts of capital, and stretch over different time periods. The return per period is the average profit or cost saving per period divided by the average capital required. From the various product concepts the one with the highest expected rate of return is chosen. Similar to the previous evaluations, the different concepts should have similar life spans to make the method valid. Because of this requirement, for similar capital requirements, this method leads to identical results as the previous two methods and therefore does not give better decision support (Kruschwitz 1993).

Payback calculations

- With payback calculations, the time it takes for the original investment amount to be recovered is calculated. In the rating of the product innovation concepts, the one having the shortest payback period is chosen. Care has to be taken however to only compare objects having similar life spans, as the result would otherwise be skewed because of depreciation and other allowances which are often based on the life of the products and have a huge impact on the payback period. Because of this, the payback criterion should never be used as the only criterion for an economic decision (Kruschwitz 1993).

Dynamic profitability calculations and techniques

In general, static economic calculations have the disadvantage that the time value of money (through differences in the time periods of in- and outflows of money) is not taken into account at all, or if so then incomplete. To fully take this into account, dynamic profitability calculations need to be done.

Net Present Value (NPV)

- The *net present value* discounts all future inflows and payments of money to a reference period, normally the present, at a rate "given" or used within the company, referred to as the minimum attractive rate of return (MARR). The difference between these future flows and the required capital investment gives the net present value. If this is positive, then the investment is regarded as being worthwhile. Practically, this NPV can be regarded as the amount of money available over and above the original investment that can be used or consumed without requiring any additional investments over and above those generated by the project itself.

Internal Rate of Return (IRR)

- The NPV of an investment depends on the interest rate being used in the calculations, referred to as MARR above. Therefore, the NPV of an investment declines with higher interest rates being used, and vice versa. The internal rate of return calculation on the other hand does not fix the interest rate beforehand, but tries to find that rate where the NPV of the investment is equal to zero. This rate gives the effective return achieved with the capital tied to the investment. Should this rate be higher than the MARR, the investment from a profitability perspective is worthwhile.

Annual equivalent

- The annual equivalent is the third dynamic profitability calculation. It is deduced from the NPV method, but the achieved annual profits and/or losses are distributed equally over the life span of the investment. This means that the actual in- and outflows of money are transformed into equivalent (identical nominal cash values), equi-distant (identical time intervals), and uniform (identical amounts) cash flows.

Technology calendar

In order to obtain information required for these calculations, the technology calendar (cf. Chap. 4.9) is a support tool allowing the medium- to long-term estimation of manufacturing costs – an important aspect in determining future cash flows.

Result: selected product concepts

The result of the concept evaluation are selected or chosen product concepts, where the technical and economic feasibility has been established. This then allows the start of the realization in the form of a product and market development process (Brandenburg 2002).

Concept evaluation at Center Positioning Systems Ltd

... This morning Mr. Stone wanted to start with the concept evaluation phase. Therefore he reviewed the innovational strategy and the innovation-related objectives which they had set up before while defining objectives. "Mr. Stone! We have to broaden our product range; we must identify new markets ... and then strike! That's how it works!" Smiling, Mr. Stone remembered the visionary speech given by his boss during the strategy discussion regarding defining objectives. Apart from entering new markets Mr. Stone had noted *technology leading* and *diversification* in their product range as innovation strategy.

Input is taken from preceding phases

"What do you suggest? The QFD method again?", Mr. Stone asked surprised while he was reading the course of action for valuing concepts. Being skeptical in the beginning he soon realized that it would be a good idea to analyze customer demands using the QFD method "...and right there it is explained how to use the method", he realized contently.

This afternoon, Mr. Stone invited all BU managers for another meeting to discuss the results of the "idea detailing" phase. Therefore he had taken the rating and evaluation system from the "idea evaluation" phase as well as the additional information gained in the "idea detailing" phase and transferred them into the model. Today's meeting should be about comparing the technical, now fully developed product concepts with customer demands, which had been specified as well using the House of Quality.

The discussion went on according to Mr. Stone's expectations. His colleagues worked well and evaluated the product concepts according to the criteria set up in the rating and evaluation system. Compared to customer demands the "IntEnd" was taking the lead again. Mr. Stone thanked everyone for their eager cooperation and went back to his office. Despite late time he wanted to prepare the next planning step.

Validation of technical realization

"Good morning, Charly!", Mr. Stone said slightly blinking while entering the tool and mold department. He had met Dr. Palmer the evening before to discuss the results of yesterday's meeting with him. Obviously enthusiastic about the systemic work of his staff chief, Dr. Palmer approved the realization of an "IntEnd" prototype. Based on this prototype the technical characteristics should be validated.

"IntEnd" prototype

Charly Builder had lead the creation of prototypes at Center Positioning Systems Ltd for more than 15 years and brought forward quite a few innovation ideas as prototypes already. This task was a challenge even for Mr. Builder though; he had to create something completely new and they had very little experience in the medical field. "This could take a while ...", Mr. Builder was thinking, "we can't built that with our conventional tools."

Five weeks later Mr. Stone had the first model for a micro positioning system to be used in an endoscope on his desk however. He examined the tiny hybrid construction made from polyethylene and titan very carefully. "It's hard to believe that this system actually works", Mr. Stone thought.

In the previous week he had made an appointment with the leading professor for internal medicine at the university clinical center of Aachen, where he would like to present the product innovation for the first time. Carefully he put the prototype back in his casing. "Now the system only has to be built into the endoscope shaft", he was thinking loudly. Therefore he had spoken to several producers of endoscopes two weeks ago already and finally decided on a basic model the micro positioning system should be built into. The complete system should be ready in three days, so he could show it to Dr. Palmer.

Reapply portfolios from valuing ideas

Until then Mr. Stone wanted to prove the technical and economical realization of the system. He once again reviewed the portfolios he had set up during valuing the ideas. The company benefit portfolio remained unchanged; the technology-market portfolio could be adjusted further though. The new insights gained during idea detailing as well as the new criteria taken from discussions with the BU managers could lead to a slightly different classification of product ideas within the portfolio. It was possible to present the *customer benefit* in more detail, to assess the *substitution security* more accurately and to identify the *synergy effects in R&D* and the *competencies in production* with more certainty. Relating to customer benefit the *influence of subjective preferences* was eminent. As there were more technological criteria available by now, an assessment was not that reliant on personal preferences anymore, which might lead to a shift in the portfolio's classification. Regarding *substitution security* they gained new information referring to patent analysis and customer researches. "A system like 'IntEnd' cannot be copied very easily!", Mr. Stone nodded confidently. They had also achieved detailed information on the R&D synergy potential.

Mr. Stone and his team weren't aware of similarities between the development of micro positioning systems and the development of larger systems for the tooling industry – it was just a different scale. Resulting from these new insights the *technology potential* had to be adjusted with "IntEnd" taking over the lead as a product alternative.

Analyzing economic efficiency

Mr. Stone had analyzed a product's economic efficiency many times before; still he asked his young colleagues from controlling for the most sensible cost account. He knew of course that a dynamic cost account led to more specific results than a static one, still an update on the latest methods seemed like a good idea. Afterwards Mr. Stone decided to calculate the cost-effectiveness and combine it with an amortization account.

"We won't get that cheap", Mr. Stone thought, "still a cost-effectiveness of 35%, estimated roughly, is a very good argument to implement our innovation plan. And the payback will only take three years!", he nodded. "I would do it!", he decided for himself and started to repeat his calculations for the other product alternatives…

3.9 Implementation Planning

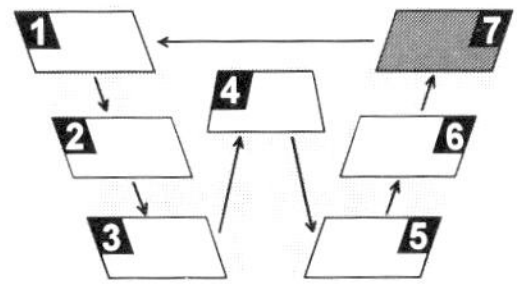

In the planning phase "implementation planning" the insights gained and results obtained from the previous planning steps are aggregated in a so-called InnovationRoadMap (IRM). This roadmap is used to look at the interactions between the environmental and market requirements on the one hand with the technological product solutions on the other hand, over the short, medium and long term. In developing the InnovationRoadMap the following planning steps need to be done:

- Classification of future projections and innovation tasks,
- Relating the market-based with the technology-based innovation tasks,
- Deducing company-specific implementation activities.

The ultimate aim of the implementation planning is the systematic representation of specific company activities for the developed product ideas and concepts by using the InnovationRoadMap.

InnovationRoadMap showing the results of the innovation planning process

In this last planning phase of the InnovationRoadMap methodology all the singular results from the previous planning phases are combined and aggregated. The graphical illustration of the singular results is done with the aid of the InnovationRoadMap, which maps the prognoses of future market and technology developments with the enterprise-specific product innovation planning.

Mapping of market-related requirements and product ideas

The InnovationRoadMap provides the user with a tool through which company specific market requirements or potentials can be compared with or related to the relevant product ideas in a systematic, repeatable and easily understandable way over a medium- or even longer-term time horizon (Brandenburg 2002).

Deduction and synchronization of company-wide product innovation activities

An important goal of the InnovationRoadMap-methodology is the deduction of product innovation activities and their synchronization within the planning horizon. To do this, the positions and described characteristics of the market and technology groupings are being used.

The basic concepts of the RoadMap-methodology are described in detail in chapter 4.9. In order to fully exploit and integrate the additional information or knowledge obtained during the various planning phases, the original roadmapping techniques have to be expanded. The required input information for the InnovationRoadMap is based on the methods tool kit of the implementation planning (Fig. 3.25).

Performing the three planning steps requires only the roadmapping instrument itself, with no additional methods or techniques. In this last planning phase all the information gathered throughout the IRM-process are entered into a roadmap, which then becomes the InnovationRoadMap. The developed and modified structure of this InnovationRoadMap is shown in figure 3.26.

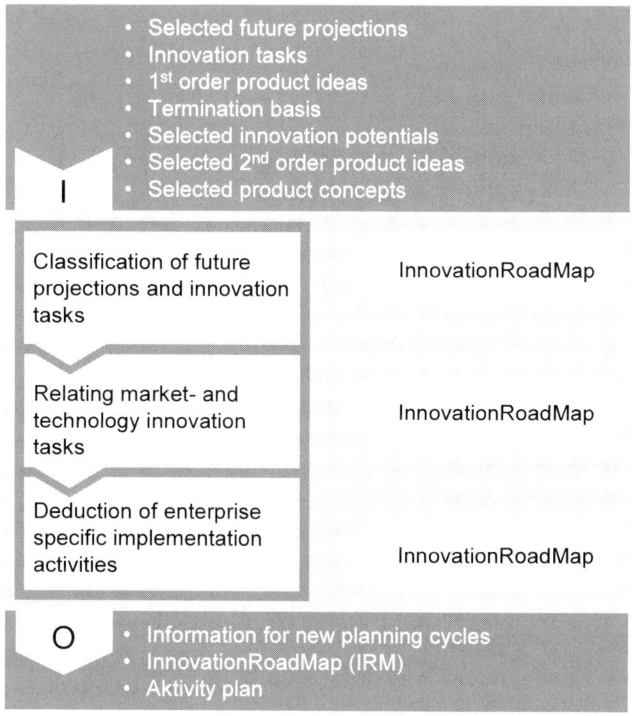

Fig. 3.25 Methods tool kit for "Implementation planning" (cf. Brandenburg 2002)

3.9.1 Grouping and structuring of the future projections and innovation tasks

Three-leg of market, product technology and time

The structure of the InnovationRoadMap is "three-legged", consisting of the axes *market*, *product technology* and *time*. The placement of the projections and tasks within this structure is based on the knowledge gained during the previous planning phases.

Strategic window

In choosing the time to enter the market, one pre-supposes that the technological product innovation allows the manipulation of future market developments by being the innovation leader. However, the earliest possible entry is not always the strategically optimal time (Brandenburg 2002). ABELL refers to a "strategic window" within this context, where a logical and effective entry into the market has a specific time frame. He is of the opinion that there is only a very brief time period during which the market requirements on the one hand and the technological innovations of an enterprise on the other hand actually agree with each other. It is only during this brief time window that market entry will be successful (Abell 1978).

The area "market" on the Innovation-RoadMap

The *market area* of the InnovationRoadMap consists of the axis *market* and *time*. Here the future market developments are shown through innovation tasks (problem ideas). These are deduced mainly from the results of the future analysis (planning phase 2), the market-based research into idea generation (planning phase 3), and idea detailing (planning phase 5). It is also possible to prioritize and/or concretize these according to increasing requirements as time progresses. The timing of the innovation tasks is done not only over the planning horizon of the future projections, but also takes into account priorities based on norms and values or even laws and regulations that might have an impact (Fig. 3.26). A possible criterion for the placement of the innovation tasks along the "market" axis could for instance be the future weightings of the innovation tasks themselves (Brandenburg 2002).

Placement of the innovation tasks along the market and time axes

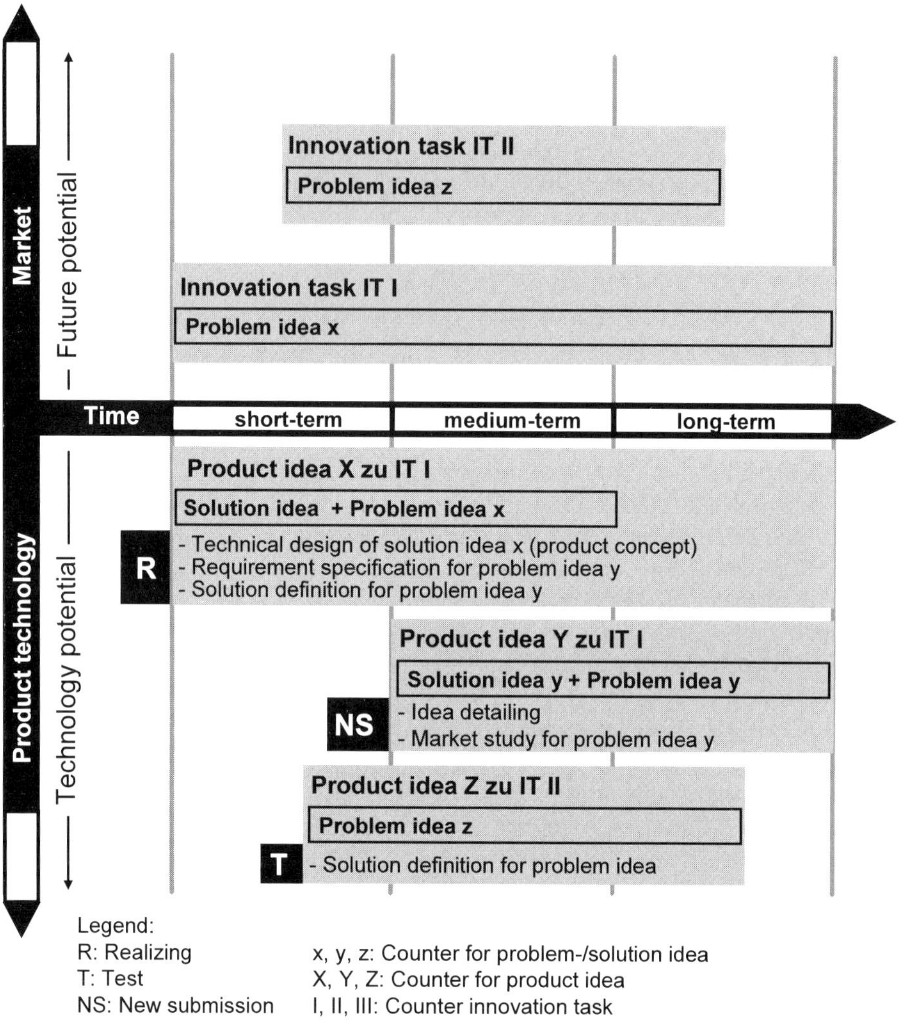

Fig. 3.26 Structure of the InnovationRoadMap (cf. Brandenburg 2002)

3.9.2 Correlating market-deduced and technology-based innovation tasks

The area "technology" of the InnovationRoadMap

The "technology" area of the InnovationRoadMap has as its two axes technology and time. Here the results from the idea generation (phase 3) and idea detailing (phase 5) phases as well as the insights gained in the future analysis are shown as future product ideas, based on the relevance of the required technology to the enterprise. The specification of the ideas and potentials within this area of the roadmap are based on multiple criteria. Prioritization of the various product ideas and their timing is done according to the criteria *implementation*, *testing*, and *resubmitting*. Differences also exist in as far as the manner of the activity between market and technology relation is concerned, as well as the status of the activity in the planning process. A further grouping or definition within this area is the way or manner in which the R&D can be used. By coupling at least one problem idea from a market perspective with at least one product or solution idea, the two areas of the roadmap are linked. Placement along the time axis depends on the urgency of the product idea on the one hand and the required development time on the other. Placement along the technology axis has to be based on the evaluation of the technology potential (Brandenburg 2002).

Placement of product ideas along the "product technology" and "time" axes

After all the information and results from the previous planning phases are entered into the InnovationRoadMap, in this implementation phase the company-relevant and -specific innovation activities required for implementing the chosen product ideas have to be worked out.

3.9.3 Deduction of company-specific implementation activities

With the visualization of the InnovationRoadMap the user is able to get an overview over which product ideas can and/or should be realized over the short-, medium- and longer-term planning horizon with the aid of the analyzed innovation tasks. By looking at the various market and technology areas over different time frames, an integrated and synergistic evaluation is guaranteed. Based on the placement of the results in the InnovationRoadMap the following strategic and operational decisions can be made (cf. Brandenburg 2002):

Making strategic and operational decisions

- A chosen product concept has to be realized through either an internal or an external development.
- The technological realization of product ideas has to be supported by building demonstrators and/or prototypes.
- If an expanded technology development is required, a development co-operation agreement with possible suppliers has to be initiated.
- If innovation potentials or product ideas have been identified, concrete market studies can be undertaken.

- To refine or find obviously missing problem or solution ideas, idea generation and idea detailing workshops have to be initiated.
- Market entry strategies have to be planned.
- With product ideas having a high innovation potential, structured patent analysis / patent applications need to be formalized.
- For identified product ideas with a high innovation potential new markets and business segments need to be developed.
- For product ideas or concepts that have been classified as "resubmit", responsibilities for their monitoring have to be assigned.

Identifying existing synergies

By showing all the different innovation activities in a summarized format, synergies between the various activities become visible and therefore usable. By showing both the company specific technology potentials and the market and environmental situation, a two-way creative exchange of ideas is possible. By correlating or connecting product ideas and innovation activities, a transparent integration of the market and technology views is supported (Brandenburg 2002).

Advantages for the user

Besides the advantages to be gained by the visual representation of the modified roadmap, there are further gains to be made through the application of the InnovationRoadMap-methodology. For instance, technology gaps can be identified, and specific solutions found to close these. Also, resources can be better allocated to refining product ideas according to the priority and the placement within specific time frames (Brandenburg 2002).

Regular repetition

The InnovationRoadMap is a dynamic planning instrument. Therefore it is advisable to repeat the exercise regularly, the time interval being based on the pace of change in the environment, the market, the industry branch and the enterprise itself. Through this cyclical repetition, it provides not only a once-off technical support for a specific planning project, but rather a continuous support of innovation activities within a wider innovation control environment (Brandenburg 2002).

Implementation at Center Positioning Systems Ltd

... The important day to finally present results was coming closer. But Mr. Stone was not worried by it, on the contrary: he was excited about presenting the results of his innovation planning. "The last weeks were quite exhausting but we came up with some extraordinary results", he thought while looking at the checklist for the final implementation phase. In the meantime Mr. Stone started to write down the necessary input, being quite experienced by now. "We already have all the information required!", he realized contently and took the necessary files from his shelf. As before, the required input information had been produced in preceding phases already. This time he was going to use the "selected future projections" and "innovation potentials", "innovation tasks", "basis for scheduling" and the "1st order product ideas" acquired in the second planning phase as well as the "selected 2nd order product ideas" and the "product concepts" which they had acquired during the evaluation of ideas and concepts.

Classification of future projections and innovation tasks within the InnovationRoadMap

First Mr. Stone classified the future projections and innovation tasks within the InnovationRoadMap. Therefore he used the basis for scheduling to sketch every single future projection and innovation task along a timeline. Doing so he draw a bar for the future projection "Medical technology" and labeled it "FP: Medical technology". The bar was overlapping the short- and medium-term area. Afterwards he tended to the innovation tasks. After looking at his schedule he marked the time areas accordingly, e.g. for the "endoscope positioning" task. Furthermore he added the according problem ideas to the InnovationRoadMap.

Connection of technological product ideas and innovation potentials

After writing down all innovation tasks he tended to the product technology side of the IRM. Once again he drew a timeline to schedule product concepts. The boxes created in such a way were filled with technology-based product ideas and some measures necessary to realize the product.

First sketch of the IRM

It was already noon when Mr. Stone looked at the first sketch of the IRM on his desk (Fig. 3.27). During the afternoon he wanted to finish the IRM in order to present it. "This will be easier than expected!", Mr. Stone thought looking at the finished IRM a couple of hours later.

Business-specific product innovation potentials

The next day Mr. Stone was looking to the last transfer step, the derivation of business-specific product innovation potentials. Therefore he once again took his collected planning documents to review the product concept assessments made by his BU management colleagues. It was obvious that the "IntEnd" concept was most promising for Center Positioning Systems Ltd. It had high synergies between market and technology, yet could be realized in a medium-term scale. Mr. Stone's recommended course of action was therefore an internal development request. Furthermore the contacts to endoscope specialists should be intensified. They also had to think about collaboration between medics and Center Positioning Systems Ltd engineers. After writing these recommendations into the IRM he closed his presentation file and started his weekend in a good mood.

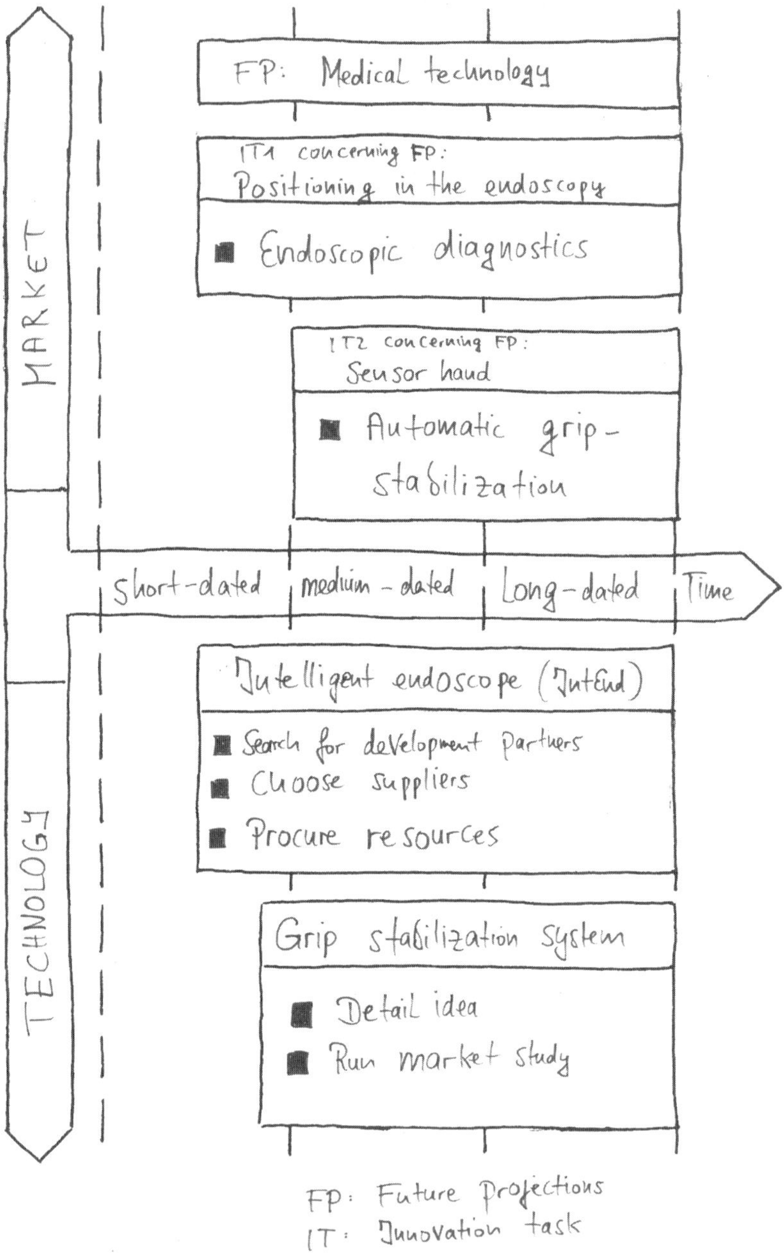

Fig. 3.27 Sketch of the InnovationRoadMap (Case study: Center Positioning Systems Ltd)

Final presentation

The following Monday Mr. Stone presented his results in front of the business management. The systematic approach in planning and the excellent documentation of all steps led to convincing and well-founded arguments. Based on these Dr. Palmer approved the development of "IntEnd" only two days later. "You have really done a great job", Dr. Palmer said while slapping the shoulders of Mr. Stone "we'll proceed exactly as you have suggested!"

3.10 Conclusions for the IRM Methodology

A periodic repeat of the planning is important

At the end of the implementation planning and with the successful application of the IRM methodology, one has product concepts which are described and evaluated with regard to the required development task right up to the planned market entry. With this, the planning cycle on which the IRM methodology is based is complete. In order to keep pace with the dynamically changing or developing markets, technologies and enterprises, the main activities of the goal forming, future analysis and implementation planning should be repeated periodically (Brandenburg 2002). The IRM methodology therefore provides a platform that increases the innovation capacity of an enterprise.

The IRM-methodology is a tool facilitating systematic planning, not an innovation machine!

It would be wrong to assume that the use or application of the IRM methodology will automatically result in innovative, successful products. The creativity of the people involved is still the most important ingredient – this however can be supported through this methodology. The structure of the methodology does guarantee a redundant-free and systematic way of systematically applying the planning steps required for successful product innovations.

Applying the IRM methodology is flexible

The concrete use of the methodology can be different from case to case, and it is adaptable. As shown in the case example used, the application of the various planning steps can be done with different techniques and inputs. These differences are normally based on or the result of financial, time, or content constraints. For example, budgetary considerations might determine the extent of a market research, time constraints might limit the level of detail of a planning phase, while highly complex product ideas might necessitate further, or even extreme, levels of detail already in the analysis and evaluation phases.

The IRM methodology as a complete instrument or tool

The method described here within the context of innovation planning is very goal-oriented; it supports the generating of qualitative and detailed ideas, allowing exploration of latent as well as future client and market requirements. The company strengths are exploited optimally; the development of complex products is more controllable through transparency, which results in objective, logical and repeatable idea or solution choices. The methodology makes it possible to live with uncertainties, to synchronize the development of markets and technologies; it is completely transparent and stimulates creativity.

The IRM methodology is an instrument or tool that supports the systematic planning of product innovations.

4 Methods Description

WALTER EVERSHEIM, THOMAS BREUER,
MARKUS GRAWATSCH, MICHAEL HILGERS,
MARKUS KNOCHE, CHRISTIAN ROSIER,
SEBASTIAN SCHÖNING, DANIEL E. SPIELBERG

For the successful planning of product innovations, it is not only important to know what has to be done, i.e. that information within the innovation process must be generated, but above all one must know how this can happen. For that purpose, several methods suitable for implementation were already mentioned in chapter 3, and their application briefly described.

A more intense look at the method "landscape" will be discussed in this chapter. Some important and comprehensive methods of product planning will be described in detail. The focus will be on the most important basics of each of the methods. Furthermore, the possibilities for the implementation in practice will be highlighted. These are in particular the application areas, efforts (time, cost, …), possibilities or advantages, disadvantages, as well as method restrictions. Furthermore, their application within the InnovationRoadMap methodology (IRM) will be described.

An overview about the methods and their allocation to the phases of the IRM methodology is shown in figure 4.1.

Scenario management has the purpose of developing alternative future possibilities. Using the rules of technical evolution, technological developments will be predicted. The portfolio analysis as an evaluation method will be used in the phases of idea and concept evaluation, but also in the future analysis. For the solution of technical problems, as well as for the development of product ideas in the phases of idea generation and idea detailing, the QFD methodology, intuitive- and analogy-based solution finding, TRIZ (Theory of Creative Problem Solving), laws of technical evolution, as well as bionics will be recommended.

In order to estimate the acceptance of future products by the customer, the conjoint analysis will be presented. By means of technology roadmapping future technology developments will be predicted, analyzed, and visualized.

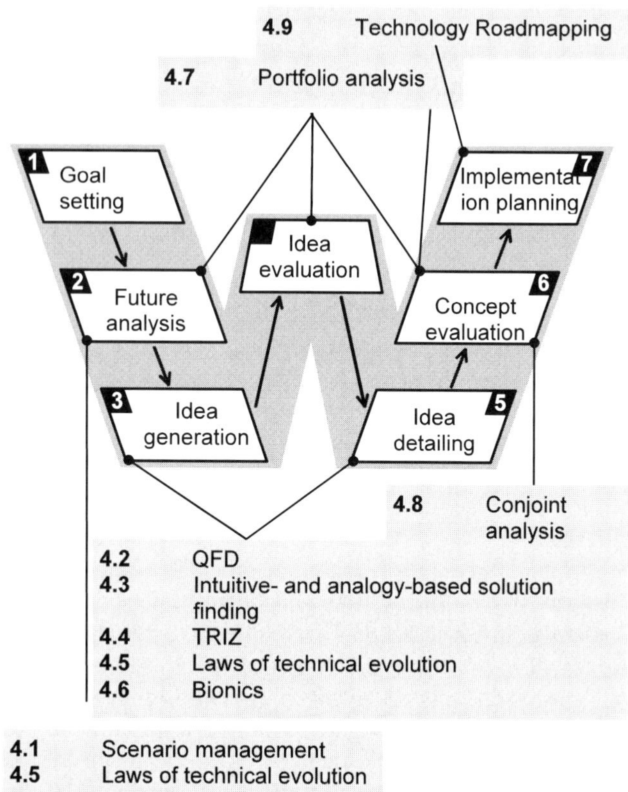

Fig. 4.1 Assignment of the methods to the process phases

Further methods which find a use within the framework of the IRM methodology are captured in an appendix of the Methods Data Forms.

4.1 Scenario Management

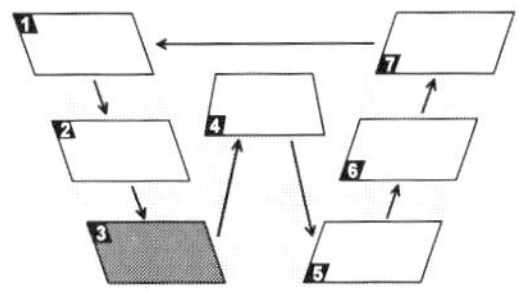

The scenario management aims at developing the so-called future robust lead figures, goals, and strategies.

- In a five-phase process the development of essential influencing factors of the consideration frames will be forecast and summarized in future scenarios.
- The scenario management is a relatively difficult process within the future analysis. With its help a comprehensive picture of possible developments of a consideration area will be designed.

Basic thoughts of the scenario management

The scenario technique, as implemented in the military field, was transferred into economic sciences by KAHN and WIENER[17]. They defined a scenario as a "hypothetical sequence of events constructed for the purpose of focusing attention on casual processes and decision points" (Kahn, 1967).

Using this basic thought as point of origin, the scenario technique or scenario analysis for the industrial implementation in a process plan, the so-called scenario management, was incorporated, which acts as a guideline for the development of scenario projects.

Aim: Develop future robust strategies

Scenario management aims at developing the so-called future robust guide pictures, goals and strategies. With this, the core thesis is that, with the aid of the scenarios, not a general prognosis will be developed, but further development possibilities will be considered. With the help of such alternative future pictures the risk for an enterprise to make a bad speculation shall be avoided especially during turbulent times (Gausemeier, 1996).

Scenario management supports the future analysis

The primary task of a scenario project is the support of corporate decision-making. The decisions can be related to an enterprise, a product or a technology. In the framework of the Innovation-Roadmap methodology, scenario management can above all support the future analysis.

For the application of the scenario technique several process models exist in the literature. They differ mainly in the exact outlining of the several steps, as well as through the process-specific orientated vocabulary of the authors. The

[17] HERMAN KAHN and ANTHONY WIENER developed "scenario writing" in the sixties at the Hudson Institute, USA.

content spectrum of the suggested structures is essentially the same (Mißler-Behr, 1993). As an example, the scenario management in accordance with the phase model of Gausemeier will be described here, which is widely spread in German-speaking countries. The model contains five phases, as shown below (Gausemeier, 1996).

Phase 1: Scenario preparation

In the *scenario preparation,* the forming field, as well as the scenario field, which surrounds the forming field, will be defined. Within the IRM methodology the forming field is known from goal setting. The forming field will be analyzed and described in its current situation.

Phase 2: Analysis of the scenario field

The aim of the *scenario field analysis* (2nd phase) is to identify key factors. For that purpose the scenario field will first be structured into impact areas. Within these areas the influencing factors will be determined. The relevant influencing factors will be summarized as so-called key factors in an influence factor catalogue.

Phase 3: Drafting future projections

The *scenario forecast* is the core phase of the scenario management. It starts with the preparation of the key factors, which will be described in their current situation. Within the formation of the future projections the "outlook into the future" will take place. For each key factor several projections can be drafted and summarized in a project catalogue.

Phase 4: Scenario forming

The aim of the *scenario forming* is to summarize the future projections into projection bundles, and to compress, in a 2nd step, this project bundles to rough scenarios. Tools of these process steps are the consistent analysis and the cluster analysis. By means of the consistent analysis, the bundled future projections will be checked regarding their compatibility (plausibility check) and the possible controversial future bundles will be eliminated. At the same time, with the cluster analysis, most similar projection bundles or groups will be respectively summarized until the envisaged number of rough scenarios (in general two to four) is achieved. The end of the scenario formation is the creative interpretation and visual description of the rough scenarios, so that clear future pictures are made available (Gausemeier, 1996). The phases of the scenario field analysis, the scenario forecast, as well as the scenario formation are summarized in figure 4.2.

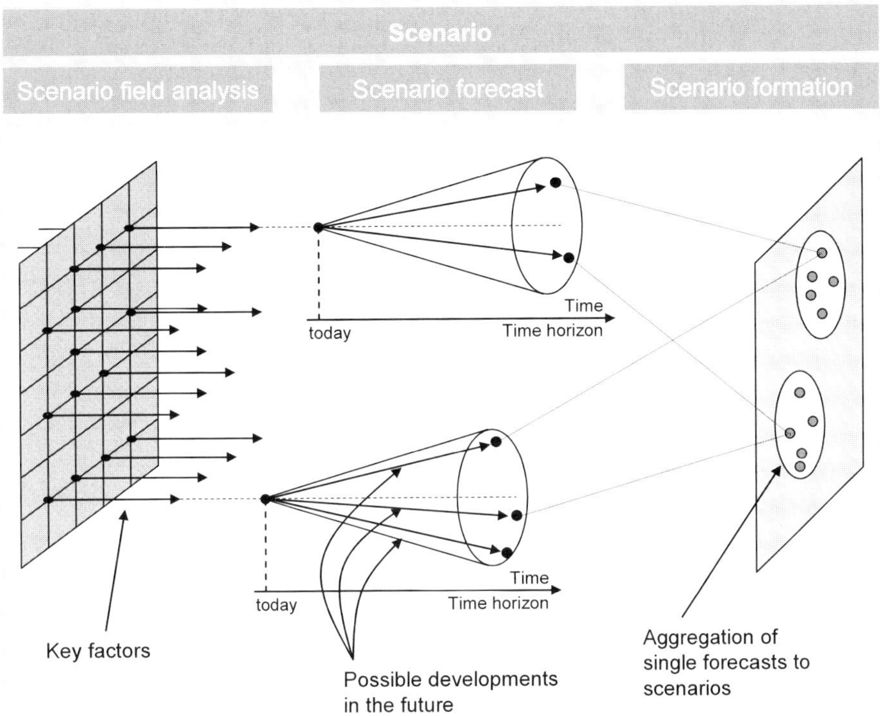

Fig. 4.2 Three phases of the scenario development (compare Gausemeier, 1996)

Phase 5: Scenario transfer

The *scenario transfer* starts with an influence analysis. Here, the impact of the scenarios on the formation field will be determined, in order to plan possible measures for chance utilization and risk avoidance. The aim of the scenario transfer is the deduction of innovation potentials which result from the developed scenarios.

Advantages and disadvantages

An advantage of the scenario technique is the consideration of the general interdependence of the influencing factors. Alternative development trends will be clarified in several directions. Especially in complex prognosis problems which do not allow a simple trend extrapolation, and when the question arises about the motivated explanation of future perspectives, the scenario technique offers valuable decision-making aids.

The work outlay for carrying out a complete scenario project is substantial. If, for instance, the influence or the key factors have to be motivated, a large amount of information must be gathered, or the opinions of experts, who are specialists in the different areas of consideration, must be taken. Furthermore, the handling of a large number of influencing and key factors, in the range of 30-40, is time-consuming. Besides that, analysis programs are necessary for scenario formation.

Branch scenarios

On the grounds of which are necessary for the complete carrying out of a scenario project, scenario management is suitable for big businesses or corporations. The latter use the scenario technique for the development of branch scenarios. This corporative use offers the advantage that future developments by the participating partners (companies, customers, suppliers, association representatives, institutions, etc.) will be explained from different perspectives which leads to more objective estimations.

Combining existing products: Linking SM and QFD

In order to use the scenario technique for the (further) development of specific products, the combination of methods with the QFD methodology is recommended (Eversheim et al, 2001a).

In companies, it is often that a product exists, which resulted from earlier projects of customer specifications. This serves as input information for QFD application. The existing analyses contain the current customer requirements and the relative weighting of the customer requirements. Based on the customer requirements, the technical characteristics are defined, and their importance for the fulfillment of the customer requirements is calculated. Based on the technical characteristics the product is designed (Chap. 4.2). Due to the differing importance of the technical characteristics the product design is determined by the most important technical characteristics.

Capturing changes in customer requirements

If the essential customer requirements are changing over time, the requirements on the product design will also change. In this case the scenario technique can be used for prognostic purposes for the changing of the customer requirements in the future (Fig. 4.3).

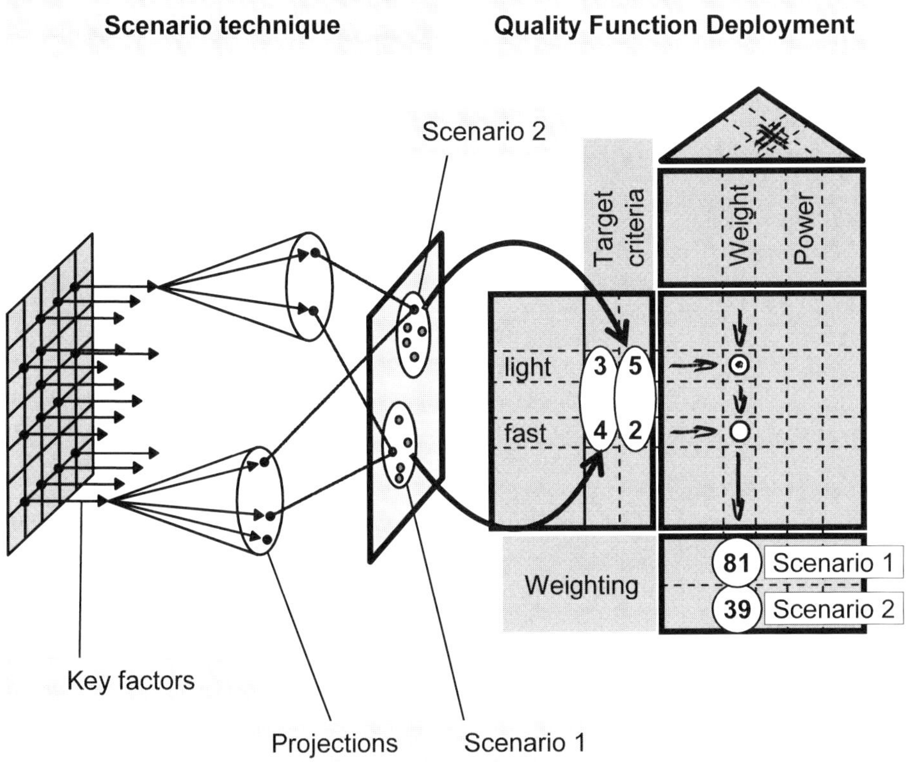

Fig. 4.3 Combination of scenario technique and QFD (Eversheim et al, 2001a)

Starting from the forecast scenarios, the future customer requirements can be derived. The key factors identified during the scenario analysis can be used due to the fact that they often stay in a direct or indirect relation to the customer requirements with respect to a product. Depending on the foreseen development of the key factors, the importance of the customer requirements changes (Fig. 4.3).

4.2 Quality Function Deployment (QFD)

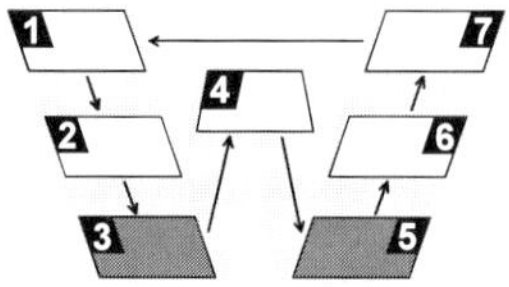

With the methodology of the QFD, freely formulated customer requirements are systematically translated into a technical language through which the evaluation and weighting of different customer requirements can be undertaken. The QFD methodology can be implemented in earlier or later development phases for different purposes. In innovation projects, the QFD methodology is used for the identification of controversial requirements, which later can be executed in problem solving, for example by using the TRIZ methodology.

History of the QFD methodology

The concept for quality development known as *Quality Function Deployment* (QFD) was first presented in the late sixties by Yoji Akao in Japan and in 1972 implemented in the ship plant of the Mitsubishi Heavy Industries, Kobe. However, this concept became known under this name only at the end of the seventies. The QFD is a methodology which allows a process design up to its crucial production steps. It is mostly based on non-technical and complex formulations of customer requirements. With the QFD methodology, the customer wishes are translated into a technical language so that the development of requirement-related products is supported throughout all process steps – planning, development, design – (Akao, 1992). For that translation, several translation matrices are used. They are marked as quality tables or the House of Quality (HoQ). In a simple structured equation the translation takes place in four phases (Fig. 4.4). During each phase, matrices are used for deriving visualization as well as evaluation of the interactions. In comprehensive implementations of the QFD methodology, up to 30 matrices are processed (King, 1994).

House of Quality

The four phases of the QFD methodology

During the first phase the customer requirements are put together and then transformed into product characteristics. Then, in the second phase, the developed product characteristics are further translated into assemblies and single parts. During the third phase the most important characteristics for execution, process steps and parameters are derived. The fourth phase serves for the translation of operation rules into production instructions or quality insurance measures (Akao, 1992).

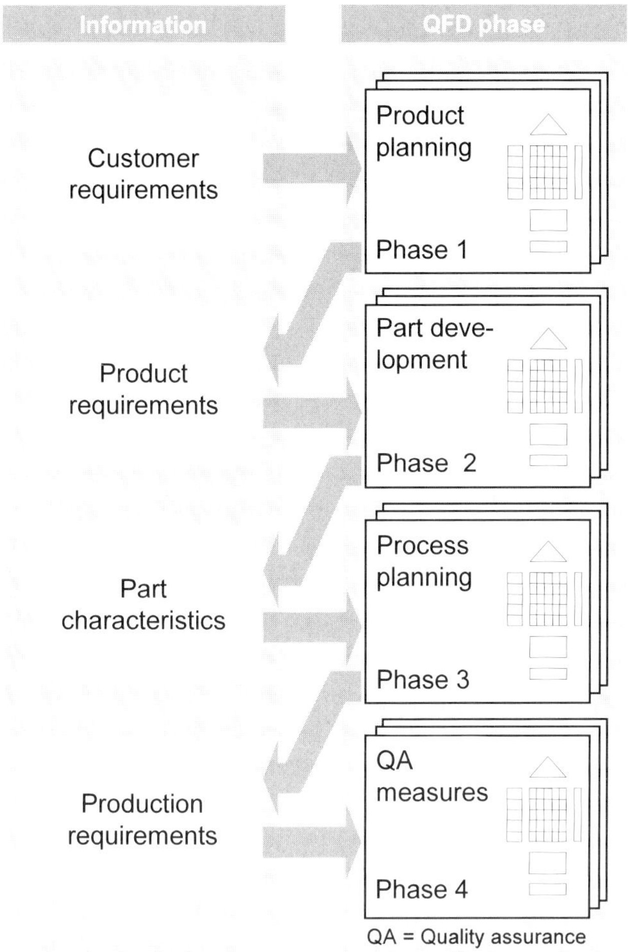

Fig. 4.4 Phases of the QFD

The principle approach for the design of a House of Quality is shown in figure 4.5.

QFD in earlier phases: Deduction of customer requirements and their translation into product or performance characteristics

Especially in the earlier phases of the product planning, the first phase of the QFD methodology is very suitable. During this phase the customer requirements are translated into solution-neutral product or performance characteristics (Geisinger, 1999). The customer requirements are determined, structured and weighted. For weighting the requirements, GEISINGER recommended using the conjoint analysis (Geisinger, 1999).

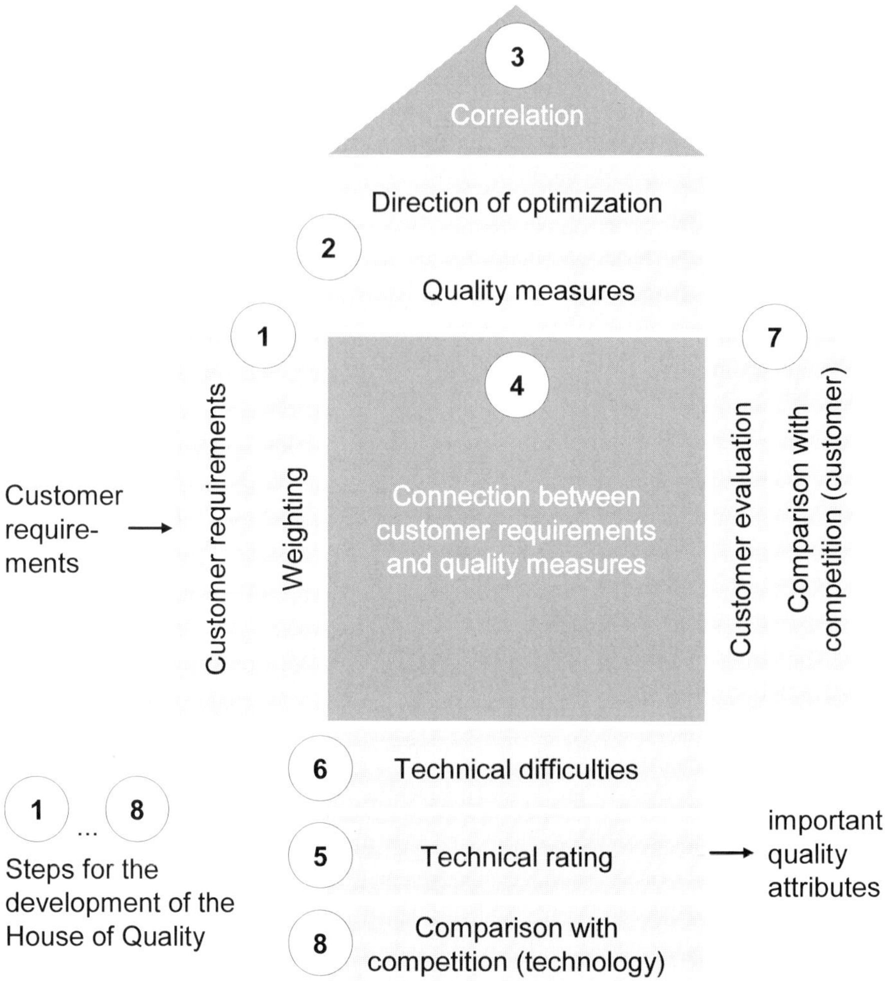

Fig. 4.5 Building the House of Quality

Furthermore the weighting can be carried out through pairwise comparison (Wengler, 1996). For the structuring of the customer requirements the *Kano-model* offers a good approach (Fig. 4.6). With this model, the customer satisfaction is reflected depending on the fulfillment level of each requirement (Kano, 1995)[18].

[18] The Kano-model was named after Noriaki Kano, a Japanese professor and business consultant (Kano, 1995).

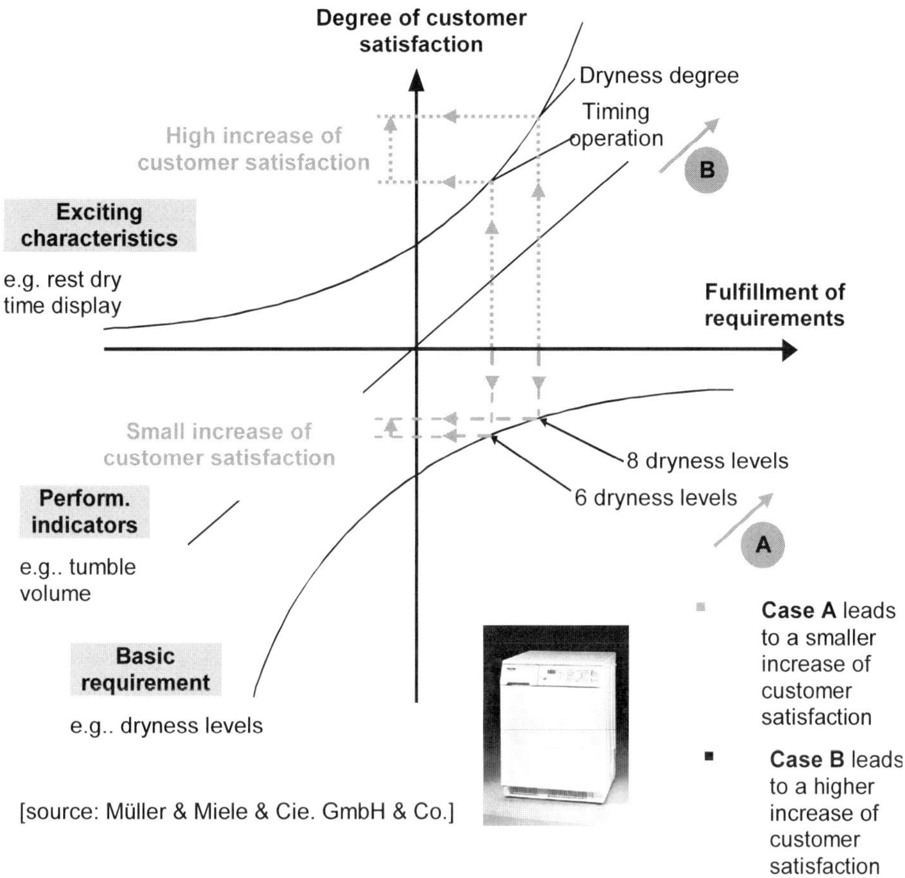

Fig. 4.6 Kano-Model for the tumble dryer

Kano-model: three types of requirement characteristics

KANO distinguishes three types of requirement characteristics. The *basic requirements* are pre-judged by the customers as self-explanatory. The *performance requirements* contribute proportionally to customer satisfaction. The *exciting characteristics* are not expected by the customer, contributing however to a higher degree to his satisfaction. The main understanding is that in questionnaires the customer expresses only a certain portion of his requirements (King, 1994). Here, one must remember that due to time-related dynamics, exciting characteristics, which are established on the market later, lead to performance characteristics and even to basic requirements. For the success of a product the balance of these three factors is crucial (Teufelsdorfer & Conrad, 1998).

Advantages and disadvantages of the QFD methodology are compared

For the transformation of customer requirements into objectively measurable technical parameters the production characteristics are identified during the second step of the building of the House of Quality (Eversheim, 1994; Hartung, 1994). For this, TEUFELSDORFER & CONRAD suggest a functional analysis (Teufelsdorfer & Conrad, 1998). Thereafter, the correlations between the characteristics are calculated and written into the roof of the House of Quality. For the estimation of the correlation, an optimization direction is given for each product characteristic. The connection between customer requirements and quality characteristics is presented in the connection matrix, from which the weighting of the quality characteristics is deduced (Akao, 1992). Apart from these obligatory process steps, the HoQ can reflect competition comparisons from the customer's point of view (right in the rows) as well as from a technical point of view (in the columns).

Advantages and disadvantages of the QFD methodology are compared

The execution of the QFD methodology requires a substantial workload. The crucial point is set on a strong technical operational level and reaches far into the phases of the series development or production planning. An advantage is the clear documentation structure which supports a function-overlapping communication as well as a consensus agreement. The systematic market-orientated deduction of the right product characteristics and the connected goal transparency can be translated onto the innovation planning of technical products.

4.3 Intuitive- and Analogy-based Solution Finding

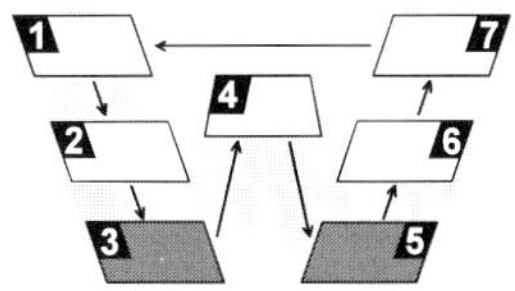

In this chapter the principle of the "intuitive solution finding" based on practice-relevant creativity techniques such as brainstorming and brainwriting is described. Furthermore, it is shown how analogies can be used when dealing with synectics, bionics and the TRIZ methodology. Main guidelines for the application of creativity techniques will complete this chapter.

Intuitive solution finding

Trial-and-error approach

The intuitive solution finding corresponds to the natural approach of mankind related to problem solving. In the so-called trial-and-error method the different possibilities for solving a problem are checked until a satisfactory solution is achieved (Spielberg, 2002) (Fig. 4.7).

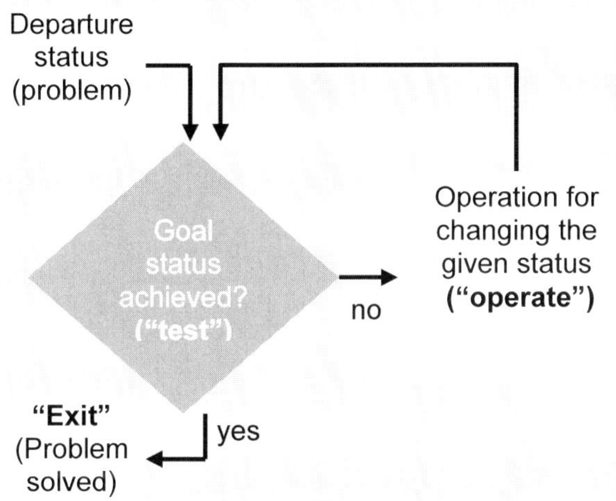

Fig. 4.7 TOTE-cycle (Test, Operate, Test, Exit) as an intuitive problem solving process (Müller 1990)

Minimal-systematic procedure

For the support of this minimal-systematic procedure, a number of creativity techniques were developed in the past, which support idea finding in the problem solving process (Spielberg, 2002). In the literature, these methods are known as intuitive creative methods. In contrast, techniques that are supported by a solution search through a focused problem analysis and a systematic approach are labeled as systematic and analytic methods (Schlicksupp, 1992). In order to collect ideas from different persons with different backgrounds by using comparatively little effort and simple tools, many techniques of both categories are applied in a workshop lasting several hours with mostly interdisciplinary teams. In the literature, a number of methods and instruments for idea generation are listed (Fig. 4.8) (amongst others Geschka, 1986; Haberfellner, 1999; Hauschild, 1997; Schlicksupp, 1992; Walter, 1997).

Creativity techniques: brainstorming and synectics

From the large number of creativity techniques developed mainly in the business environment, which are applied not only for solving technical problems, the methods of *brainstorming* and *synectics* are selected and explained here as significant representatives. The methods can serve as a representative for numerous variants and specific further developments and can contribute to the fundamental understanding of other creativity techniques. Based on brainstorming, the intuitive-related solution search, and based on synectics, the analogy-related approach for solution search during the phases of idea generation and idea detailing are described.

Analogy-based approach for idea generation and idea detailing

For a detailed description of other creativity techniques, the references shall be consulted.

4.3.1 Brainstorming and brainwriting including the 6-3-5 method

Brainstorming

Brainstorming is a method for free and therefore intuitive idea generation in a team. Only a few main rules are valid. So, the solution ideas for the specific task should be freely expressed; a critical discussion does not take place yet. In fact, the ideas suggested by the other participants should consciously be picked up and developed into new ideas. The most important leading thought in this form of idea generation is the principle of quantity before quality and association. In order to minimize competition-related hurdles, the team should be hierarchically homogeneous, but interdisciplinary. For the generated ideas no copyrights are accepted.

Analogy method	Solution of technical problems through investigation of prototypes (examples) from other departments (cf. Synectics)
Bionics	The transfer of biological structures, mechanisms and systems into technical solutions
Brainstorming	Free idea or solution generation in a creative environment
Delphi method	The use of expert knowledge for solution or forecast development
Force-fit method	Generation of new solution ideas by bringing together two different notions through a creative thinking process
Function analysis	Consideration of several possibilities of function fulfillment during the choice of different product functions
Heuristics	Support and guidance during the search for solution on the basis of heuristic principles
Method 6-3-5	Pick up and further development of ideas
Mindmapping	Cartographic representation of thought contents and the resulting idea streams
Morphology matrix	Systematic development of new ideas through a direct confrontation of different attribute specifications
Secondary field integration	Development of solution concepts by taking into consideration different constraints
Problem solving tree	Graphic representation of complex interactions and activities
SIS Methode	Bringing together different single solutions to form a complete solution
Synectics	Intensification of activities for the solution search (cf. Analogy method)
TILMAG method	Determination of new solution ideas through a multi phased association process
TRIZ	Theory for creative problem solving

Fig. 4.8 Methods for idea generation (cf. Eversheim et al. 1996)

Method 6-3-5

An important variant of brainstorming is the so-called brainwriting, which brings the sketched process of free mutual stimulation into a written procedure, which is partially also separated into space and time. Here, the *6-3-5 method* shall be mentioned as an example: in five minutes, six participants write down three ideas on a given topic. Thereafter, the form is given to the next participant and the whole process is repeated five times. Through this approach the first participants steer the thoughts of the next participants, as these can read the already written ideas and can further develop them (Brankamp, 1997; Schlicksupp, 1992).

In the industrial practice the basic principles of these methods are well known. Since brainstorming has been recommended and successfully applied since the sixties already, the mentioned rules of creative working have been established in many companies as a natural component of the company culture. In a certain way, these are universally valid for other creativity techniques (Schelker, 1976).

4.3.2 Synectics

Synectics

The method of synectics was developed with the aim of supporting idea generation in problem solving processes in the direction of unconventional solution concepts (Spielberg, 2002). Here, the idea generation, which is completely free during brainstorming, is expanded by a focused analysis of facts not connected to the problem, i.e. the solution finding is steered. The aim is to discover analogies whose transfer onto the considered problem could bring new solution concepts. In a determined process, the considered problem is so far abstracted that analogy building, e.g. in the area of nature, technology or society, is possible. The identified ideas are transferred onto the problem (Fig. 4.9).

Discover analogies

Focused steering of the creativity

The advantage of this methodology lies in a focused steering of the relativity. On the one hand, it produces a stronger problem focus and on the other hand, it generally provides new and surprising solution concepts. Simple questions such as "Does this problem exist in nature, and how does biology solve it?" can lead the search into promising directions.

Synectics is substantially more complex than brainstorming and is not that common in practice. The results are in general estimated as being valuable (Schlicksupp, 1992; Schelker, 1976).

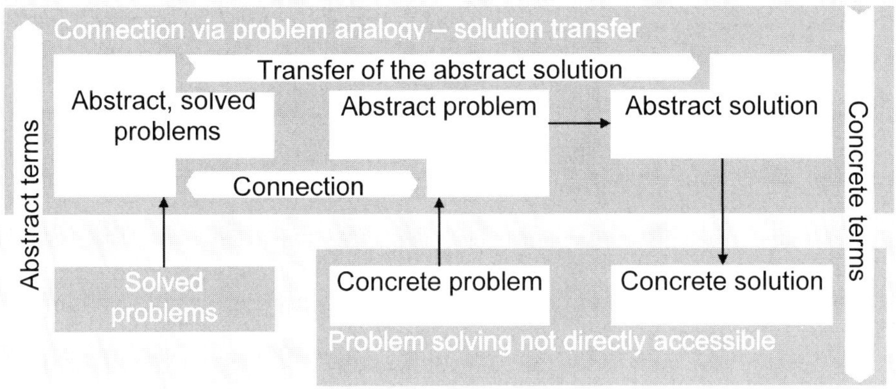

Fig. 4.9 Principle of the analogy-based problem solving

Synectics can be seen as a representative for the methodical concept to find a solution based on analogy setting and transfer. An essential elaboration of this fundamental concept took place in bionics (Chap. 4.6). Through comprehensive fundamental studies in the field of biology, bionics can fall back on suitable analogies that can be transferred onto the technical field and help the user developing superior solutions. In the past, this method was successfully applied in the field of aerodynamics (implementation of the so-called winglet; sharkskin surfaces on aircrafts); in light building (sandwich procedure), or in surface technology (self-cleaning lotus surfaces). HILL has made an essential contribution to the systematic research of problem solving with design principles of nature. As a point of origin he uses the concept of the principles of controversial-oriented problem solving (Chap. 4.4.5). By doing this – e.g. through the cataloguing of structures – he succeeded to open up bionics for the systematic problem solving process (Nachtigall, 1998; Hill, 1999) (Chap. 4.6)

Problem solving through analogy setting and transfer

Bionics

Contradiction-oriented problem solving

Analogy setting is an essential part of the TRIZ methodology (Chap. 4.4). GENRICH ALTSCHULLER (the founder of the TRIZ methodology) recognized that many creative problem solutions are based on an analogue process. As a starting point, ALTSCHULLER uses the fact that similar problems were already solved in another context – for the purpose of this methodology there are technical connections. These solutions must be found and transferred to the concrete problem. For the purpose of building analogies and transferring solutions to a problem, the TRIZ methodology

TRIZ methodology

offers efficient tools (methods) which steer the solution search (Altschuller, 1984; Herb, 2000).

Basic rules for applying creative techniques

Basic rules for applying creative techniques

For an efficient and successful application of creative techniques a few basic rules have to be considered. These rules are known from brainstorming and are valid for most creative techniques (Schlicksupp, 1992; Herb, 2000):

- If possible, the group potential is used for problem solving (expertise of several people).
- Criticism during idea generation and killing phrases are not allowed.
- Ideas are taken up in a group in order to initiate associations.
- Even the most unusual concepts and suggestions are given space for development.
- As many ideas as possible are put together – their evaluation will be undertaken later.

Result

The selection of suitable methods depends on the aspired solution level

A variety of methods are available to support solution finding during the phases of idea generation and idea detailing. The type of creativity stimulation, the systematic approach, and the work effort of the different methods differ a lot. For the situation-specific choice of a suitable method the available time, budget and staff resources are compared to the aspired solution level. It has to be considered that in general, with greater effort invested, solutions at a higher level can be expected.

4.4 TRIZ Methodology

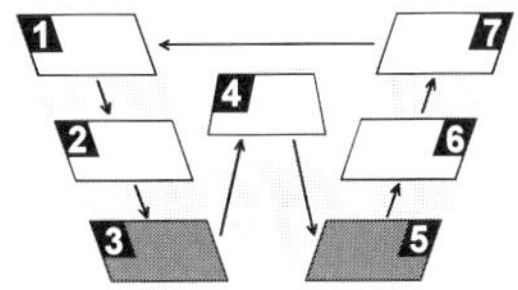

The TRIZ methodology offers the opportunity to abstract technical conflicts that often occur within the operative planning phases, and to channel them into a solution. The methodology can best be used during the phases of idea generation or detailing, since during these phases ideas are generated which often lead to a technical contradiction that can be solved. The highly systematic methodology supports the user in his quest of keeping the realization effort low. The essential element of the TRIZ methodology is the contradiction matrix, which is based on an analysis of more than 2.5 million patents[19]. Parallel to this aid the TRIZ methodology offers further tools which simplify solution finding.

The TRIZ methodology

TRIZ is a methodology which was developed in the former Soviet Union for the technical problem solving process. TRIZ is the Russian, by now internationally used acronym for "Theory for the Solution of Inventive Problems". The basis of the methodology lies in the works and empirical studies of the Russian scientist GENRICH ALTSCHULLER. He assumes that the way to an invention follows certain laws and rules. In order to prove this assumption ALTSCHULLER analyzed a number of patents and came to the following conclusions:

- The detailed description of a problem often leads to creative solutions.
- Many problems have already been solved under different names in other fields and areas, but are comparable from a content point of view.
- The contradiction is the central, innovation producing element of numerous patent scripts.
- The further development of technical systems follows certain basic rules.

Analogies are identified and used

The general approach of the TRIZ methodology is based on the identification and use of analogies and follows a basic pattern of four steps. First, the specific problem is analyzed and abstracted. In the abstract form of the problem description analogies to earlier problem statements are used in order to

[19] The data about the number of analyzed patents vary. Altschuller himself is claimed to have analyzed about two hundred thousand patents. During the further development of the methodology, up to 2.5 million patents are claimed to have been incorporated into the TRIZ knowledge base.

transfer the solution principles of these problems to this particular problem (Fig. 4.10).

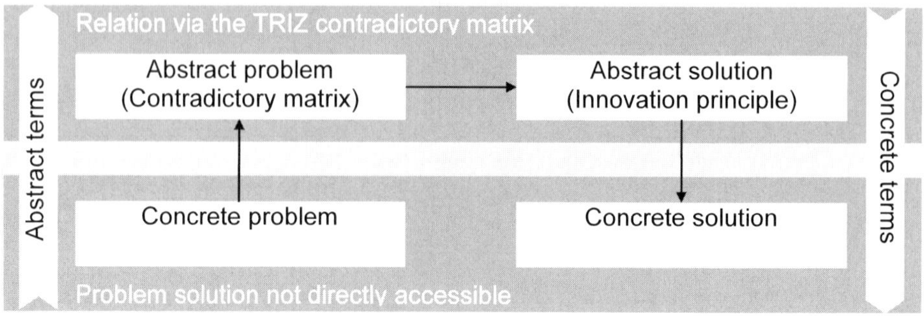

Fig. 4.10 Problem solution process in the TRIZ methodology

TRIZ methodology is a set of tools

For methodical support ALTSCHULLER and his colleagues developed a number of tools, which are presented below (Fig. 4.11). Basically, these tools/methods can be classified into the categories systematics, knowledge and analogy, which represent the three fundamental columns of the TRIZ methodology.

Systematic tools

The systematic tools support the developer in the analysis and detailing of the problem situation. They are for example used for the development of basic concepts or the identification of functions to be optimized.

Knowledge based tools

The knowledge-based tools of the TRIZ methodology contribute a wide spectrum of knowledge of the areas of chemistry, physics, mechanics, and thermo-dynamics and make it accessible through suitable methods.

Analogy tools

On the one hand, the analogy tools are based on the patent analyses, which ALTSCHULLER started and which are still continued today; on the other hand, they are supported by the long-time experiences from research, development and application.

Primarily, the tools of the systematics column are used for the formulation of a concrete problem, while the knowledge- and analogy-based tools are mainly applicable to problem solving. The following presentation of the TRIZ tools follows the representation as shown in figure 4.11. The TRIZ methodology does not require a certain sequence in the application of the tools, nor does a specific procedure have to be followed. Thus, the tools listed below are just a small representation of many other possibilities.

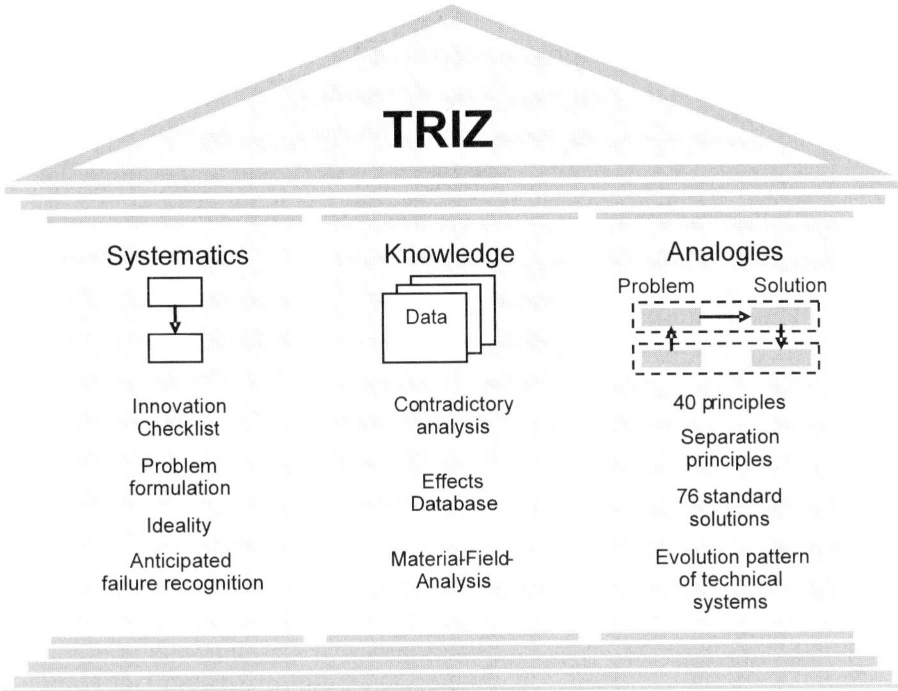

Fig. 4.11 Tools for the TRIZ methodology

4.4.1 Innovation checklist

With the *innovation checklist* the problem formulation is supported by a structured listing of system- and situation-related aspects (Terninko et al, 1998). This list contains information which can be classified into six topics from a content point of view (Fig. 4.12).

Innovation checklist

First, data are collected about the environment and the system environment to be developed or improved. In order to do so, the system identification is determined, the primarily useful function of the system is formulated and the current or desirable system structure is presented. On this basis the functioning of the system when executing the primarily useful function can be described.

Available resources

An exact analysis of the system environment is necessary for the identification and the use of the available resources. For example, free available heat can be used as an energy source for a sub-system. With regards to content, the identified resources of the system or system environment can be classified according to material-related, field-related, functional, informative, time-related, and space-related resources.

Innovation Checklist

Information regarding the system to be improved and its environment ☑
"What is it about?"

Available resources ☑
"Which resources can I use?"

Information regarding the problem situation ☑
"Why is there a problem at all?"

Changing the system ☑
"What are the boundaries regarding the problem solving?"

Selection criteria for the solution concept ☑
"What do I want to achieve with the new product?"

History of solution attempts ☑
"Has the problem already been solved somewhere else?"

Fig. 4.12 Thematic content of the innovation checklist

Analyze the problem situation

Accept system changes

Formulate selection criteria for solution concepts

Document history of solution attempts

The next topic of the innovation checklist to be discussed is the information about the problem situation. Here, the data regarding the aspired improvement of the system or construction as well as possible disadvantages which should be avoided are collected. A mechanism or *modus operandi* of the "bad solution" is highlighted and the development history of the problem is analyzed. In a further point of the checklist the change of the system is discussed. It is determined whether and in which magnitude and in which boundaries changes of the system are acceptable. Then, the selection criteria of the solution concepts are formulated, i.e. the technical and economic properties as well as a related time plan are specified. Furthermore, the expected new type of the system is documented. In a last thematic focus of the innovation checklist the history of past solution attempts is dealt with. Here, earlier attempts for problem solving as well as systems that describe a similar problem are investigated.

4.4.2 Problem formulation/Effect structure analysis

Operation structure analysis

After a first concretization of the task as well as of the system to be developed and its environment, the general problem will be divided into different part problems during the *effect structure analysis*. The goal is to split complex interactions into simple manageable connections in order to be able to better recognize the problems.

Cause-and-effect diagram

An important sub-step is the formulation of a *cause-and-effect diagram*. Elements of this diagram are damaging (DF) and useful functions (UF), as well as connections between single functions (Fig. 4.13). The construction of the diagram starts "either with the primarily damaging functions (PDF) or (with) the primarily useful functions (PUF)" of the whole system (Terninko et al, 1998). In a second step connections can be made based on specific standard questions.

Figure 4.13 shows a cause-and-effect diagram for a melting furnace. The primarily useful function of the melting furnace is the *extraction of metals*; the primarily damaging function is the *explosion danger* due to water leakage. The water leakage could occur due to cracks in the pipe as well as high water pressure. Through this systematic presentation the chain of causation can be reconstructed from the PDF to the PUF, the extraction of metals.

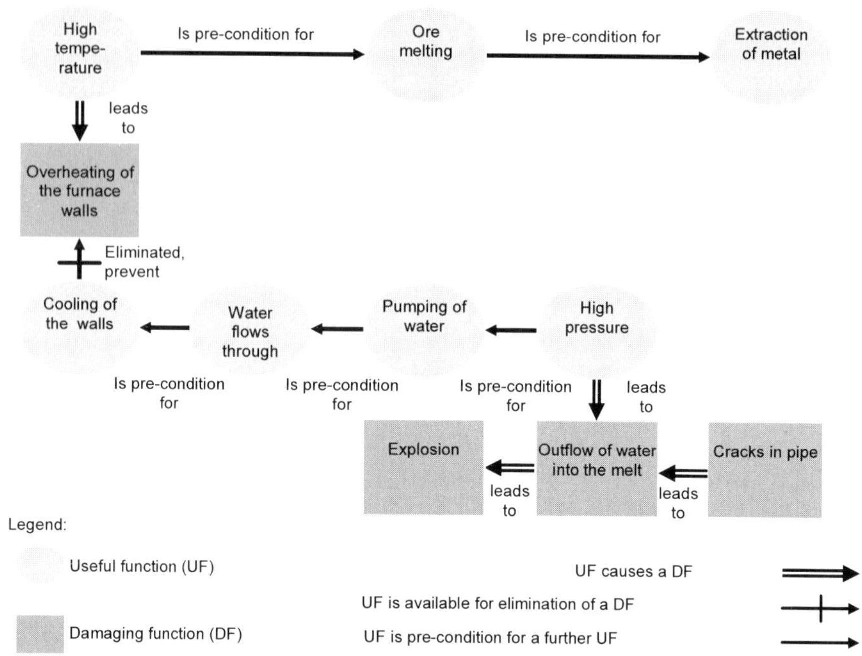

Fig. 4.13 Cause-and-effect diagram for the example of a melting furnace (Terninko et al. 1998)

Contradictory development

Through specific questions to each of the single functions, part problems can be defined. The questions are: "Can a benefit follow out of a DF?", "Can a UF be improved?" If, out of a function, two interactions result, e.g. a function leads to another function and at the same time is a pre-condition for a third one, then a contradiction occurs, which must be solved.

Combining QFD methodology and TRIZ

For systems which, due to their high complexity for instance, allow many possibilities for cause-and-effect diagrams, it is recommended to use the QFD methodology (Quality Function Deployment) (Terninko et al, 1998). The application of the QFD methodology leads to the representation of contradictions in a so-called correlation matrix in the roof of the House of Quality (Fig. 4.14).

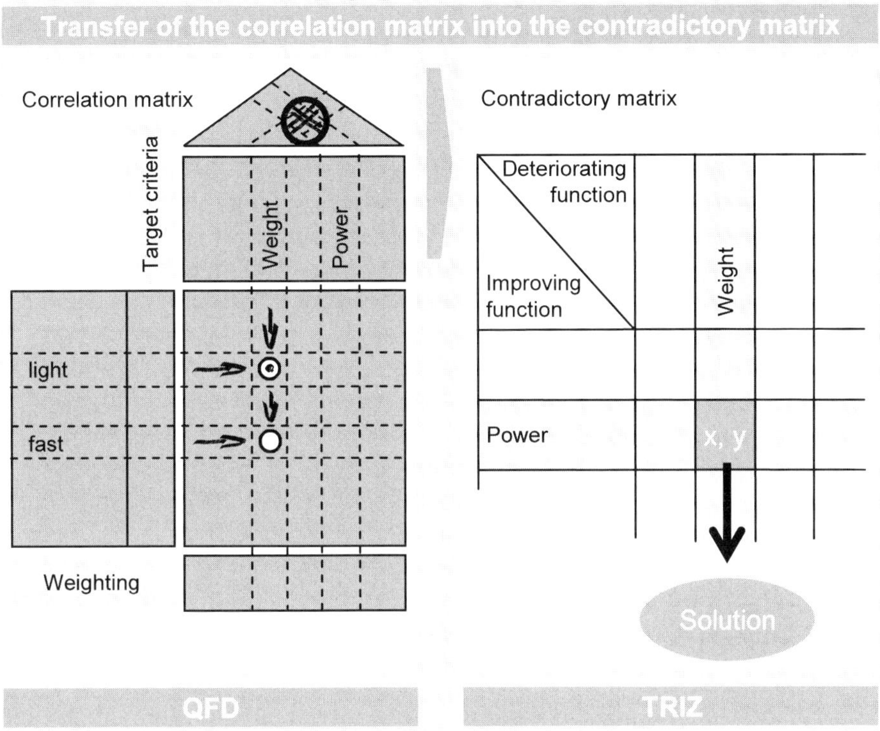

Fig. 4.14 Combining QFD and TRIZ (Eversheim et al. 2001a)

As with the QFD methodology customer requirements are translated into technical specifications, it is assured that the contradictions exist in the form of technical incompatibilities which can be solved by using the contradictory matrix (Fig. 4.15). The systematic combination of the QFD methodology and the TRIZ contradictory matrix can be methodically supported (cf. Eversheim et al, 2001a; Teufelsdorfer & Conrad, 1998).

4.4.3 Ideality

Ideal System

An *ideal system* is achieved when there is no system anymore but when its function is still applied (Altschuller, 1984). In general, such a result is utopian. However, it is applied as a mental game in order to abolish thought barriers.

Ideality is the ratio of the sum of all useful functions to the sum of all damaging functions

Ideality is defined as the ratio of the sum of all useful functions (UF) of a system to the sum of all damaging functions (DF). The formulation of the ideal product should support the user during the goal definition. He can orient his work towards the ideal and in this way work in a goal-oriented and efficient way and find new solution principles (Altschuller, 1984). By using this approach, weaknesses of the existing system are recognized and functions are made apparent which do not contribute to the fulfillment of the actual purpose of the machine, but became necessary because of compromise solutions of part problems. For example, heat pumps are installed in machines in order to divert excessive heat, although this does not correspond to the actual purpose of the machine, namely for instance the milling of components. By focusing on the ideal product new solution principles can be discovered. With regard to the above-mentioned example it is advisable to try to prevent heat generation at all.

Ideal end result

The first step when using the tool of ideality is the formulation of the "ideal end result" (Altschuller, 1984). The following ideas are used to reach this goal:

- No machine exists; the required effect is, however, achieved.
- There is no time or energy consumption, the required effect is, however, achieved.
- No material is available; its function is, however, fulfilled.

The following hints (Terninko et al, 1998) can help when trying to achieve the best possible fulfillment of ideality:

- *Elimination of supportive functions:* help functions do not directly contribute to achieving the required tasks and can therefore often be replaced.
- *Elimination of components:* the task of functional components might be taken over by available materials or functional resources or fields.
- *Recognition of self-supplies:* in many systems there are functions which could additionally be used for other functions. For that purpose the single functions of a system must be completely disclosed.
- *Replacement of single components:* by using models, simulations or cost-effective replicas substantial costs could be saved.
- *Change of the functional principles:* other processes or work systems might better fulfill the desired function than the existing do.

- *Utilization of resources:* this hint is partially contained in the previous ones; it is specifically aimed at using the resources for the problem identified by the innovation checklist.

4.4.4 Anticipative error detection

Anticipative error detection

The anticipative error detection is also known as subversive[20] fault analysis (Terninko et al, 1998). The following question is fundamental: "What can we, and what can Mother Nature do to make our product or our process fail?" It is sought for damaging or inefficient characteristics of the product or of the process, which could result in the maximal error case – the destruction of the product or the process. In general, there are nine steps in the anticipative error detection, which Terninko describes as follows (cf. Terninko et al, 1998):

1. *Formulate the original problem:* in this first step the primary utility functions as well as the central problem are defined. It is also tried to list conditions and reasons through which the problem could have been caused. The elimination of this problem should not negatively influence utility functions or create new problems.
2. *Formulate the inverted problem:* In contrast to the first step, here, the problem is formulated as a primary utility function and in this way is reversed. The problem is defined as a requirement.
3. *Intensify the inverted problem:* the inverted problem is exaggerated. New effects which further intensify the problem are searched for and applied on the problem.
4. *Search for obvious solutions for the inverted problem:* through analogies out of everyday life it is tried to find realization possibilities for the inverted problem. The solution should include the previously formulated intensified effect.
5. *Identify and use resources:* it is tried to identify useful resources out of the surrounding environment or system, which could contribute to the solution of the problem.
6. *Search for useful effects:* besides physical, chemical or geometric effects, human, animal or psychological effects are observed for the solution of the inverted problem.

[20] Subversive (lat.): overturning, demolishing

7. *Search for new solutions:* in this 7th step it should be tried to realize the described effects, i.e. to solve the inverted and intensified problem by taking into consideration limited resources.
8. *Re-inversion and verification:* based on the found solution, hypotheses are formulated and tested.
9. *Develop methodologies for error prevention:* the problem is inverted again in order to formulate the solution for the elimination of errors.

4.4.5
Analysis of technical conflicts: contradiction matrix

Conflict of goals

39 system parameters or technical standard parameters resp.

Contadiction matrix

40 innovation principles

A challenging technical problem is characterized by a not (yet) solvable conflict of goals, i.e. by a minimum of two parameters that have to be optimized but that cannot be realized simultaneously by the developer with the available technical means, or does not allow an acceptable compromise (Altschuller, 1984). For the description of contradictions ALTSCHULLER numbers 39 system parameters or technical standard parameters, which enable the description of most contradictions when they are sufficiently abstracted. These standard parameters are opposed to each other in a contradiction matrix (Fig. 4.15). When searching for a solution, the one parameter that should be improved is selected from the row. From the column, the parameter that will be impaired is selected. ALTSCHULLER identifies 40 innovation principles that were used for problem solving in the patents analyzed by him and allocates them to the corresponding contradictions in the contradiction matrix. There are fields which contain more principles but there are also fields that are empty, i.e. up to now, no sufficiently clear and abstractable patent for the solution of the particular conflict has been found (cf. Terninko et al, 1998). The order of the innovation principles in a field corresponds to the frequency of the patents which were solved by using the corresponding innovation principle.

Example of conflict analysis

The application of the contradiction matrix can be explained by the following example: an enterprise produces drills for different applications. The goal of the optimizing measure is to accelerate the drilling process. This could be achieved by increasing the drilling performance, which is represented by the rotational speed of the drill. A disadvantage is that by increasing the drilling performance, more heat is generated during the drilling process, which can lead to friction welding, broken die or material distortion. Thus, the increase of the drilling speed presents a technical conflict vis-à-vis the increasing temperature. For the solution of this conflict the contradiction matrix suggests the following innovation principles – *28: Replace mechanics, 30: Flexible shells and films, 36: Phase transition* and *2: Disconnection,* the latter being identified as the most common principle solution for this kind of problem.

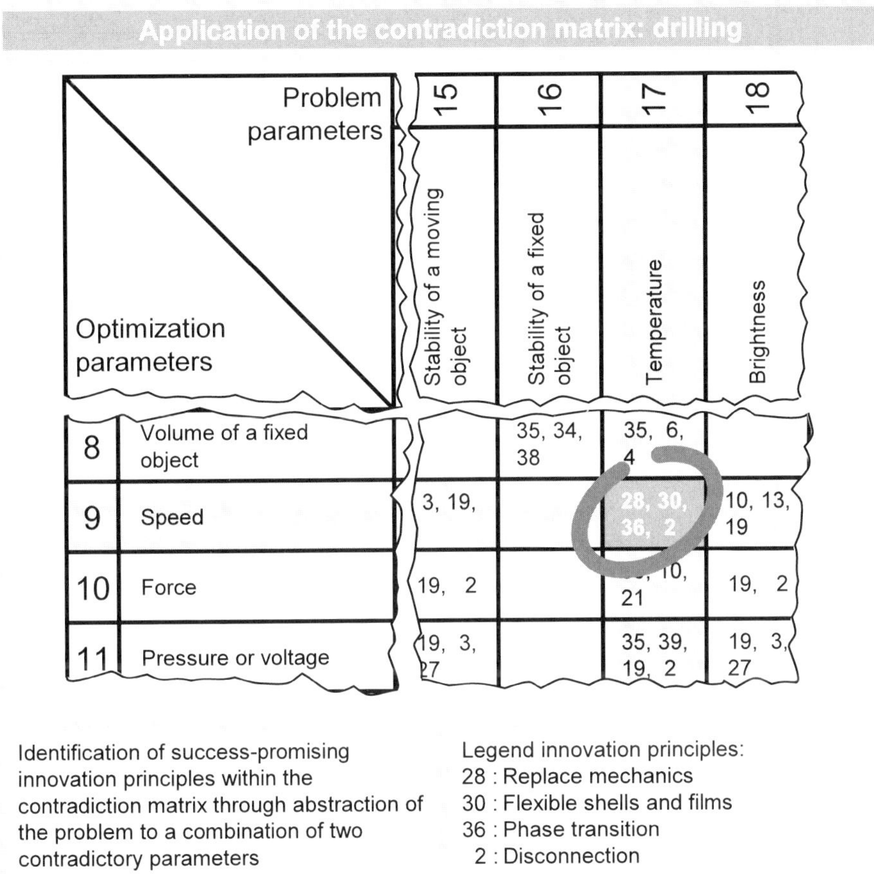

Fig. 4.15 Excerpt from the contradiction matrix

In the list of the 40 innovation principles identified by ALTSCHULLER, the following specifications of the named principles can be found:

- *Replace mechanics:* the conflict will be solved through the replacement or exchange of systems (mechanical, optical, acoustical, etc.) or fields (electric, magnetic, stationary, periodic, structured, etc.). Example: in order to increase the holding force of a metallic coat on a thermoplastic, the coating process is carried out in the presence of an electro-magnetic field, which increases the holding force.
- *Flexible shells and films:* the solution principle consists of replacing or isolating usual constructions by flexible shells or thin films. Example: in order to reduce water loss from plants, their leaves are treated with polyethylene spray. The polyethylene solidifies and leads to a better growth of the plant because the oxygen can easily pass this protecting film; the water vapor, however, passes through more difficultly.
- *Phase transition:* in the example of the drilling tool, it is this innovation principle that leads to a solution. In general it says: use the effects created during the phase transition of a substance - volume change, heat generation or absorption. In the example, the drill was coated with paraffin, a substance that dissolves during heating and takes out the heat in form of steam (Fig. 4.16).
- *Disconnection:* this innovation principle suggests that through the removal or cutting off of the interfering component of an object, or through the separation of the necessary component or the essential characteristic, the conflict can be solved. Example: the security at airports is increased by playing tapes of birds (of prey) – other birds stay away from the airport (the property "sound" is separated from the object "bird" and used in another context).

4.4.6 Analysis of physical contradictions: separation principles

Physical contradiction

A physical contradiction exists if the existence of a defined status is required simultaneously with a contrarian property (e.g. an object should be hot and at the same time cold). Basically, a technical contradiction is transformed into a physical one by identifying the parameter that influences both, the desired as well as the undesired result. The physical contradiction is then defined by exactly this parameter (Terninko et al, 1998):

- *Technical conflict:* The heating of component "A" improves "A", at the same time, however, destroys component "B".
- *Physical conflict:* At the same time, "A" must be hot and cold.

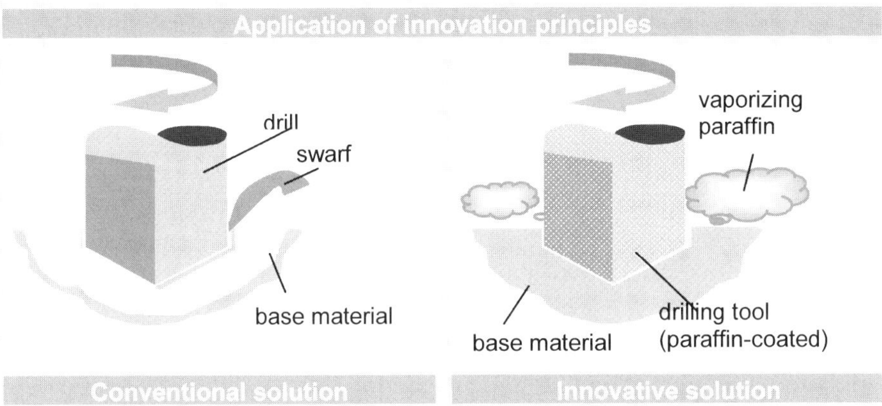

Fig. 4.16 Solving the contradiction with innovation principle 36: Phase transition

Technical contradictions are transformed into physical contradictions

The transformation of a technical into a physical contradiction can be useful especially when the contradiction matrix does not offer a suitable solution. The physical conflicts are particularly suitable to find almost ideal solutions for completely contradictory requirements on a physical level. The latter are solved through the separation of requirements. For this approach there are four *separation principles*:

Four separation principles

- *Separation in space:* Solving the problem through the local separation of components or the fragmentation of a component into further single components which in sum give the same result. Example: spatial separation of a charging station and a cell phone.
- *Separation in time:* the functioning of a system is segmented with regards to time, so that conflicting requirements, functions or conditions are needed at different times, i.e.: a process is subdivided into several subsequent sub-processes; a process which, however, does not influence the desired function. Example: a chair lift, where the contradictory functions of *fast transport during the run* and *slow movement for safe access and exit* have to be realized. The conflict is solved through a separation in time: at the access and

exit points, the chairs are disconnected from the hauling rope (Herb, 2000).

- *Separation within an object and its parts:* if a system has to fulfill contradictory functions or if it has to work under conflicting conditions, the system is subdivided into sub-systems and one of the contradictory functions is carried out by a different sub-system. Example: a structured window glass which combines the functions *translucent* and *protected against views*.
- *Separation through condition change:* the conflicting requirements are separated through modification of conditions under which simultaneously a useful and a damaging process are taking place. The system or the environment are to be modified in such a way that only the useful process is carried out. Example: drugs that could be life-threatening when taken excessively could be covered with a sick-making substance, so that vomiting is forced when taken excessively (Terninko et al, 1998).

4.4.7 Effects database

Describe the purpose and overall function

Present energy, material and signal flows

In order to use the *effects database*, function structures for the planned object are identified. For that purpose the overall function or respectively the purpose is presented in a requirements list containing all energy, material, and signal flows as a black-box (Fig. 4.17). Then, the overall function is subdivided into part functions. Afterwards, the part functions are structured and combined (VDI, 1993).

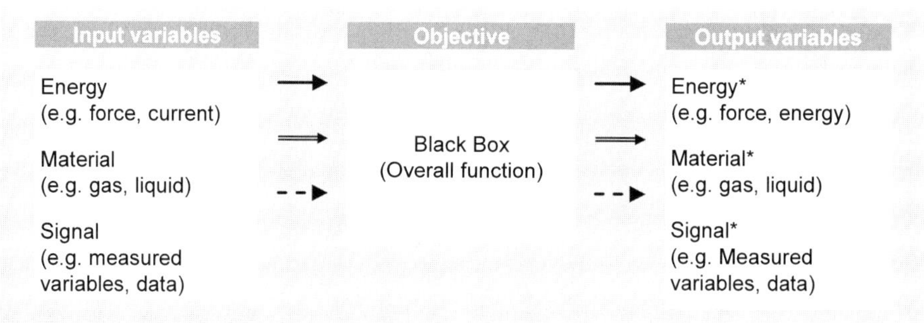

Fig. 4.17 Black Box

A function is the complete description of an activity of a technical object that is available or still to be designed; a technical object can perform one or more functions (activities) (compare Koller, 1994).

Definition: Function

When functions cannot be sub-divided into different activities anymore, Koller speaks about *elementary functions.* For these elementary functions he provides tables which are sorted according to energy and material operations, energy and material combining operations, as well as data and information using systems. From these tables, functional structures with an integrated appearance can be derived without suggesting a certain solution. KOLLER (Koller 1994) and LINDE (Linde and Hill 1993), for instance, suggest such systematics of physical effects. Moreover, the software *TechOptimizer* provides computer-assisted support (Invention Machine 1998).

Elementary functions

4.4.8 Substance-Field (S-Field) analysis

The *S-Field analysis* is based on the assumption that each basic element of a technical system has three components: an energy, a field (F), which influences a system component (S1) in such a way that another system component (S2) experiences a modification (Altschuller 1984). In most cases, the S-Field analysis is applied in order to model problems of existing technical systems and, based on this, to analyze and compare the solution ideas – coming from the effects database for instance (Terninko et al 1998).

S-field analysis

With the Substance-Field analysis technical systems can be easily modeled. The method will be most successful if a systematic problem definition is conducted beforehand. Once the problem has been defined, different solution concepts can be generated by resorting to standard solutions – ALTSCHULLER mentions 76 (Chap. 4.4.9) - and some elementary rules (Altschuller 1984).

76 standard solutions

First, the cause-effect relationship between fields and substances must be identified; then, in a second step, the Substance-Field model is built. A minimal functional technical system consists of two substances and one field. It is investigated whether the system is complete, shows a damaging function which has to be eliminated, and/or is efficient enough. If the investigation shows that there is an obvious need for improvement, new solution concepts can be determined by means of the 76 standard solutions (Terninko et al. 1998).

This approach is shown in figure 4.18 with the example of an air hammer. The task here is to shatter a boulder. This should be done by using a pneumatic hammer that performs a pulsed mechanical force on the boulder.

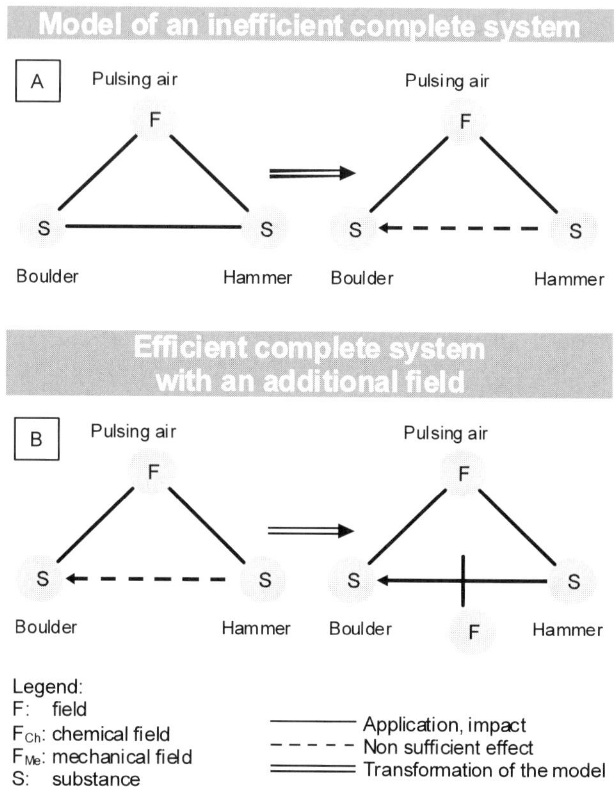

Fig. 4.18 Substance-Field analysis

The system is completely described with the components (air) hammer, pulsed air and boulder. In a next step the efficiency of the system can be analyzed and exemplarily represented. It could for instance turn out that the power of impact is not sufficient to destroy the boulder (Fig. 4.18, part A). In part B of this figure a suggestion for improvement is developed, using standard solutions (Chap. 4.4.9). The adaption of each standard to the specific problem leads to new innovative ideas and solutions. Starting from the inefficient complete system, a new complete efficient system is generated by adding an additional energy, a chemical field. The field can for instance be realized by using an aggressive liquid (chemical), which chaps the boulder and makes it easier to destroy.

4.4.9 76 Standard solutions

Physical effects and phenomena

Currently, the collection of standard solutions compiled by ALTSCHULLER counts 76 "physical effects and phenomena". It was developed with the purpose of using knowledge from science and technology outside of the specific disciplines in which it was discovered. The compilation of effects should help the user look for solutions outside of his discipline and to break through psychological (thought) barriers. Since the effects are presented on an abstract level, the S-Field analysis can be used for preparing the problem (see above). The standard solution applied in the example above is "add a new substance". Further examples for standard solutions are "use magnetic liquids" and "use resonance" (Altschuller 1984; Herb 2000).

4.4.10 Laws of Technical Systems Evolution

Standard development patterns of technical evolution

During his investigations, ALTSCHULLER discovered that technical systems follow the so-called *standard development patterns* of the technical evolution (Chap. 4.5). Such a standard development pattern describes, in the sense of a generally applicable rule, a sequence of changes which congruently occur during the further development of many different systems. The knowledge and application of this standard development pattern allow already today to find further technical development for the future. TERNINKO ET AL. (1998) distinguish the following development patterns:

- *Stepwise evolution:* each technical system goes through several development stages which describe the maturity level of a design or the quality of the system as a function of time.
- *Increasing ideality:* each system accomplishes useful and damaging functions. The general evolutionary further development of systems to increased ideality is based on the improvement of the ratio between all positive and all negative functions.

- *Uneven development of parts of the system:* a product mostly consists of different system elements, which - seen isolated – are all in a different phase of their lifecycle. The component that first leaves the maturity phase slows down the whole system. An under-developed part too limits the whole system until it is completely developed. The key to further development results from the understanding of the overall function as an interaction of many part functions.
- *Increased dynamic and controllability:* increased dynamic means the development of a static system to a dynamic system and further to an improved dynamic. Increased controllability means the development of a system that is not influenceable from outside at its lowest development phase; in the next phase the system becomes controllable and finally becomes a system that integrates the properties steering and controlling.
- *First increased complexity, then simplification:* the further development of a simple (mono) system leads via a two-phase system to multi-component systems which, in a further step of the synthesis, fulfill all these functions as a monolithic system of a higher ideality.
- *Change of symmetry and asymmetry:* attempt of solution finding by which single system parts are systematically designed suitable or non suitable, in order to avoid undesirable effects, and to improve the allover performance.
- *Miniaturization and increased use of fields:* this pattern describes the development of big, rigid systems towards immaterial systems (fields) via smaller, lighter systems.
- *Less human interaction:* the development of systems in which humans are performing just monitoring and pure intellectual activities in systems.

Application of the TRIZ tools

In order to integrate the modules of the TRIZ methodology into a systematic process plan for problem solving, the so-called *ARIZ*[21] *algorithm* was developed. The approach, which at the beginning consisted of only four steps, has become much more complex from 1959 until today, so that an updated version according to *Ideation International Inc.* consists of approximately 100 steps (IWB Software 1997). The TRIZ

[21] ARIZ: Algorithm of Inventive Problem Solving (Abbreviation according to the Russian acronym (Altschuller, 1973)).

algorithm is a very challenging way of problem solving: it should only be used if the application of the other TRIZ tools did not lead to a result (Terninko et al. 1998).

4.5
Laws of Technical Systems Evolution

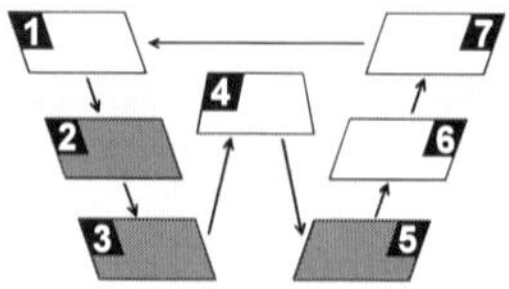

The laws of technical systems evolution can be applied in two different ways during the development process. On the one hand solution ideas can be developed by applying evolution anticipated by using the knowledge of evolutionary principles.

Looking into the future of markets and techniques

The laws of technical evolution describe an idealized process that technical systems become subject to because of further developments and inventions (Pannenbäcker 2001). The knowledge about these laws not only holds big innovation potentials but also supports the user in deciding between various product alternatives and innovation goals. The laws of technical evolution enable a look into the future of the market and product development (Terninko et al. 1998). This tool identifies the evolution phase of a system and from this deduces the need for action. The core element is the knowledge about life phases and lifelines of technical systems (Altschuller 1984).

The laws of technical evolution can be applied in both, the phases of idea generation and idea detailing for finding a solution, as well as in the phases of future analysis for innovation planning. The work effort as well as the team size strongly depend on the situation.

Although the laws of technical evolution have been known for a long time (Altschuller 1984), there yet exists no complete, harmonized procedure which could implement this knowledge (Pannenbäcker 2001). In the following, eight evolution principles are suggested which enable the developer and decision maker to design new solution possibilities and to anticipate technical developments.

The principles listed here are essentially based on different approaches of the TRIZ methodology (Altschuller 1998; Herb et al. 2000; Linde and Hill 1993; Livotov 2002; Pannenbäcker 2001; Terninko et al. 1998; Petroski 1992; Petrov 2001; TriSolver Group 2002; 2nd Innovation Workshop 2002) (Chap. 4.4). They were simplified for the practical application. The evolution principles cannot be clearly separated from each other and influence each other.

Principle structure and approach

The structure and the procedure are identical for each of the eight principles. The evolution of technical systems is put over time and subdivided into phases. The system to be considered and its components are analyzed and the actual development status is then assigned to a phase according to the respective principle. In accordance with this positioning, it is searched for future developments. For the methodologies of synectics (Chap. 4.3) and TRIZ (Chap. 4.4) the described principle of analogy is applied:

The concrete technical situation is abstracted and compared with the phases of the different evolution principles. Through the general description of this phase as well as of the following phases that the system will probably pass through, the user can, through analogy analyses, draw conclusions about both, the current situation as well as the future development of the specific technical system.

Steps of the evaluation principles

The application of the evolution principles is divided into three steps depending on the effort and the level of the target results. Depending on the target-setting, one of the three steps is selected. In the first step the evolution principles that demand a comparatively low effort are applied on an existing technical system in order to quickly improve its performance characteristics and reduce the production costs. The second step allows a more detailed prognosis. Here, not only the technical system is investigated but also its components. The results are for instance design sketches, specifications of components, and the selection of suitable manufacturing processes, which can be realized on the medium run. In the third step evolution principles are comprehensively applied in order to develop solution ideas which in general can only be realized well in the future and bear a high innovative potential. In this step, the evolution principles are applied on technologies or branches and are accompanied by patent search, market and trend analyses (TriSolver Group 2002).

Application is software-assisted

The application of the evolution principles as described below can be software-assisted. The programs of "Invention Machine" and TriSolver describe comparable evolution principles, give ideas and support the user in his systematic search for solution (Invention Machine 1998; TriSolver Group 2002).

Evolution principle 1: Lifecycle curve

S-curve

The life phases of both, technical as well as biological systems can be represented by an S-shaped curve, the lifecycle curve (Fig. 4.19). This fact is based on the systematic analysis of technology history (Brockhoff 1999). The fundamental idea of all life-cycle models and the resulting concepts is based on generally observable biological processes: objects have a limited durability; they come into existence and disappear. In analogy, models for market, branch, enterprise, product and technology cycles were developed and to a certain extent integrated (amongst others Brandenburg 2002; Bullinger 1994; Pümpin 1992; Servatius 1985; Pfeiffer 1982). In the following, only the lifecycle of technical systems or respectively products is considered. Since individual products do not necessarily follow the idealized form of the S-curve, but product or technology families, however, do so, the lifecycle curve can be used as a support for the decision concerning the development and application of technologies (amongst others Brandenburg 2002; Perillieux 1987; Wolfrum 1984; Bullinger 1994; Michel 1987).

Life phases

In the graphic representation the maturity or quality of a system over time is applied (Terninko et al. 1998). The curve can be divided into four areas. In the first phase - the childhood or the creation of the products of a product family - the development of the system is still slow. Then, the product family enters the phase of growth and is rapidly improved. In the phase of maturity the development pace stagnates; in old age, the products are displaced by a new product family or integrated into super-systems. Due to the high cost pressure in the last phase the quality of the products often decreases (Altschuller 1998). Furthermore, the lifecycle curve gives an idea about the turnover that can be achieved with a product or a product family depending on time.

Curves serve the forecast

Based on such curves it can be anticipated at what time what kind of products will be displaced by other products and no more profit can be expected on the market. Enterprises can use this information in order to launch new products at the right time and to optimize or replace old products on time (Brankamp 1971).

By using the S-curve representation (Fig. 4.20) the considered technical system, e.g. a product of the own company, can be positioned on the lifecycle curve for strategic decision making. Since, with regards to quality, the S-curve correlates with the curves “number of inventions”, “invention level” and “product success”, the current position on the S-curve can be determined if the position on one of the other curves is known. If the development of a system can only be observed over a small period, which in general is the case, only the curve progressions in this interval can be described. It is thus only possible to know whether the curve rises or falls. Several sections of the corresponding curves may be assigned to this identified progression. As the progressions correlate with each other, the positioning can be concretized if the sectional progressions of several curves are known.

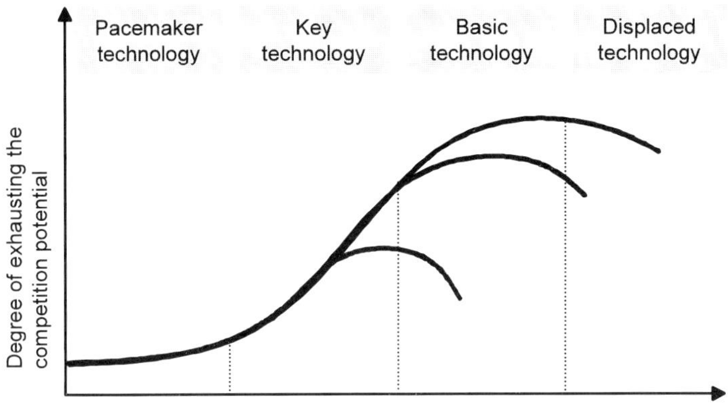

Indicators	Creation	Growth	Maturity	Age
Uncertainty about technical capabilities	high	average	low	very low
Investment in technology development	average	maximum	low	very low
Broadness of the potential application fields	unknown	big	established	decreasing
Type of development requirements	scientific	application-oriented	application-oriented	cost-oriented
Impact on the cost-performance ratio of the products	secondary	maximum	marginal	marginal
Number of patent applications	increasing	high	decreasing	none
Type of patents	concept-patents	product related	process related	
Access barriers	R&D potential	personnel	know-how, licenses	application know-how
Availability	very restricted	restructuring	market oriented	high
Duration of the competitive advantage	long	average	average	very restricted

Fig. 4.19 Lifecycle phases of technologies (Little 1981)

The progression of the number of inventions can be determined by a patent search. In order to assess the level of inventions, it is necessary to perform an analysis and an evaluation of the inventions and patents for the specific time period. A market analysis about this period can give information about the success of a product. If, for instance, the number of applied patents related to the product decreases as well as their level of invention, while simultaneously success is growing, the product will be positioned in the first half of the growth phase (Herb et al. 2000).

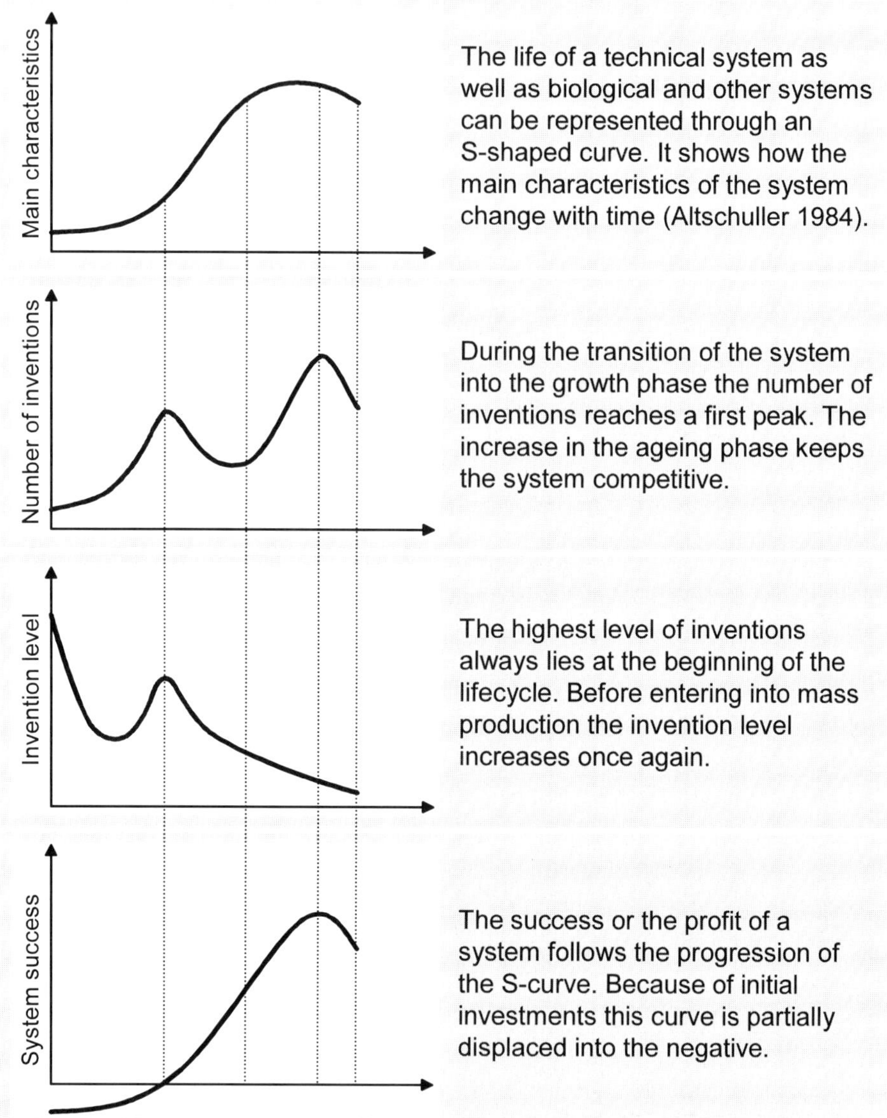

Fig. 4.20 Lifecycle curve and other curves correlating in time progression (Altschuller 1998; Herb et al. 2000; Teufelsdorfer u. Conrad 1998)

Evolution principle 2: Increasing ideality

Technical systems are developing towards increasing ideality. This means that the number of useful functions increases while the number of damaging functions decreases. The ideal end product, the aim of the evolution, fulfils a function without existing itself (Altschuller 1984; Herb et al. 2000; Petrov 2001; Terninko et al. 1998).

Development and evaluation of product concepts

The principle of ideality can be used in order to develop new and better product concepts (Chap. 4.4) as well as to evaluate product concepts. For that purpose existing product concepts are analyzed by identifying the positive and negative functions. New solutions are being provoked by the goal of increasing positive functions and reducing damaging functions or by the aim of developing the ideal product. The TRIZ methodology provides solution approaches which support the developer in idealizing product concepts. New product concepts can be evaluated concerning their ideality level.

Example: the first wireless telephones were used as so-called field telephones in the military domain. Due to their weight and size, the devices could just be carried by a person. Later, the devices of the C-network were available for civil use in Germany, which already had the size and weight of a small suitcase. The small, comfortable devices of the latest generation have clearly developed towards ideality. In the future micro-telephones may be implanted so that they can fulfill their task without being noticeable.

Evolution principle 3: Uneven development of the systems and their parts

Analyzing sub-systems

Each technical system consists of subsystems (e.g. assemblies and components) which can be understood as technical systems on its own. These subsystems have their own lifecycle and accordingly follow the laws of technical evolution.

ALTSCHULLER recommends checking the systems for the completeness of the necessary subsystems and the energy flow through these (Altschuller 1998).

If subsystems are developed in an unbalanced way, the development of the whole system can be slowed by an underdeveloped subsystem and the improvement of another subsystem can become almost superfluous. This fact can be used in order to identify and further develop the subsystems that impede the improvement of the super-system. An analysis of the subsystems can show that the entire system can be essentially improved if the right subsystem is optimized.

Example: For a long time, the aircraft development was stagnant because the developers focused on the improvement of the engine power instead of improving the aerodynamics of the aircraft. After the development of modern streamlined aircraft shapes, the aerodynamic improvements stagnated again. This made revolutionary improvements on the engines necessary, which led to the replacement of the piston engines through newly developed turbo jet engines. Currently, work is in progress for the improvement of aerodynamics by means of bionic principles (winglets and structure coats).

Evolution principle 4: Increased dynamics and application of fields

Increase degrees of freedom

In the course of time, technical systems become always more dynamic, flexible and diverse – the degrees of freedom increase. The development follows three sub-principles (Herb 2000; Terninko et al. 1998; Petrov 2001):

- Systems are increasingly sub-divided into mobile components.
- The use of fields increases.
- The information density within a system and its parts increases.

Figure 4.21 symbolically illustrates this principle with the example of a steering-wheel.

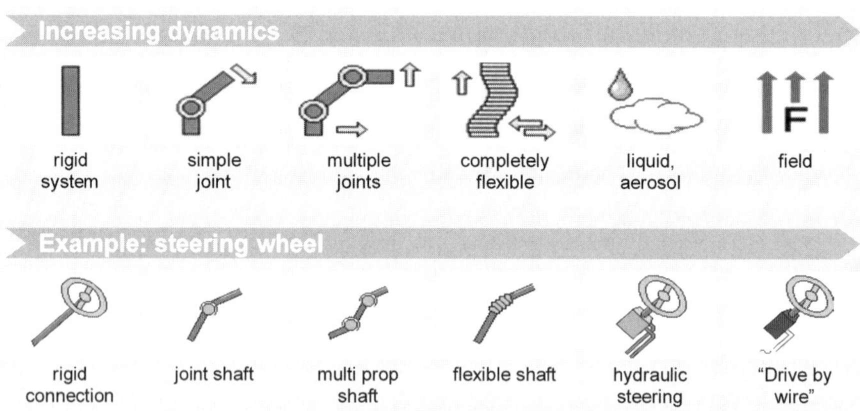

Fig. 4.21 Increased dynamics on the example of the steering-wheel (Invention Machine 1998)

As another example, the development of the "wheel" as a subsystem of mobile "vehicles" will be described in the following:

During the first evolution step, the massive wooden wheel was replaced by a spoke wheel. The use of additional substances which supported the running surface led, via the change from wooden to metal wheels, to the development of air-filled rubber tires. In air cushion vehicles, the wheel is completely replaced by air-cushions. According to the pattern of the evolution principle the next step will bring about the use of fields as currently tested for the magnetic levitation train Transrapid.

Evolution principle 5: Through complexity to simplification

Development to simplification through complexity

Technical systems first develop towards increased complexity to become ingeniously simple afterwards. The development changes from a system existing on its own, which with time becomes always more complex, via several systems existing in parallel, to a combined system incorporating the systems existing in parallel. This combined system contains the functionality of the previous systems and can be seen as their super-system. The further evolution takes place at the level of the super-system (Altschuller 1998; Herb et al. 2000; Terninko et al. 1998).

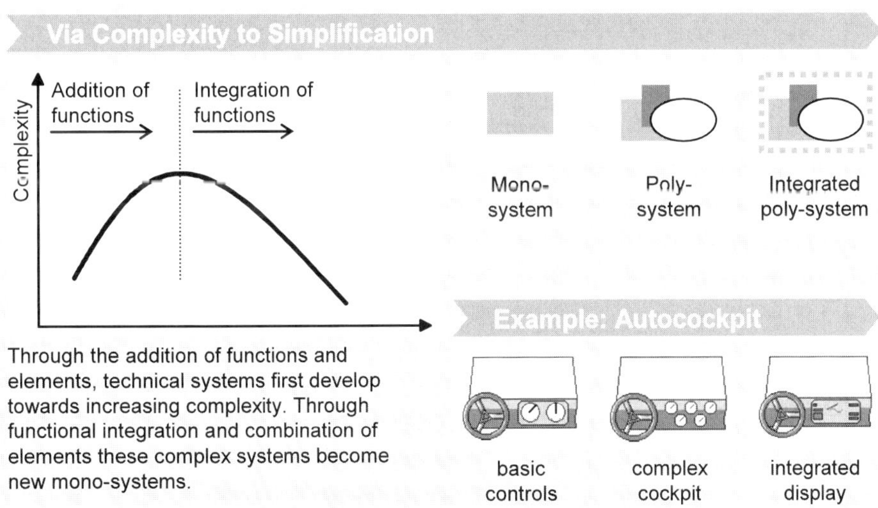

Fig. 4.22 Evolution of technical systems via complexity to simplification on the example of the car-cockpit (Mann 2001; Invention Machine 1998)

Figure 4.22 shows the development of a system over time on the example of the car-cockpit. The progression of the system development can be divided into two phases: in the first phase the complexity increases. More functions are made available and the number of subsystems or systems existing in parallel increases. The second phase is the phase of simplification. Here, functions and systems are integrated, components are reduced and processes simplified.

Example: the first cars had only few functions. Accordingly, only few controls and displays were necessary. With the development of additional functions and comfort increase new controls and displays had to be added, which led to a high complexity of the cockpit. In new higher class cars the different functions are combined in a few control elements so that few elements are necessary to operate the existing functions. For instance, the "controller" of the iDrive-operation concept of the BMW 7-series allows to control up to 700 functions with one single element.

Evolution principle 6: Increased coordination

Coordination increases

This principle can also be characterized as "Law of the rhythm coordination of the parts of a system". It says that in the course of the evolution of a technical system, actions and rhythms must be increasingly co-ordinated and integrated (Fig. 4.23). The evolution principle can be realized in practice by applying three measures (Altschuller 1998; Herb et al. 2000; Terninko et al. 1998):

- Systematic use of breaks and intervals.
- Optimization of systems through focused congruence of the system elements. Use of symmetry and resonances.
- Optimization of systems through focused non-congruence of the system elements. Use of asymmetry and avoidance of resonances.

Example: the use of a candle or a simple bulb can be understood as an uncoordinated action because the illumination of a room is neither focused towards a certain direction, nor takes place uniformly. The light is better coordinated if, in addition, a reflector is used. In order for the light to be used more effectively and efficiently, light sources and the supporting systems were developed to form a laser, which almost emits an axis-parallel, monochromatic, coherent light. In the next evolution step the light is used in intervals (pulsed).

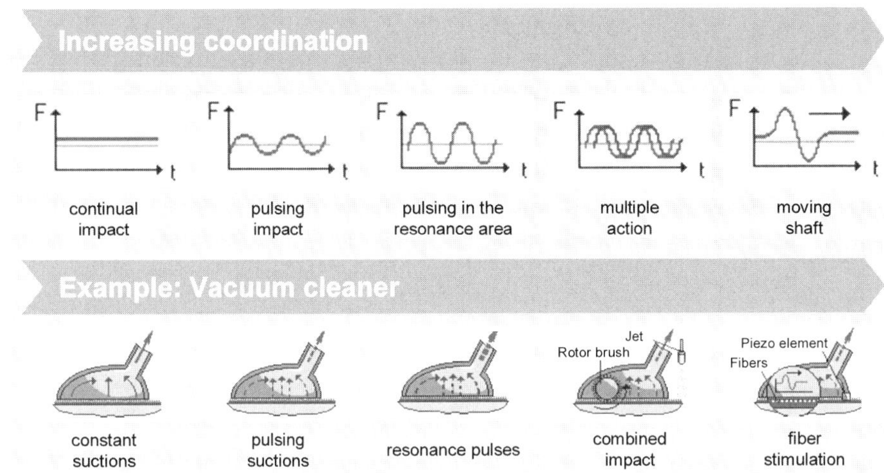

Fig. 4.23 Increased coordination on the example of a vacuum cleaner (Invention Machine 1998)

Evolution principle 7: Transition from macro to micro level

The evolution principle "transition from macro to micro level" consists of three sub-principles (Altschuller 1998; Terninko 1998; Herb et al. 2000; Teufelsdorfer and Conrad 1998):

Development first happens on macro level and then on micro level

- The evolution of a system goes towards miniaturization.
- At first the macro, i.e. the entire system is improved, then improvements take place on micro level, i.e. in subsystems.
- Increasingly smaller elements are used to fulfill the desired function. Via the use of simple parts and material structures, the development goes towards an atomic level. In the next development step fields are applied; this level can be characterized as the sub-atomic level.

Example: the first watches needed so much space that they were installed in big buildings such as church towers. The development went via long case clocks and pocket watches up to wristwatches.

Evolution principle 8: Increasing controllability, self-control and automation

Controllability, self-control and automation are increased

On the one hand the controllability of systems increases in the course of the development process, on the other hand control is more and more carried out by the system itself. In this way the effectiveness of the system is increased and the human is relieved from routine tasks. Realization, control and decision-making by humans are replaced in this order (Herb et al. 2000; Linde and Hill 1993; Mann 2001; Terninko et al. 1998).

Example: formerly, a fireplace was used for heating the living area. It was necessary to regularly add wood in order for the fire to be kept alive or to control the temperature. Today, this task is carried out by heating systems. They control the input of heating material, keep the desired temperature and automatically switch on and off.

4.6 Bionics

Bionics can contribute to the solution of technical problems in many ways, e.g. through the direct use of biological systems, the copying of biological structures, the realization of biological principles, as well as by using ideas from biology for the development of technical solutions.

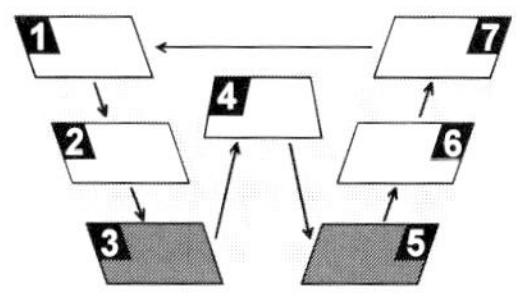

Learning from nature

People have always been learning from nature. The first written document on this subject comes from LEONARDO DA VINCI (1452-1519). Amongst others, he studied the anatomy of birds and their flight behavior in order to collect ideas for the design of flying apparatus (Hill 1999).

Take biology as a model

Today too nature often serves as a model for technology. Over millions of years the natural evolution process has created a huge amount of optimized adaptations to the existing circumstances. It is assumed that on earth there exist about 1.5 million animal species, from which about half are insects, and about 0.5 million plant species. This incredible diversity can serve as a virtually inexhaustible pool for technical innovations. Nevertheless, nature does not provide solutions for every technical problem. The wheel, for instance, was invented by humans, without any biological model (Hill 1989).

Bionics

NACHTIGALL (1998) defines bionics as follows: Bionics as a scientific discipline systematically deals with the technical realization and application of constructions, procedures, and development principles of biological systems. This also includes the co-operation of being animated and non animated parts and systems as well as the economic-technical application of biological organization criteria. NACHTIGALL distinguishes two areas within the scientific discipline: biotechnology or technical biology and bionics. Technical biology deals with biological fundamental research with the aim of using technology to gain knowledge about procedures and constructions of nature. Bionics, in contrast, uses exactly this fundamental knowledge as stimulator for technical inventions (Fig. 4.24) (Nachtigall 1998).

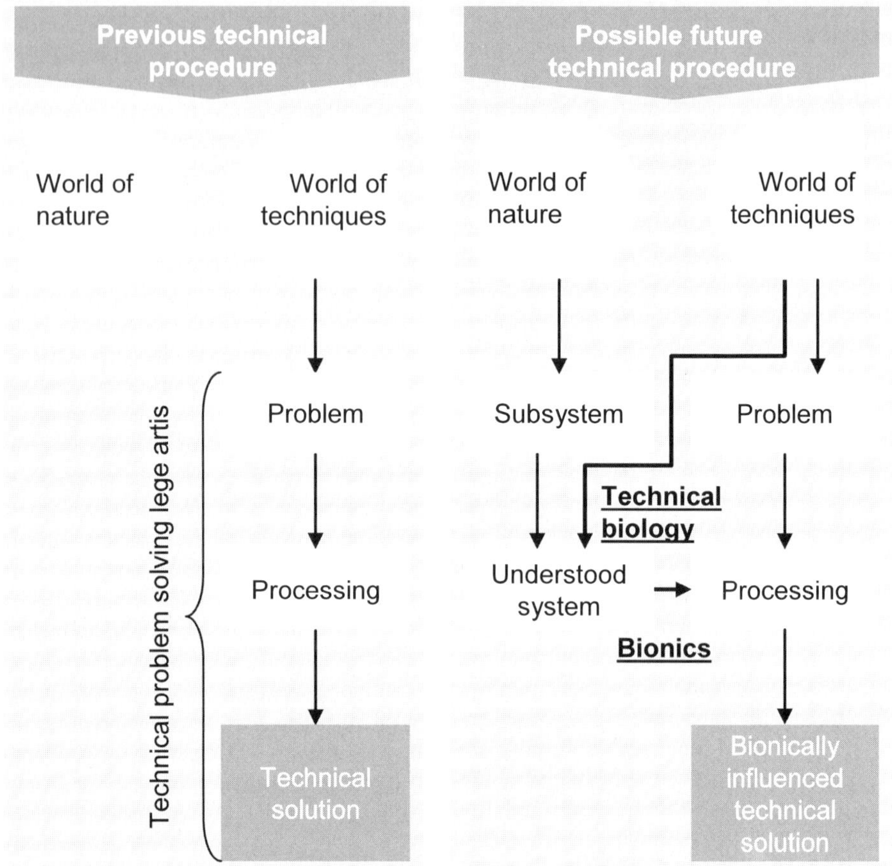

Fig. 4.24 Interaction between technical biology and bionics as possible future technical procedure (Nachtigall 1998)

Sub-disciplines of bionics

Technical biology as well as bionics include a number of sub-disciplines in which topic-specific research as well as topic-specific search for technical solutions is done. This introduction is discussed in detail at the end of the section "Bionics" in chapter 4.6.4.

According to HILL, the knowledge about nature can be used for techniques in many ways: biological systems or sub-systems can be directly used or can bring about, via abstraction and association, new technical solutions. Figure 4.25 shows two further possibilities for the bionic procedure, which are situated between abstraction and direct application: biological structures can be simulated through technical materials and biological principles can be realized in technical solutions.

Below, the four mentioned possibilities are described in more detail.

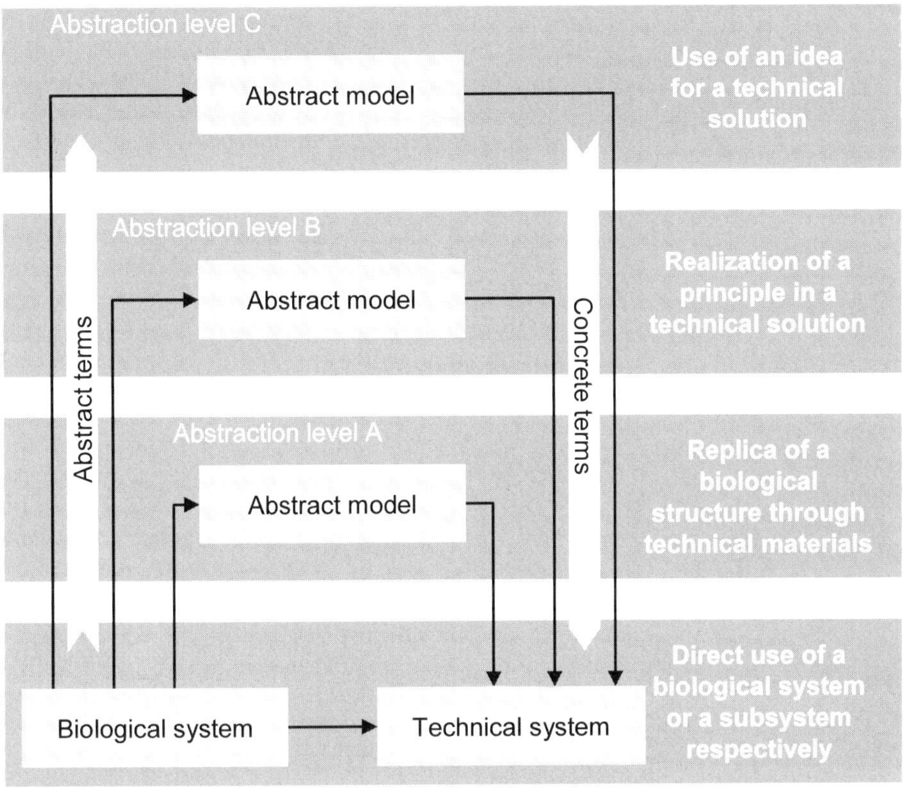

Fig. 4.25 Possibilities of the bionic procedure (Hill 1999)

Bionics is comparatively complex and should ideally be carried out by an interdisciplinary team of biologists and engineers. High level innovations are to be expected. The work effort can be reduced through tables of biological structures and principles.

In the following it is described how bionics can be used in the phases of idea generation and idea detailing, starting from a problem, for developing solution ideas following nature's example.

4.6.1
Direct use of biological systems

Use of biological systems

The direct use of biological systems shows different advantages such as the renunciation of environmental pollution and consumption of raw material. Besides, such systems are available or easy to cultivate. In mining, for instance, flies are used as an efficient and reliable sensor for mine gas due to their exceptional sense of smell. If the concentration of poisonous gases in the air increases, the flies emit characteristic impulses which can be registered by electronic analytical devices, which in turn immediately activate an alarm (Hill 1997). In these cases, we speak of biotechnical linked systems (Hill 1999).

4.6.2
Biological structures

Copying biological structures

Not only the imitation of nature leads to creative performances, but also the abstraction and transmission of functional principles (Nachtigall 1986b). On the level above the direct use the structural analogy is situated (abstraction level A). Here, biological structures are completely or partially realized in form of technical structures (Klaus and Liebscher 1976). As an example for the structural analogy Nachtigall uses the comparison of a TV-tower with a blade of rye (Fig. 4.26) (Nachtigall 1986a).

Copying biological functions

Functional analogy is the congruence of the functions of a technical and a biological system. The kind of elements and the structural constitution can differ from each other (Klaus and Liebscher 1976). "The (conclusion by analogy) results from the theoretical transmission of functional characteristics of the still unknown, not clearly formulated search object (technical system as target system) to the characteristics of the analogical object (biological system as starting system)" (Hill 1999).

In order to make systematic solution finding possible, Hill (1997) provides structural catalogues which are organized according to the fundamental functions forming, changing, transmitting, recording/blocking, separating/connecting as well as supporting/carrying of substance, energy and information. Furthermore, he provides effect catalogues for the effects forming (increasing/decreasing) and changing.

The catalogues can very well be incorporated into the constructions systematics according to VDI 2221. According to WÜSTENBERG (1998), solutions for the individual sub-problems can be found in HILL'S catalogues if the system to be developed is first described by a black box and this allover function then divided into partial or elementary functions.

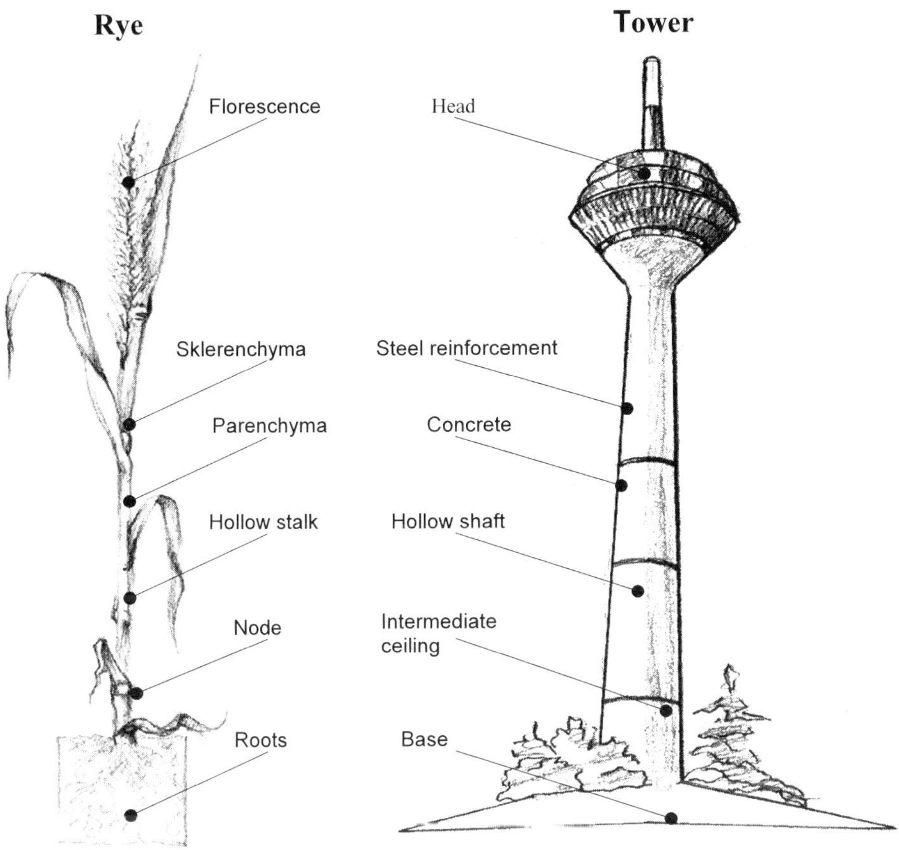

Fig. 4.26 Example of structural analogy: blade of rye and TV-tower (Nachtigall 1998)

4.6.3 Laws of evolution and other biological principles

Applying biological principles

At the abstraction level B biological principles are transformed into technical solutions. For that purpose HILL presents a few principles of biological functions that the constructing engineer can use as an inspiration for improvement and the new development of technical systems. For a better understanding and as an inspiration, the principles are listed and briefly characterized.

- *Minimum-maximum-principle:* "In biology, a maximum of performance and stability is achieved by using a minimum of material and energy" (Hill 1998).
- *Principle of multi-functionality*: This principle can for instance be seen with bird claws. It serves not only for grabbing, but also for transportation, for defense, for locomotion, and for building the nest. Through the principle of multi-functionality the material and/or the manufacturing effort can be reduced.
- *Principle of structure and function specification: A*vailable system elements are slightly redesigned in order to fulfill additional functions. The antenna cleaning apparatus of the honey-bee for instance has long and short brushes. Similarly to some toothbrushes, the long brushes are for preliminary cleaning and the short brushes are for polishing. Another example are the human teeth with incisors and molars.
- *Principle of third party and environmental energy use:* Using this principle, birds, when flying in V-formation, use the air flow of the birds in front of them in order to reduce their energy consumption by approximately 23%. Moreover, environmental energies such as water, wind and solar energy are used in many ways.
- *Principle of dynamics:* Many biological systems can adapt to different conditions in a dynamic way. The plumage of birds for instance opens and becomes permeable to air when they move their wings upwards. When they move their wings downwards, the plumage closes and the bird is buoyed.
- *Principle of the optimal arrangement:* Using this principle, trees adapt to difficult environmental conditions such as wind or snow. The areas of the tree which experience higher tensions are strengthened. Similarly, an increase in bone mass can be observed at especially strained points and a decrease in bone mass at unstrained points. Short living systems are optimized according to the strain. This principle is used in the CAO-method (Computer Aided Optimization) in order to optimize the progress of stress in technical components through simulated adaptive growth (Mattheck 1998).

- *Principle of closeness of process sequences:* Nature lives according to the principle of recycling in order to optimally use the available resources. Following this principle, the forest continuously renews itself: dead vegetal and animal remains are incorporated in new creatures.
- *Principle of function-structure-interrelation:* Basically, the structure of a biological system adapts itself to the natural circumstances and the required functions. The pecker of a flower bird for instance is adjusted to the form of calyx so that it can extract the nectar from the bottom of the calyx. In this context, it is asked in technology which structure would best fulfill the required function.

For each principle HILL (1998) provides a number of examples as well as a list of questions on the basis of which the engineer can check whether a specific principle would improve the system (Hill 1998). For further principles HILL offers principle catalogues for solution finding with examples. These are in particular biological function and structure principles, biological organization principles and ecological formal principles (Hill 1997).

The basic principle of nature which brought about the biological diversity and abundance is the evolution. Biological systems continuously improve in order to optimize themselves and to adapt to the changing conditions. HILL filed a selection of evolution laws, steps, trends and stages, which can be transferred to technology. Through the transmission of these evolution techniques to the state of the technology, it is possible to determine development trends, to disclose development reserves and to identify development directions (Fig. 4.27). Through the application of evolution laws effectiveness factors can be found, development goals can be confirmed and first blurred solution ideas can be discovered. According to REICHEL, for instance, one evolution step is "self-monitoring and protective function". This step finds an application in the safety valve of a pressure cooker: the valve opens when overpressure occurs; in this way, it guarantees a constant pressure and thus prevents an explosion of the pot (Hill 1999).

KURSAWE and SCHWEFEL describe the evolution laws mathematically and from this develop procedures to solve optimization problems. For their approach "only a minimum of information regarding the task to be solved is needed". With this approach highly complex problems can be processed (Kursawe and Schwefel 1998).

4.6.4 Inspiration from nature

Receiving inspiration from biology

The search level C is the highest level of abstraction. Here, inspirations from biology are used for technical solutions. Principles and functions must not directly be transformed into technical solutions; they rather should stimulate and steer the creativity of the inventor. In this case, bionics can rather be assigned to synectics and so to the intuitive lateral methods, which is often the case in literature. Through associations with analogies from biology and the connected alienation of the problem to be solved, new thought patterns are created, which can lead to the development of new solution concepts (Peples 1999).

Concerning the systematic application of bionics, also by using catalogues, HILL makes it clear that a strategic approach increases the probability of finding an innovative solution but that it is, however, no absolute guarantee for an invention. Nevertheless, solution finding is supported and the intuition is steered towards a promising direction (Hill 1999).

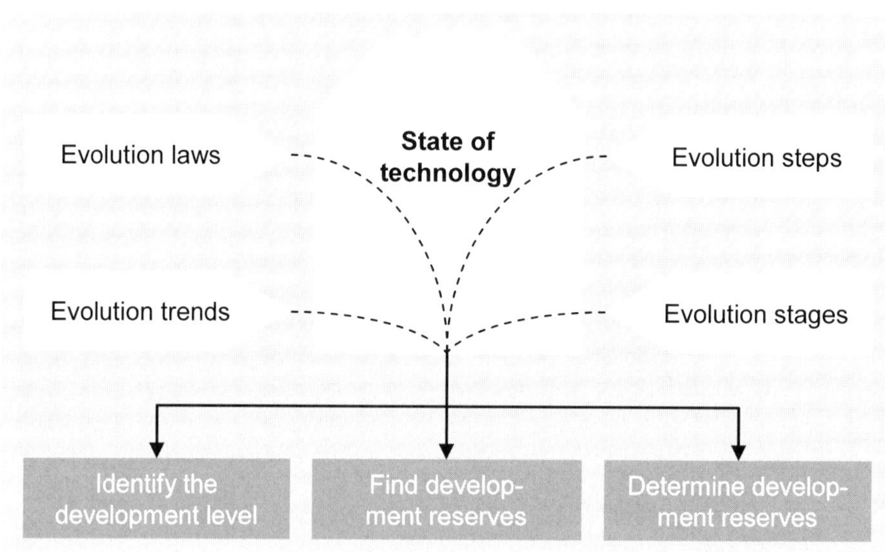

Fig. 4.27 Evolution laws for goal determination (Hill 1999)

Sub-disciplines of bionics

Structure bionics into sub-disciplines

In his detailing (1998), NACHTIGALL suggests to structure bionics into sub-disciplines. If the problem to be solved falls into one of these subject areas, the developer can find ideas in the associated chapters and the included examples. It might also be advisable to do some own research in the corresponding area. In the following, the sub-disciplines are listed and briefly explained (Nachtigall 1998):

- *Historical:* This subject area describes the development of bionics.
- *Structural bionics:* In this field biological structural elements and materials are investigated concerning their solution potential for technical problems.
- *Construction bionics:* In this sub-discipline, ideas can be won from biological light-weight constructions. But also a return to traditional building materials can lead to successful solution concepts.
- *Climate bionics:* The study of natural constructions or traditional buildings can bring about new ideas especially in the areas of passive ventilation, cooling and heating.
- *Construction bionics:* Biological construction elements and mechanisms serve as a model for technical developments. For instance, the burr served as a model for the Velcro fastener.
- *Movement bionics:* In this area the movement forms of walking, swimming and flying are considered, especially with regard to flow adjustments and driving mechanisms.
- *Device bionics:* This sub-discipline correlates with the areas of structure bionics and construction bionics. Here, the technical use of biological constructions is investigated.
- *Anthropo-bionics:* This area includes human-machine intervention, ergonomics, and robotics.
- *Sensor bionics:* This sub-discipline deals with the collection of physical and chemical signals. Special interest is paid to the positioning, orientation and recognition of chemical substances.
- *Neuronal bionics:* The development of neuronal networks and simultaneous computers are two examples for data analysis and data processing which took the brain as a model from nature.

- *Process bionics*: Nature is characterized by total recycling and waste avoidance. This ecological aspect makes process and implementation bionics interesting for process engineering as well as for the entire industrial society.
- *Evolution bionics*: NACHTIGALL especially cites the mathematical formulation of evolution techniques and strategies as a useful optimization method for complex systems and processes, when computer simulations cannot be carried through because of the complexity of the problem.

Conclusion

At the end of the chapter "Bionics" it can be said that nature can serve as a model for ecological technology (Von Gleich 1998). High hopes are put on process bionics. In nature, the recycling technique and photosynthesis have been brought to perfection. Concerning the organization of individual organisms as well as the interactions in both, small and big ecological systems, biology is by far superior to all systems created by humans (Nachtigall 1998).

In the preceding presentation of bionics, special focus was put on the sub-categorization of the invention method. The elements "Direct use of biological structures", "Biological structures", "Biological principles", "Inspiration from nature", and "Sub-disciplines of bionics" can be used independently from each other as tools for solving technical problems during the phases of idea generation and idea detailing.

4.7
Portfolio Analysis

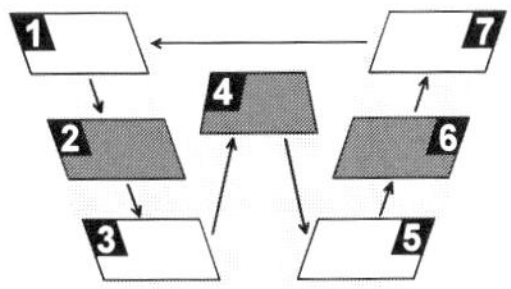

The portfolio analysis is used during the planning phases "Future analysis", "Idea evaluation" and "Concept evaluation" of the W-model.

- In this context it is used for the evaluation and selection of innovation tasks or respectively product ideas.
- Besides these special portfolios of the innovation process some other fundamental portfolio techniques are used in innovation management.

History

Portfolio analyses are part of the most important methods of strategy development and formulation. The term portfolio was taken from the fiscal sector. The first portfolio concept that can be allocated to financial management was suggested by MARKOWITZ in 1952. This portfolio consists of a planning method for composing bond bundles ("Portefeuille") which, assessed according to certain criteria, should bring an optimal interest rate on the capital invested at the stock-exchange. Here, it was essential to bring the contradictory aims (return, risk and liquidity) of the magic triangle in the portfolio management into an optimal ratio to each other. The idea of the portfolio was later transferred to other economic fields such as real investments (cf. Gabler 1997).

Holistic way of posing a problem

At the beginning of the seventies, the portfolio principle was applied onto holistic problems in diversified enterprises. A pioneer in this process was the American concern *General Electric*. Using a portfolio developed by the consultancy *Boston Consulting Group,* a chance/risk-balanced product and market program was determined.

Since this first industrial application the portfolio analysis has been modified a lot. It counts among the most commonly used analysis and planning instruments and can be found in many versions in different management disciplines.

Portfolio types

In the strategic development of product programs on enterprise level, the *market portfolio* is used in order to determine the aspired positioning of the business fields in the portfolio of the company. On the basis of these business field portfolios, which are based on the assumption that product and production technologies develop relatively constantly to each other and thus must not explicitly be considered, other portfolio versions were developed. In so-called *technology portfolios,* individual aspects of the technology development are integrated. Together with the business field portfolios they represent the second type of portfolios that can be incorporated into the innovation process (Fig. 4.28).

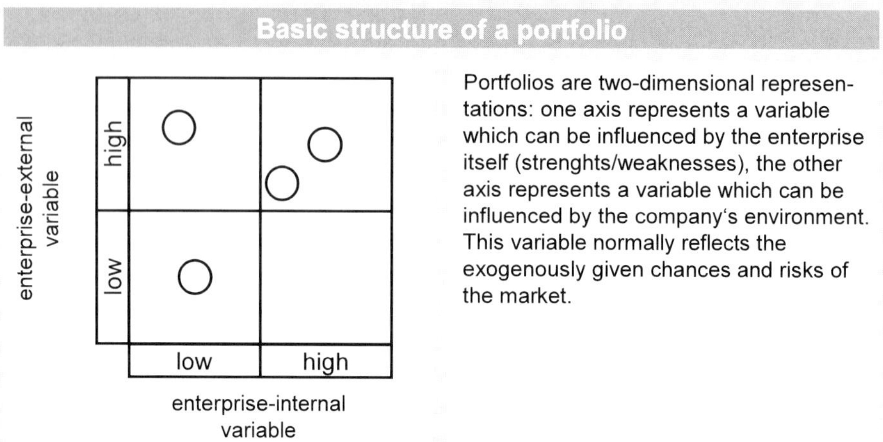

Fig. 4.28 Basic structure of a portfolio

Furthermore, other types were developed for special applications. Here, only the *process portfolio* shall be mentioned as an example. This portfolio is used for identifying core processes as well as for deriving process strategies based on process efficiency and effectiveness (cf. Eversheim 1996).

Application in the IRM

In the W-model, the portfolio technique is used in the phases "future analysis" and "idea evaluation". In the first phase, the portfolio technique is used to evaluate and select innovation tasks. In the second phase, the portfolio acts as an analysis instrument for representing and evaluating product ideas. In both cases, the advantages of the portfolio are easy understanding, good visualization and transparency (cf. Brandenburg 2002). Their use thus supports the innovation process. Besides these specially developed portfolios that have already been described in the corresponding chapters, the market and the technology portfolio are relevant for the innovation process, whose applications are described in the following sections.

4.7.1 Market portfolio

Product/market combinations

The aim of the market portfolio is the identification of the optimal product-market combination. The question to be answered is the following: which product on which market can effectively (strategically) ensure the company's success? This is the task of the strategic corporate planning and requires the division of the company into different strategic business units (SBU)[22].

Strategic Business Unit

The market portfolio is a method for the strategic planning of the single SBUs. With this method, the SBUs of a company are assessed with regard to their current market position and their development possibilities. The portfolio allows not only a comprehensible presentation of the results but also the deduction of standard strategies for the different SBUs (cf. Hinterhuber 1996).

There exist different types of market portfolios. The two most important representatives are the market share/market growth-portfolio of the *Boston Consulting Group* (known also as BCG matrix) and the branch attractiveness/business field strength-portfolio of McKinsey (Pfeiffer et al. 1982). For better understanding of the market portfolio types, the BCG matrix is presented below.

4.7.2 Portfolio of the Boston Consulting Group

Lifecycle model

The BCG matrix is represented through a market-related and an enterprise-related axis (Fig. 4.29). In this regard, the market growth represents the market factor and allows referring to the lifecycle model, which is closely connected to the BCG portfolio. In the first as well as second phases of the product lifecycle, a high growth of the turnover or revenue is generally expected. In the two last phases, rather little growth or even shrinkage is observed (cf. Kotler and Bliemel 1999). Concerning the market factor, two sections, high and low market growth, can be distinguished.

22 A strategic business unit is the grouping of preferably equal product and market combinations to form a plannable unit.

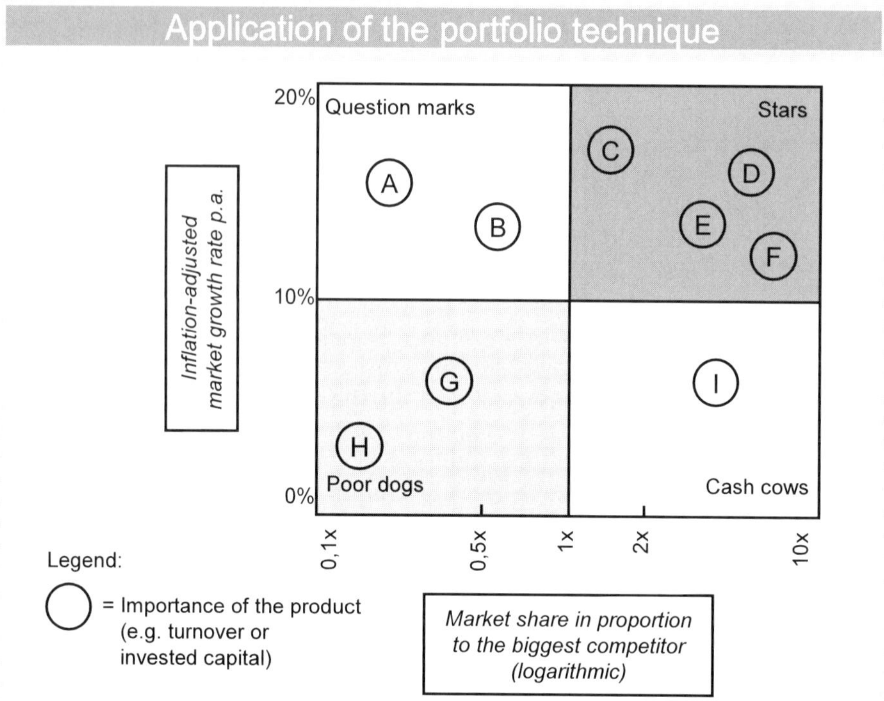

Fig. 4.29 Market portfolio of the Boston Consulting Group

The concept of the learning curve

The second, enterprise-related axis is the axis of the relative market share of the company. The deduction of this dimension is based on the learning curve concept. The relative market share should indicate what lead on the learning curve and thus what cost advantages the company has. Through a higher cumulative production volume a company reaches scale-related cost advantages, which manifests the connection between high market shares and advantageous cost-situation (cf. Hinterhuber 1996; Nitzsch 1998). Just like the market growth, the relative market share is divided into two areas, high and low.

Market share

The operationalization of the relative market share axis follows from the comparison of the company share with the share of the biggest competitor in this product field. If the relative company share is bigger than the one of the biggest competitor, the market share is high. If it is smaller, the share must be considered as low. Since it concerns a ratio, it is recommended to use a logarithmic scale. If the market growth rate is higher than 10% the market growth is considered as high. If it is lower than 10% the market growth is low (cf. Hinterhuber 1997; Nitzsch 1998).

In its actual sense, the BCG matrix refers to business units. Strategic business units can be characterized through products or product groups (Nitzsch 1998).

From the division of the portfolios follow four basic matrix fields, to which the terms *Poor dogs, Question marks, Stars* and *Cash-cows* are assigned. The matrix fields as well as the resulting standard strategies are explained below (Kotler and Bliemel 1999; Nitzsch 1998; Pfeiffer et al. 1991):

Poor dogs: low market share, high growth rate

Products that are located in the matrix field of the so-called *poor dogs* must be seen as a problem for the company. Neither do they present a good cost structure, nor is it likely to realize a profit in a non-growing market in the future.

Question marks: low market share, low growth rate

A company with products that are assigned as *question marks* is situated in a fast growing market with a low market share. As the product is still in the introduction phase, the experience in manufacturing the product is still low. This often results in a bad cost situation since there are no cost-reducing learning effects.

Stars: high market share, high growth rate

Products are characterized as *stars* when a relatively high market share is achieved and thus a cost beneficial situation occurs. The market growth is positive and promises to earn profits. Generally, capacities must be created in such a case in order to respond to the demands of the market.

Cash cows: high market share, low growth rate

Products with a relatively high market share and low market growth are so-called *cash cows*. For those products the company has achieved corresponding cost digression effects through experience and can achieve high profit margins on the market. As the required capacities are available the investment needs for this type of product is low.

To each field of the BCG matrix or to each type respectively, a standard strategy (categorical behavior pattern) can be allocated (Kotler and Bliemel 1999; Pfeiffer 1999).

- *Expansion:*
  The company has decided to expand the market share of the product. The approach is recommended for *question marks* in order to become a star by gaining market shares.
- *Maintain:*
  In this strategy it is important to preserve the current market level. This strategy is suitable for the lucrative cost situation of a *cash cow*.

- *Harvest:*
  In this case the company is interested in withdrawing liquid funds from the product field in order to reduce or even stop its engagement in the middle or long term. This approach is recommended for weak *cash cows* which are situated rather at the end of this phase of life as well as for not success-promising *question marks* and maybe *poor dogs*.
- *Rejection:*
  If a product or product group is rejected the resources set free with this act can be used for other products. This tactic is suitable especially for *poor dogs* but also for uncertain *question marks*.

Standard strategies

With regard to the BCG portfolio, it must be critically noted that basically, the standard strategies or categorical behavior patterns are logically comprehensible; they can, however, not be seen as a dogmatized recommended course of action. Specific environment constellations can require another behavior, which contradicts the strategic recommendation. Moreover, this portfolio – just like all portfolio versions – does not allow representing the time-related development of the products. This planning instrument can not be made dynamic (cf. Pfeiffer et al. 1991).

4.7.3 Technology portfolio

Technology development

The above-presented market portfolios are – from a historical point of view – the portfolio variants which were developed first. They are based on the assumption that product and process technologies develop relatively consistently and therefore should not explicitly be considered (cf. Bullinger 1994). Due to this generally missing development in many branches and markets new forms of special technology portfolios were developed as a decision-making support for the strategic technology management and technology planning (cf. Bullinger 1994; Servatius 1985; Pfeiffer et al. 1991).

Technology attractiveness

In *technology portfolios* too, external and internal recording parameters are compressed into two dimensions (cf. Pfeiffer et al. 1991). As external parameter the technology attractiveness records the sum of technical and economic advantages which can be achieved through this technology.

Resource strength

The resource strength as internal parameter, in contrast,

represents the technical and economic control of the technology field.

Besides the "pure" technology portfolio (cf. Wolfrum 1994) there exist further developments of the method, which establish the relationship to market- and technology-strategic aspects. It is in particular the concepts of the consultancies that pursuit this extended approach (cf. Schmitz 1996). In figure 4.30 the described concepts are put together again and compared with regard to the recorded dimensions and characteristics.

Authors	Dimensions	Characteristics
McKinsey (Krubasik 1982)	- Relative technology position - Technology attractiveness	- Integrated market technology portfolio - Focusing on technologies - Based on the S-curve model
Pfeiffer et al. (1982)	- Resource strength - Technology attractiveness	- Pure technology portfolio - Product and process technologies - No integration in the overall planning
A. D. Little (Servatius 1985)	- Relative technology position - Position within the technology lifecycle	- Integrated market technology portfolio - Based on technology lifecycle - Strategy dependent on PLC- and TLC-phase
Booz, Allen & Hamilton (Gerpott 1991)	- Relative technology position - Technology importance	- Isolated technology consideration - No synchronization with the market planning - Criteria for technology importance unclear
Michel (1987)	- Relative innovation strength - Innovation attractiveness	- Planning object: innovation fields - SBU- and technology specific portfolios - High complexity in application
Wildemann (1987)	- Technology priority - Market priority	- Technology portfolio - Orientation towards the current market situation
Porter (1986)	- Company advantage - Technology attractiveness	- Pure technology portfolio - Fuzzy-based evaluation method - No separation between internal/external variables
Pelzer (1999)	- Future orientation - Technology mastering	- Product-neutral technology evaluation - Long-term orientation

Fig. 4.30 Comparison of technology portfolio concepts (cf. Wolfrum 1994)

Recommended actions

It should be mentioned that, according to PFEIFFER, the technology portfolio is only an aid within the entire portfolio method. The portfolio method includes the steps "technology identification", "position determination", "transformation of the current into future state", and "deduction of recommended actions" (cf. Eversheim 1996). Recommendations for action are generally deduced directly from the position of a technology in the portfolio and therefore are also characterized as standard strategies (the instruction for action is "standardized" for the various fields). The suggested strategy should, however, be questioned because the specifications of the particular branch have not been considered and often a rigid realization acts as a self-fulfilling prognosis (cf. Wolfrum 1994). In this context, WOLFRUM distinguishes two major variants of technology portfolios (cf. Wolfrum 1992): first, pure technology portfolios, in which exclusively technological aspects are recorded and processed. Second, portfolios in which a connection towards market- and technology-strategic aspects is made.

In the following, two technology portfolio types are presented, which are suitable mainly for being used in innovation management. First, the technology portfolio according to PFEIFFER (Pfeiffer et al. 1991) is presented; then, the potential portfolio according to PELZER (Pelzer 1999), which was developed at the Fraunhofer IPT, is explained.

4.7.4 Technology portfolio according to PFEIFFER

Generation and observation cycle

PFEIFFER's approach considers both, the generation cycle, which takes place before the market cycle, as well as the observing cycle for the strategic analysis process regarding the two dimensions of technology attractiveness and resource strength. The approach is based on the assumption that with a tendency towards expanding generation cycles and at the same time contracting market cycles, the innovator can always reach a substantially higher volume of business as the imitator (Eversheim 1996). The technology portfolio analysis according to PFEIFFER takes place in several steps (Pfeiffer et al. 1991):

Identification of technologies

First, relevant product and production technologies are identified according to dissection criteria (e.g. systems, subsystems, component groups, elements, processes). The recorded processes are presented in technology lists.

Determination of technology attractiveness and resource strength

Economic and technological advantages

The technology attractiveness describes the economic and technological advantages which can be achieved strategically through further development in a particular field. With the dimension of resource strength, the funds necessary for the realization of the technology potential, which are already available in the enterprise, and which are measured in relation to the competitors, are taken into consideration. The indicators of the dimensions technological attractiveness and resource strength as well as the general structure of the portfolio are shown in figure 4.31. Based on the listed criteria the determined technologies can be classified in an as-is portfolio of the company.

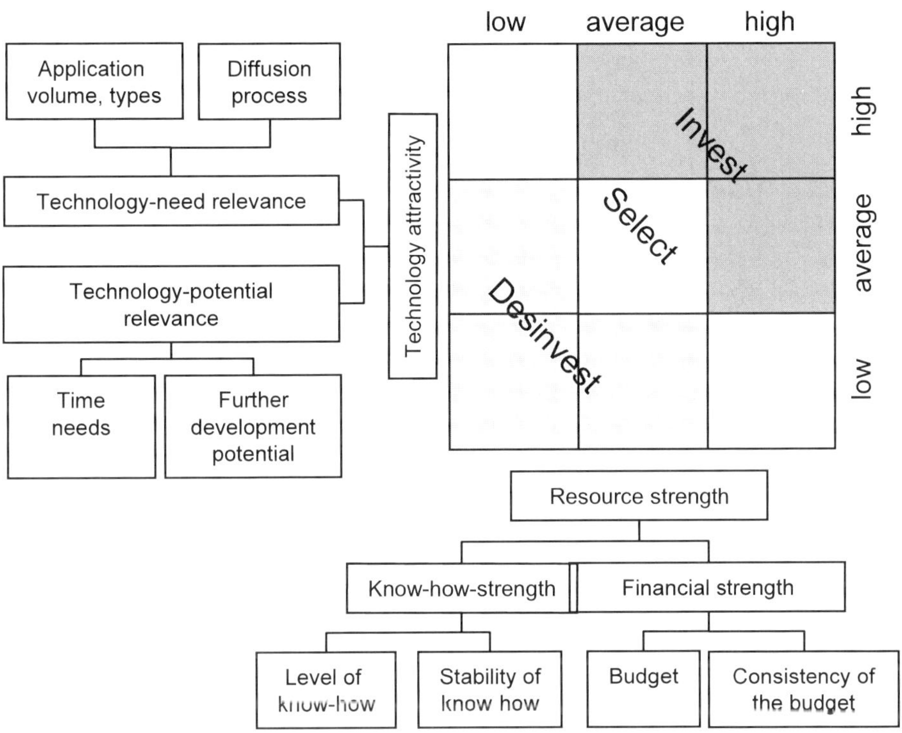

Fig. 4.31 Technology portfolio according to PFEIFFER

Transformation of the current technology portfolios in a to-be portfolio

Current and future portfolios

In order to show future technological chances and risks of a company, the current situation manifested in an as-is portfolio has to be transformed into a future one. For that purpose, potential competitive technologies must be identified and be positioned in the portfolio.

Deduction of recommended actions

Standard strategies

Through the confrontation of current and future technology portfolios final recommendations for actions must be deduced. This portfolio concept also provides standard strategies, which can be used for decision-making and support in action. The strategies to be distinguished are the one of *investment*, of *selection* or respectively *differentiated consideration*, and the one of *disinvestment*.

PFEIFFER's approach supports the decision-making process if it deals with the allocation of resources for specific projects of the product development (cf. Eversheim 1996).

4.7.5 Potential portfolio according to PELZER (Fraunhofer IPT)

Technology Push

With the methodology of the potential portfolio new development suggestions are to be generated, which especially underline enterprise-internal strengths and potentials. Here, the initiative for production innovation comes out of the enterprise. Customer and market requirements are only taken into account during the evaluation of alternative product ideas. The process is thus based on the *principle of the technology push*. In order to support the planning of new products on the basis of company-internal technological potential, the Fraunhofer Institute for Production Technology IPT Aachen developed the potential portfolio method.

Identification of potentially strong techniques

In this version, the application of the portfolio technique aims at identifying potentially strong technologies. The company is given the opportunity to focus on those technologies, which on the one hand are well-mastered internally, and on the other hand show substantial advantages compared to alternative technologies in the future. In order to be able to meet these requirements, the potential portfolio (Fig. 4.32) has the axis *established technologies* and *promising technologies*.

Mastering technologies

The *established technologies* reflect the company-internal strengths and weaknesses. These can be divided into the sub-criteria material means, application performance, and development know-how. Classifying a technology on the axis of *established technologies* takes place through the comparison of the productivity of the concerned company with the best-in-class technology user.

Future orientation

The axis of *future orientation* describes the company-independent chances and risks, which are connected with the realization of the particular technology. It is a measure of the competition relevancy of this technology. A company-neutral evaluation of the technology compared to other technologies with regard to the sub-criteria cost leadership potential, differentiation potential, further development potential and image-improvement potential takes place.

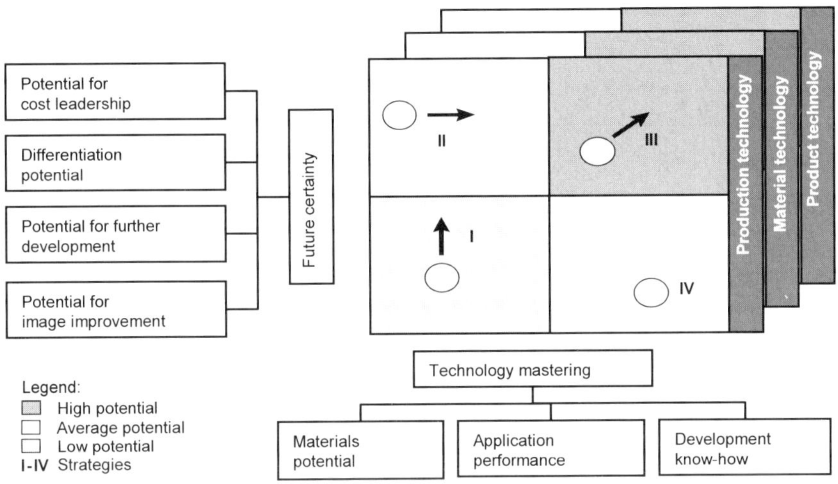

Fig. 4.32 Potential portfolio according to PELZER

Evaluation criteria

For the technologies (production, material, and product technologies) the mentioned evaluation criteria are equal for the shown abstraction level (Fig. 4.32), and differences result only at deeper detailing levels.

The actual methodology of the potential portfolio is subdivided into five phases (cf. Pelzer 1999):

1. *Potential analysis*
   First, the boundaries of the scope of analysis are set. On this basis, the potential types to be considered (material means, funds, information, staff) are selected and the relevant potentials are established through checklists and interviews.
2. *Potential evaluation*
   The established potential types are evaluated through the potential portfolio which allows a differentiation in product, production and material technologies. The

differentiation simplifies the free combination in the phase of idea generation. The positioning of the potentials in the portfolio is based on the fuzzy-set theory.

3. *Creation of search fields*
   For technologies with a distinguished potential, single product characteristics are deduced by means of a product model. These support the development of product ideas during the following idea generation.
4. *Product idea generation*
   Success-promising product ideas are generated during workshops using creativity techniques.
5. *Idea evaluation*
   To guarantee conformity with the market, the methodology concludes with the idea evaluation, i.e. the prioritization of the product innovations by means of company-internal and -external criteria.

4.7.6 Portfolio use on the example of Hilti AG

STEFAN NÖKEN, *Head Corporate Engineering, Hilti AG*

The Hilti Group is the worldwide leading supplier of high-quality system solutions for professional customers in the building industry. Besides highest quality and direct customer relations, outstanding innovation counts among the key strengths of the company. Amongst others, a differentiation on the market is achieved through innovative products and superior technology.

Strategic planning

Thus, Hilti attaches highest importance to the use of state-of-the-art production technologies. During the accurate strategic planning, currently used technologies are regularly checked with regard to their future relevance. On the other hand, the benefit potential of new, not yet used technologies is assessed.

Strategic attractiveness

The evaluation of the benefit potential of the different technologies as well as the deduction of strategies takes place by means of *technology portfolios* (Fig. 4.33). As a parameter, the strategic attractiveness of the technology is used over the technology competence. The *strategic attractiveness* comprises the sum of the technical and economic advantages which can already be made available or realized through the application of the technology. Here, the differentiation potential, the cost-reduction potential, the influenceable sales volume, the synergy potential, and the technology maturity are evaluated in detail.

Technology competency

Technology competency, in contrast, is a measure for the company-internal control of the technology field. The rating takes place in comparison to the competition and comprises the subcriteria application experience, equipment, partner network, know-how stability, and the further development know-how.

As-is-to-be portfolio

Based on the mentioned evaluation criteria an as-is portfolio and in a second step a to-be portfolio is established. Both, the target position for the technologies already in use as well as new technologies are positioned in the portfolio. From the confrontation of the current and the future technology positions the medium- and long-term recommendations for action are deduced.

The application of the *technology portfolio* provides an important basis for the deduction of basic strategies, e.g. the investment or disinvestment in different technology fields. Moreover, the portfolio forms the basis for the substitution planning of technologies and the construction of know-how in manufacturing technologies that will be relevant in the future.

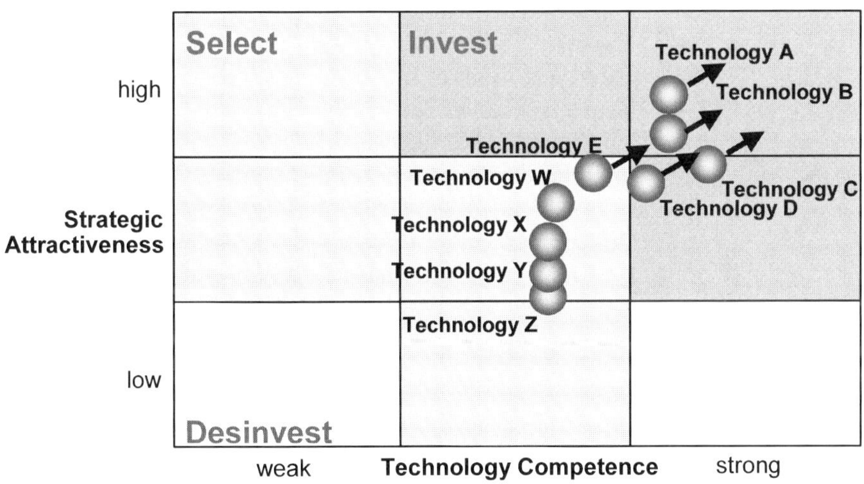

Fig. 4.33 Technology portfolio of the Hilti AG

4.7.7 Critical appreciation of technology portfolios

High aggregation

In a dynamic environment, it generally seems problematic to get to an adequate decision using trivial geometric solution patterns (cf. Schmitz 1996). Furthermore, one could criticize the very high aggregation of the individual assessments on only two aspects of the evaluation during the positioning (cf. Werners 1993), which in practice often requires resorting to detailed data (cf. Dang and Lenz 1992). The isolated allocation of standard strategies to particular technologies also contradicts the initial portfolio idea, since the existing interdependencies between them are not considered (cf. Robens 1986). Besides the mentioned critique points of the method, the problem of the classification of both, the own position as well as the technology attractiveness must be mentioned. While the determination of the technology position is a problem of the respective company, the assessment of the potentials of a technology, in particular of a new one, is a problem of all users of this methodology.

Structuring, visualization

In spite of all the mentioned critique points it must be concluded that the portfolio analysis is a suitable tool for structuring one's thoughts as well as for the visualization of complex issues. Accordingly, portfolios support an easy comprehensibility and the transparency of the planning process. Finally, the portfolio analysis provides a suitable framework and a communication basis for an intensive discussion about the current situation as well as future orientations of the enterprise or its departments; and this particularly to heterogeneously assembled committees. Thus, the portfolio methodology is used in different companies in interdisciplinary teams and helps them to get a systematic regard on the situation (cf. Herzhoff 1991). This property can contribute to the improvement of the interaction between R&D and marketing.

4.8 Conjoint Analysis

The knowledge about customer demands is a pre-condition for market success. The customer demands reflect the individual weighting of product attributes or respectively the preference for certain products. In this way, customer acceptance can already be estimated in the phase of first concepts. Besides other methods (e.g. multi-dimensional scaling) it is in particular the conjoint analysis that has established itself for that purpose.

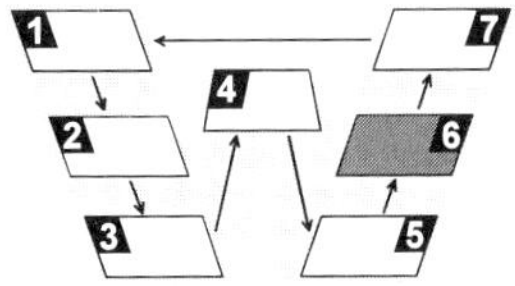

Additive customer benefit

The *conjoint analysis* is used to be able to assess the customer's acceptance of a product and its functions. In the analysis, it is assumed that the total benefit of a product is additively composed of the benefit of the individual product components. At first, all important product characteristics (e.g. weight) and their possible specifications (e.g. light or heavy) are identified. Through the combination of different specifications of the various product characteristics, several concepts are realized, which are presented to the potential customer for assessment. The customer rates the concepts according to his preferences. The number of product characteristics must be accurately planned, so that the evaluation of the different specifications can be mathematically determined from the indicated rating of the concepts.

The conjoint analysis, also called conjoint measurement or conjunct analysis, was created in 1964 in the USA by LUCE and TURKEY under the term "Simultaneous Conjoint Measurement". The authors developed a method for the simultaneous measurement of the allover effect of two or more variables on the level of interval scales by using ordinal-scaled[23] original data. The first algorithm for the evaluation of conjoint analysis data is due to KRUSKAL, who in 1965 developed the monotone analysis of variance; a methodology which still today counts among the most often used scaling methods for the conjoint analysis. In 1971, GREEN and RAO published a first paper, which brought the conjoint analysis in connection with the application area of marketing (Gerhards 2002).

New product planning

One of the most important application areas of the conjoint analysis is the planning of new products. The method is

[23] The ordinal scale allows establishing a rating by means of order statistics

supposed to answer the question "how a new product idea can be optimally designed with regard to the needs of the potential customers" (Meffert 1998).

Linking of QFD and Kano model with the conjoint analysis

For weighting of customer requirements, the conjoint analysis can be linked with the Kano model and the QFD methodology. For the early phases of the product planning, the first phase of the QFD method is of interest. In this phase, customer requirements are transformed into solution-neutral product or performance characteristics (Geisinger 1999). The relation between customer requirements and the deduced product characteristics are represented in the *House of Quality*. For the weighting of customer requirements GEISINGER recommends to use the conjoint analysis (Geisinger 1999). By means of the Kano model, the customer requirements can be structured. In the model, the customer satisfaction is represented against the degree of fulfillment of each customer requirement. The model is based on the empiric experience that in questionnaires the customer articulates only a certain part of his requirements, namely performance requirements. Basic requirements and exciting requirements are, in contrast, rarely mentioned. Basic requirements are considered as a matter of course and the customer thus does not specifically mention them in interviews. The non-fulfillment of these requirements does, however, cause strong discontentment. Performance requirements correspond to the explicit expectations of the customer and can directly be recorded, for instance by using conjoint analyses (see above), as the customer explicitly demands them. Their degree of fulfillment is linearly proportional to the customer satisfaction. Exciting requirements are not expected by the client but they are appreciated as a useful surprise. They can cause high satisfaction, not, however, discontentment (Brandenburg 2002).

Translate customer requirements into characteristics

Weighting of customer requirements

Bringing customer satisfaction into the model

Structuring of customer requirements by means of the Kano model

Application in the IRM methodology

For the concept evaluation the evaluation of product ideas or product concepts must be concretized and further developed based on the information gained during idea detailing. The following planning activities are distinguished:

1. Assessment of requirement fulfillment
2. Validation of the technical feasibility, and
3. Economic efficiency calculation.

The assessment of the requirement fulfillment deals with the evaluation of how suitable a product idea is to fulfill the concrete customer requirements. The product ideas can be compared with the requirements and technical characteristics as already formulated in the House of Quality. Furthermore, customer interviews, for instance by means of conjoint analyses, can bring the necessary information (cf. Brandenburg 2002).

4.8.1 Steps for the realization of the conjoint analysis

Five process steps

The conjoint analysis can be structured into five steps (Fig. 4.34). At first, the investigator must determine the product characteristics and their specifications and then develop a design. The third step consists of interviewing respondents; here, fictive data are assessed. Based on the obtained data, the individual part worth utility is estimated and evaluated by means of the conjoint analysis. In summary, the first step can be allocated to design, steps two and three to data collection, and the steps four and five can be allocated to data evaluation.

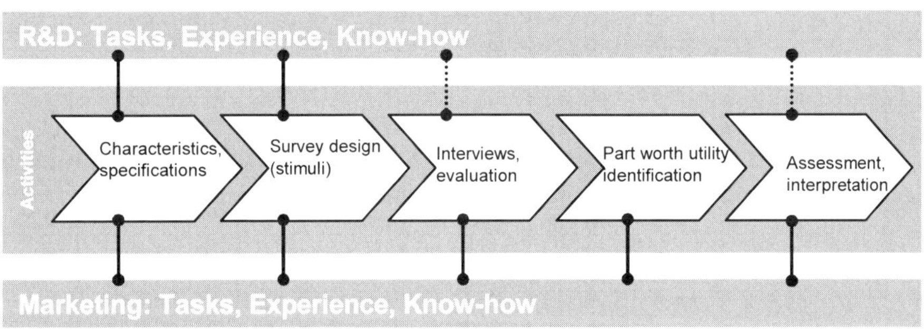

Fig. 4.34 Process steps of a Conjoint-Analysis (comp. Backhaus et al. 1996; Gerhards 2002)

4.8.2 Choice of product characteristics and the corresponding specifications

Choice of product characteristics

The part worth utility identified in the conjoint analysis refers to the specifications of product characteristics, which must be determined in advance by the investigator. The adequate selection of product characteristics and specifications is of significant importance. The selection problem is in general connected with a compromise solution, since on the one hand, as many product characteristics and specifications as possible should be incorporated into the design in order to guarantee a realistic representation of the product assessment; on the other hand, however, from a certain number of specifications, the reliability and the validity[24] of the results decrease with each additional specification. Concerning the choice of product characteristics and specifications the following aspects must be considered (cf. Backhaus et al. 1996):

1. The product characteristics for the assessment of an alternative must be relevant, i.e. the product characteristics must be selected on the assumption that they are relevant for the overall benefit evaluation of the respondents.
2. The product characteristics must be influenceable by the manufacturer in order to make the results of the conjoint analysis useable for the product design and selection.
3. The selected product characteristics must be independent. This means that the perceived benefit of a specification is not influenced by the specifications of other product characteristics.
4. The specifications must be realizable; i.e. these specifications must be technically feasible for the manufacturer.

Reliably uncover product characteristics and specifications

When generating product characteristics and specifications, not only marketing experts and designers should be involved, but also potential buyers, in order to realize a reliable disclosure of all relevant product characteristics and specifications. Possible methods for finding product characteristics are interviews with groups or individuals, as well as the elicitation interview and the repertory-grid-test. In the latter methods, the interviewed persons are presented already available product brands. On the basis of these brands, the customer specifies product characteristics important to him. If in the repertory-grid-test, characteristics resulting from the analysis of the differences of various concept alternatives are to be disclosed, the respondent is asked to spontaneously name striking characteristics of a stimulus during the elicitation interview. Compared to the process of characteristics generation, the collection of characteristics from vendor brochures and catalogues, product descriptions, analyses from test reports,

Methods to uncover product characteristics

[24] In this context it is often spoken of reliability and validity.

etc. is a much simpler method (Weiss 1992). The aim of all these methods is to obtain a compilation of potential product characteristics and characteristic specifications that is as complete as possible, and from which subsequently a subset is to be selected, taking into consideration the previously listed requirements.

4.8.3 Choice of preference model and design

Preference model

After the generation of product characteristics, a suitable *preference model* must be selected and a *suitable design* must be determined in a second step of the conjoint analysis.

Preference model

Basically, the preference model can be divided into two parts; one being the *assessment function*, which shows how each specification of a characteristic can be allocated to an assessment value, and another one is the *linking function*, which provides the information about how the assessment values of the single characteristics of the object can be aggregated to a total value (Weiss 1992).

Assessment functions

a) Assessment function:
In the conjoint analysis three assessment functions are used. Besides the *ideal vector model*, which assumes a linear relation between the height of the specification and the part worth utility, and the *ideal point model*, which is based on the assumption that there exists an ideal specification with a maximal part worth utility, it is above all the *part worth utility model* which is used.

Part worth utility model

The part worth utility model, in which each specification can be allocated to any part worth utility, can be seen as the most flexible of all assessment models, because it is the only model which admits qualitative product characteristics together with categorical specifications (e.g. color, design, etc). It must, however, be said that, compared to the ideal vector or ideal point model, considerably more parameters are to be estimated to determine the progression of the function in the part worth utility model (Weiss 1992).

Total preference value

b) Linking function:
The linking model, which links the part preference values of the individual concept characteristics to a total preference value, is usually divided into compensatory and non-compensatory models.

Compensatory models, which in many conjoint analysis applications are used almost exclusively, are based on the assumption that all concept characteristics are considered in the evaluation process and that they are interchangeable (Schubert 1991). In compensatory models, the linking of the evaluation values can be achieved in an additive or multiplicative way. Often, it is the additive model that dominates; mostly in connection with the part worth utility model, which will be used in the further realization.

In contrast to compensatory models, non-compensatory models do not allow an offsetting of the negative assessment of a concept characteristic by the positive evaluation of another one of the same assessment object. Model versions of non-compensatory type are for instance the conjunctive, disconjunctive and lexicographic models.

Form of collection

Forms of collection

Two collection forms must be mentioned in particular when speaking about the traditional conjoint analysis: the *trade-off approach* and the *full profile approach.*

Trade-off approach

When using the trade-off approach, the respondents must assess pairs of specifications according to their preference. The advantage of the method lies in the low complexity of the evaluation task.

Full profile approach

In the profile approach, which is the most common method, a stimulus consists of the combination of respectively one specification of all product characteristics. A stimulus is the combination of specifications which are given to the respondents for assessment (Backhaus et al. 1996).

The essential advantage of the profile method is the realism of the stimulus presentation; amongst others it allows the questioning of buying intentions. Since with the growing number of product characteristics and their specifications the number of possible stimuli rapidly increases, it is most often resorted to the use of a subset.

Reduction of the stimuli

Information overkill

In order not to overstrain the respondents with a too high number of stimuli, it is aimed for a reduction, in general by means of a so-called fractional factorial experimental design; corresponding information losses are accepted (Berekoven et al, 1996). Since information overkill might already occur with about 20 stimuli, most applications require the use of this method. It must be made a difference between the approach applied in a symmetric design, where the number of the specifications of all characteristics is equal, and the one of the asymmetric design, which shall be presented here.

Experimental design

For the reduction of a symmetric design the method of the Latin square is suitable. The construction of an experimental plan for an asymmetric design is based on the principle that a proportional "distribution" of the specifications occurs for those characteristics whose number of specifications is smaller (Backhaus et al, 1996). Depending on the specific problem, various design alternatives are possible for the use of a fractional factorial design. Since the complete presentation of all these alternatives would go beyond the scope of this method description, only the ADDELMAN-plan method will be referred to in the following, which is used in the applied software tool SPSS.

Approaches for big factorial designs

It has to be mentioned that for dealing with big factorial designs, basically three solution concepts were developed: the use of incomplete block designs, the consideration of only individually relevant characteristics for the conjoint analysis, as well as the use of the hybrid conjoint analysis.

4.8.4 Presentation of stimuli and interview

To select a presentation form for the stimuli, it is above all the form in which the product concepts are presented to the respondents for evaluation that is of interest. In connection with the full profile approach mainly three presentation possibilities are discussed, which can also be used in combination (Schubert, 1991):

- The verbal design,
- The visual design,
- The physical design.

Verbal description

Most of the time, the verbal description is carried out with stimulus cards. It is advantageous that almost all characteristics can be represented. The verbal description is also a simple method for data acquisition (Weiss, 1992). If the number of characteristics is too high, this can, in contrast, be disadvantageous; it also stays unclear whether this method can create ideas about real products. A visual representation of the test concepts, for instance in form of sketches, drawings, photo-mountings, etc. clearly simplifies the evaluation task of the respondents. It might, however, be problematic that under certain circumstances it is not only the test-relevant characteristics that control the evaluation process, but also other information shown in the picture or its artistic representation (Schubert 1991). The design of the test objects as physical models or products is the most realistic form of stimuli presentation. Possible forms of appearance of the objects range from three-dimensional models to a real product. Amongst others, disadvantages of this presentation form are the outlay to be made (making available all product alternatives that have to be evaluated) and – amongst others for innovation goods – the problem of availability. The interview itself can take place in written, oral, or computer-assisted form.

Visual description

Stimuli presentation

Important: Scaling of the information

During the interview with the respondents, high importance must be given to the scaling, i.e. the measuring level of the evaluation data. Basically, four measuring levels can be distinguished: the division of the non-metric scales into *nominal*[25] and *ordinal scales*, as well as the *interval* and *ratio scales*[26] as evaluation criterion of the metric measuring level (Backhaus et al. 1996). While originally, the conjoint analysis was about analyzing ordinary assessment data, today the tendency goes towards the use of metrically scaled input data (Schubert 1991). In practice, it is the method of rank distribution, which allocates a rank to each product, and the method of rank sorting, which is a sorting of the product alternatives according to the preferences of the respondents and to the evaluation of the product alternatives through metric preference values in the form of the preference value method that is most commonly used.

---

[25] Nominal scales present classifications of qualitative property specifications. An example of a nominal scale is: riding quality (sportive-comfortable)

[26] "Interval scales" and "ratio scales" are scales with equal distances; the latter has a natural zero-point

4.8.5 Selection of the estimation procedure for the determination of part worth utilities

The aim of the conjoint analysis is to identify the part worth utilities for the specifications of individual product characteristics. This happens via different stimuli and based on empirically gained holistic assessment data. It is achieved through the selection of an estimation procedure which defines the algorithm for the parameter estimation. From the gained part worth utilities, the metric utilities for all stimuli and the relative importance for single product characteristics can be deduced (Backhaus et al. 1996).

Computer-assisted analysis

Since the conjoint analysis includes the estimation of a big number of data, the evaluation usually takes place computer-assisted. The selection of a suitable software tool requires knowledge about the used statistical procedures. The estimation procedures can principally be divided into two classes, which are based on metric or non-metric algorithms and are complemented by other statistic approaches. SCHMIDT (1996) gives an overview about the most common algorithms.

Metric approaches

If the assessment data are available in a metrically scaled form, the application of an Ordinary Least Squares (OLS) regression is suggested. Alternatively, the approaches for Minimizing of Absolute Errors (MSAE) regression as well as the Analysis of Variance ANOVA[27] are available (Schubert 1991).

Non-metric approaches

The group of the non-metric approaches can be divided into ordinal and nominal algorithms.

Besides the procedures of PREMAP[28], JOHNSONS[29] and LINMAP[30], it is mainly the MONANOVA[31] algorithm developed by Kruskal that is used in ordinally scaled assessment data. The monotone analysis of variance forms an iterative procedure and is thus much more computer-

27 ANalysis Of VAriance

28 PREference MAPping

29 JOHNSON's Trade-off Procedure

30 LINear Programming Techniques for Multi-dimensional Analysis of Preference

31 MONotone ANalysis Of VARiance

intensive than the metric analysis of variance (Backhaus et al. 1996). It is based on the principle of the monotone regression: the surveyed similarity ranking is Z-transformed into metric values.

If the assessment data is available in nominally scaled form, the Categorical Conjoint Measurement (CCM) could for instance be used as estimation method. However, this approach could not establish itself in practice (Schubert 1991).

Statistical approaches

Besides the mentioned procedures used for estimating metrically and non-metrically scaled evaluation data, one can find so-called statistical approaches in the literature. They create a relation between paired comparison assessments and probability models (Schubert 1991). The approaches *Probit* and *Logit* are based on this approach.

4.8.6 Analysis and interpretation of the results

Three phases

The evaluation of the data sets collected during the conjoint analysis is usually carried out in three steps. In the first step the data of all respondents are analyzed separately (individual analysis), and then, based on all data sets or segment-specifically, summarized to aggregate conjoint solutions (aggregate analysis) (Schubert 1991). The third step consists of the presentation and interpretation of the results.

Besides the individual analysis followed by the aggregation of the obtained part worth utilities, the application of a so-called common conjoint analysis is possible.

Individual analysis

Adjusting the data sets

On the one hand, an individual analysis requires a substantial analysis effort. On the other hand it is necessary in order to sort out data sets in which the assessment of the respondents contravened the assumptions of the additive conjoint analysis. Compared to other approaches of perception and preference research, the conjoint analysis has the advantage that infractions of the assumptions and random answers of the respondents can be recognized due to the results configuration and may be excluded in order to obtain the "cleanest" data sets possible for the aggregate evaluation (Schubert 1991). Higher stress values can for instance be an indication of an "infraction" of the assumptions by the respondent.

The individual analysis does not only allow to find "incorrect" data sets, it also provides the basis for the following segment-forming and thus for a target group-specific product development (Schubert 1991).

Aggregate analysis

After the individual analysis the individual part worth utilities per specification can be aggregated by means of averaging. This procedure requires, however, a standardization of the part worth utilities for each person (Backhaus et al. 1996).

Representation and interpretation of the results

Estimated part worth utilities

The evaluation procedure of the conjoint analysis results in the estimated part worth utilities of all characteristic specifications of the investigated object. They have a metric measuring level and are directly comparable with each other (Schubert 1991).

Before the interpretation of the data, the stress value should be analyzed as a measure for the quality of the estimation. The stress value expresses the degree of congruence of the deduced ranking for the preference values with the original empiric ranking of the stimuli profiles.

Low stress values do not necessarily suggest a solution configuration rich in content, as so-called "degenerate" solution configurations, which due to a simplification of the evaluation process by the respondent result in infractions of the assumptions of the compensatory preference model, show a lower stress value too (Schubert 1991). Degenerate solutions can be recognized by the fact that relatively high part worth utilities can be observed for one or two characteristics only, while for the other characteristics the values are zero or very small. This kind of results can for instance be recognized by means of the so-called Shepard diagram (Schubert 1991).

Determining the importance of a product characteristic

After identifying the importance of the individual specifications of the product characteristics for the overall evaluation of the product concepts, the next step consists of determining which product characteristics are particularly important for the evaluation. It is the range, i.e. the difference between the highest and the lowest part worth utility of the specifications of a product characteristic that is decisive for the importance of a product characteristic for the preference change. The relative importance thus results from the weighting of the range of different product characteristic in relation to the sum of the ranges of all product characteristics (Backhaus et al. 1996).

Visualization of the characteristics

It has to be noted that for less important characteristics the preference values virtually do not change if the specifications of these characteristics are varied. For particularly important product characteristics, however, changes of the specifications result in strong changes of the overall evaluation (low willingness to compromise). During the final representation of the data in diagrams, discrete characteristics should only be visualized by means of bar or circular charts. The use of connecting lines suggests that information is also available about the specifications between the categories (Schubert 1991).

Further evaluations

Alternative concepts

Besides the estimation of part and total preferences the conjoint analysis offers further procedures for data evaluation. In the segment-specific analysis, for instance, specific target group definitions for alternative product concepts can be gained. The step for the segment-specific analysis, which counts among the most important steps of the conjoint analysis, is mainly used in the early phase of the product development. Further steps of the evaluation are market simulations, which is used for identifying the success probability of alternative product concepts, as well as the positioning of alternative product concepts by means of multiple correspondence analyses (Schubert 1991).

Advantage and success factors

Essential advantages of this method are: an exact determination of the utilities and requirements of both, internal as well as external customers and thus the optimal control of the R&D-related use of resources due to the knowledge of utility shares and areas. The impact of subjective product characteristics (benefits, imageries) on the objective product characteristics becomes transparent. By using adequate software the below-mentioned disadvantages can be attenuated or eliminated (it is to be seen as a critical success factor). Furthermore, considerable statistic-mathematical knowledge is required. The pre-selection of the product characteristics is subjectively influenced. Mostly, a past-oriented, conservative collection of product characteristics takes place.

Outlay

The method of the conjoint analysis is very time- and cost-consuming. On the other hand, there exists a number of software programs (e.g. Unicon, TradeOff, Monanova, Linmap, ACA), which support the application of the method.

4.9
Technology Roadmapping

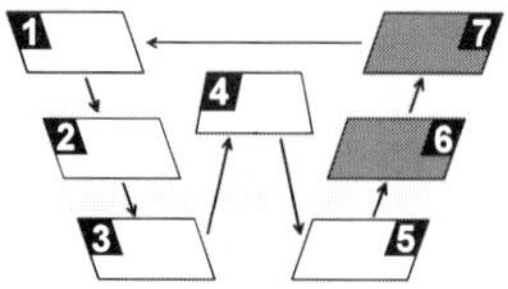

Technology Roadmapping enables the prognosis, analysis and visualization of future technology developments. The goal is the prognosis and evaluation of future developments in a certain field of activity. Roadmapping is made up of the generation of the roadmap and the actual result presentation as a roadmap.

In general, technology roadmapping allows concretizing the planning of future developments. On the one hand, it makes available a systematic-analytical and creative process for the future prognosis of product, service and technology developments, and on the other hand offers the opportunity for a sensible visualization of results (Specht et al. 2000). The goal of technology roadmapping is the prognosis and evaluation of future developments in a certain field of activity. For that purpose expert knowledge is systematically collected and diverging opinions are synchronized in group-dynamic processes. On this basis takes place the deduction of detailed options for action in an enterprise-individual framework for action.

Technology roadmapping is especially useful as an instrument for the prognosis and representation of the developments and convergences of technical products. The availability of technologies and their interaction over time are graphically represented. The process of technology roadmapping allows collecting unstructured expert knowledge and aims at stimulating the intuition and creativity of technical experts.

Advantage of the technology roadmapping

The benefits of a technology roadmap are the systematic approach as well as the structured representation of future enterprise activities. By using this method the staff responsible for planning and for implementation gets a far-reaching overview over the next strategic planning steps. Furthermore, operative steps can be deduced from the developed strategic guidelines.

Variants of technology roadmapping

In practice there exist different forms of technology roadmapping. On the one hand, they result from different reference objects; on the other hand, the goals vary depending on the different interest groups (Möhrle and Isenmann 2002). Depending on useful reference objects

Möhrle and Isenmann suggest the following classification:

- Roadmapping for central *pacesetter*, *key and future technologies*: application potentials of different application systems can be deduced
- Roadmapping for *application systems*: the objects are superior future issues (e.g. the factory of the future, the car of the future)
- Roadmapping for the *business activities of an enterprise*: current and future product programs are listed in connection with services.

The forms of technology roadmapping vary due to the participating differentiating interest groups and their different functions (Möhrle and Isenmann 2002):

- Internal control of the company's R&D units
- Coordination between different enterprise divisions
- Presentation of the competitive strategy
- Coordination of internal and external R&D activities
- Common (technological) orientation of different companies.

4.9.1 Technology calendar according to Schmitz (Fraunhofer IPT)

Technology calendar according to Westkämper

In order to connect the strategic planning horizons with each other Westkämper developed the technology calendar, an integrative approach for the department of production technology (Westkämper 1987). Due to the fact that new production concepts should be planned using the strategy of an anticipatory harmonization of product and process development, Westkämper uses the technology calendar in order to coordinate company resources such as staff, development and investments in terms of a long-term planning horizon (Westkämper 1987).

Technology calendar according to Schmitz

Based on Westkämper's technology calendar, Schmitz developed a modified technology calendar which is an extension of this technological approach in conformity with the used method. The modified technology calendar allows for a systematic, comprehensible, and enterprise-individual comparison of products to be manufactured and relevant or useful manufacturing technologies (Schmitz 1996). The axes of the modified technology calendar comprehend the dimensions “product”, “technology” and “time” (Fig. 4.35).

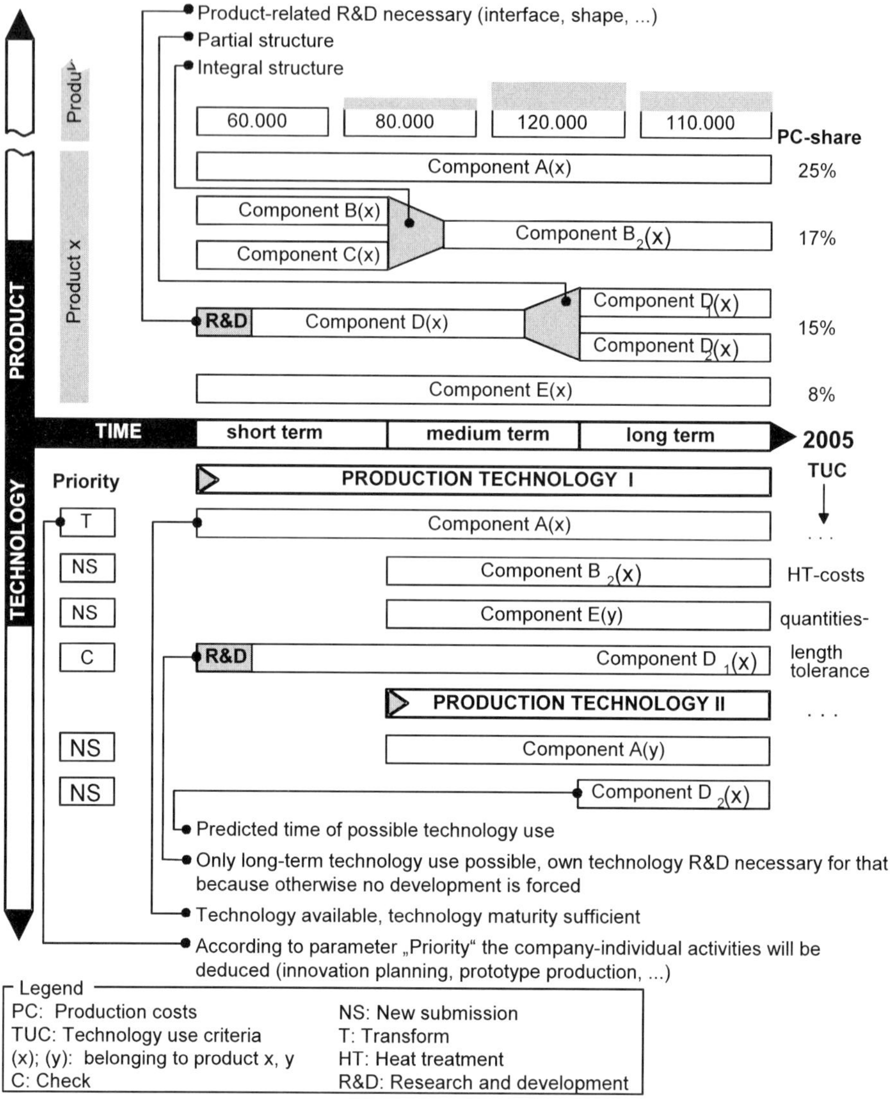

Fig. 4.35 Structure of a technology calendar (Schmitz 1996)

“Product” area of the technology calendar

In the “Product” area of the technology calendar – represented through the axes product and time – a short- to long-term sales and production plan is established for all success-promising product innovations based on predicted quantities. Underneath the quantities, the relevant components of the respective product innovations are sorted according to their share of manufacturing costs. In addition, those modifications in the design of the product structure (e.g. the introduction of an integral or partial construction) that are directly connected to the use of a particular technology are marked in particular.

Classification in time of design changes

The classification in time of the constructive changes of components mainly depends on two aspects: on the one hand, each company needs a certain time for the introduction and realization of product changes, which can for instance be due to new admission procedures or the adjustment of the tool manufacture. On the other hand, the time of implementation depends on the maturity of the manufacturing technologies which are supposed to produce the new component structures. If no change of the component structures is planned, at least those components are marked whose features must be changed in terms of construction due to the implementation of new manufacturing technologies and thus require development work in the R&D.

"Technology" area of the technology calendar

The "Technology" area of the modified technology calendar is represented through the axes "technology" and "time". Here are listed the manufacturing technologies that will be used in the future for manufacturing a component. The time of the first implementation of a manufacturing technology – and thus its starting position in the technology calendar - depends on its first *possible* concrete application on a particular component.

Priority of technology applications

For each component, the "priority" of the technology application is determined according to an established standard strategy. The classification of the components takes place with regard to technology use criteria (TUC), which are also presented in the technology calendar. If a component structure needs an own technology-related development, this particular component is marked in particular (e.g. by the comment: R&D).

Repetitive and detail planning independent of individual persons

To make sure that repetitive and detail planning can be made independently of individual persons, the planning history of both, the product as well as the technology area within the modified technology calendar is documented in a product- or respectively technology-oriented data sheet.

4.9.2
Usages of the technology calendar

Possibilities of the technology calendar

Basically, the technology calendar according to SCHMITZ gives the user an overview about which production technologies may currently or in future be used for manufacturing which components (Schmitz 1996). Based on the vertical planning of the product and technology areas on the one hand and the horizontal planning over different periods of time on the other hand, a holistic perception in the sense of a strategic planning becomes possible.

Advantages of the technology calendar

By using SCHMITZ' technology calendar, the user can, amongst others, estimate the medium- to long-term manufacturing costs of a component. Furthermore, the visualization of the results in a roadmap offers the advantage of a systematic, comprehensible, and company-individual comparison of the developed product innovations and the required production technologies. Thus, the strategic planning of future product innovations can be carried out in connection with the technological potentials and it is guaranteed that planned new product launches can be realized according to the strategic planning requirements, also with regards to technology.

Benefit of a technology calendar

The modified technology calendar is a tool which presents the following advantages for different company departments or respectively responsible persons (AWK 1999):

- Using the technology calendar, the company management can control and check the strategic alignment of the product and technology planning.
- When planning new production concepts, the production managers can predict future technology developments in a systematic form and thus take them into consideration.
- During the construction and design of product innovations both, company-internal manufacturing technologies as well as those available on the market can be considered.
- When planning investments, potentially necessary new research and development projects for the anticipated production technologies can be specifically defined.
- Using the technology calendar the purchasing department can identify the technology-specific needs in quantity and capacity and use it as decision guidance for upcoming make-or-buy decisions.

Deriving future enterprise activities

Based on the product and production technologies listed in the technology calendar; future company activities can be derived or respectively synchronized. In this regard, SCHMITZ identified some typical decisions which can be taken based on the technology calendar (Schmitz 1996):

- For the specification of short-term saving potentials detailed cost comparison calculations can be carried out.
- In the case of promising production technologies, investment calculations can be initiated.
- Against the background of the potential of identified production technologies the relevant components can be designed accordingly; this may require a change of available design drawings, process plans, order processing, etc.
- The application of innovative production technologies for future product generations can be stimulated through suitable development projects.
- For promising products, the manufacturing of functional prototypes can be commissioned.
- Development co-operations with suppliers can be initiated in order to achieve technological leaps in practice.
- Authorization and release procedures with customers or public institutions necessary for a technology change can be carried out in advance.
- Corporate planning (financial planning, qualification) can be synchronized with regard to the need of resources (capital, staff, etc.) required for the introduction or use of new technologies.
- The recheck of an apparently more efficient production technology can be determined on a later date if this technology is not yet fully available.
- Staff responsible for future production technologies can be appointed, who will be responsible for a continuous update of the technology-related standard of knowledge.

In terms of content, the activities a company initiates based on SCHMITZ' technology calendar are rather orientated towards products or production technologies. Because of this and due to the temporal impact of the measures, an interconnection of the activities can be noted.

Repetitive planning every 1 to 3 years

Based on the technology calendar the results of a dynamic planning of internal and external technology developments can be compared with future prognoses and be represented. Against this background, it is useful to repeat the corresponding planning phases every three to five years in order guarantee the necessary up-to-dateness of the planning results. Thanks to the technology calendar the production technologies which cannot yet be used for the competitive production of company-own products can already now be taken into account in the corporate planning. Based on a reduction of complexity and a systemization of the technology planning - available through the use of the technology calendar methodology - product and production technology innovations can be synchronized more efficiently. Moreover, the transparency and acceptance of the planning results can be substantially improved compared to the conventional planning methods. With the structuring of future implementation activities, the willingness to initiate necessary technology innovations in line with the enterprise strategy is increased. At the same time chances increase that these technology innovations can be transferred into practice with real success.

Advantages of the technology calendar

By using the technology calendar, the user is given information about which innovative production technologies may currently or in future be used for manufacturing which components (Schmitz 1996). Due to the temporary connection of current and planned products with available and future production technologies, a synchronization of the development processes of both areas takes place. In the course of this synchronization detailed solutions can be developed according to the strategic objective, which are situated on the level of the technological basics (Schmitz 1996).

4.9.3 ProjectRoadMap – combination of InnovationRoadMap and technology calendar

In the planning phase "concept evaluation" of the InnovationRoadMap methodology there is on the one hand the need for evaluating the developed product concepts from an economic point of view. On the other hand, it is decisive for a successful strategic planning of future product innovations to take into consideration the possibilities of available and future technology potentials and to include them into the planning.

Methodical connections

The methodical combination of InnovationRoadMap and technology calendar to form the ProjectRoadMap allows considering the compatibility of product and technology planning. Loopholes in the medium- and long-term planning of products and technologies can be identified (Vinkemeyer 1999). Furthermore, technology and market know-how can be bundled for the future and available expectations, ideas, and concerns about the current and future business of the company can be detected.

The ProjectRoadMap summarizes the company activities identified in the InnovationRoadMap and the technology calendar, with the aim of introducing a product or technology onto the market at a determined time (Fig. 4.36). For each single project a separate ProjectRoadMap is established. In this way, the complexity of the representation is reduced and responsibilities can be clearly allocated. The aim is to align the activities in such a way that an optimal target achievement and an efficient use of the available and required resources are guaranteed.

In the ProjectRoadMap the results are represented in a process plan in a product-related form by transforming the identified development needs into a project structure. All product- and technology-related activities are summarized. The ProjectRoadMap serves as a visualization tool. It complements the usual content of a project description such as process plan, time plan, and work plan. Typical decisions taken during the establishment of the ProjectRoadMap, are (cf. Walker 2002):

- the stimuli for the development of special technologies,
- the acquisition of a machine or a component,
- the cooperation with a research institute or a technology provider for the further development of a technology,
- the introduction of special machinery,
- the technology test showing the suitability for series production,
- the promotion of the product development or
- the test of available products for efficient use of technologies.

This results in the opportunity of not only aligning differences in expectations, procedures and goals (e.g. between technology and marketing) but also of communicating future visions as the basis for a jointly supported strategy (Möhrle and Isenmann 2002).

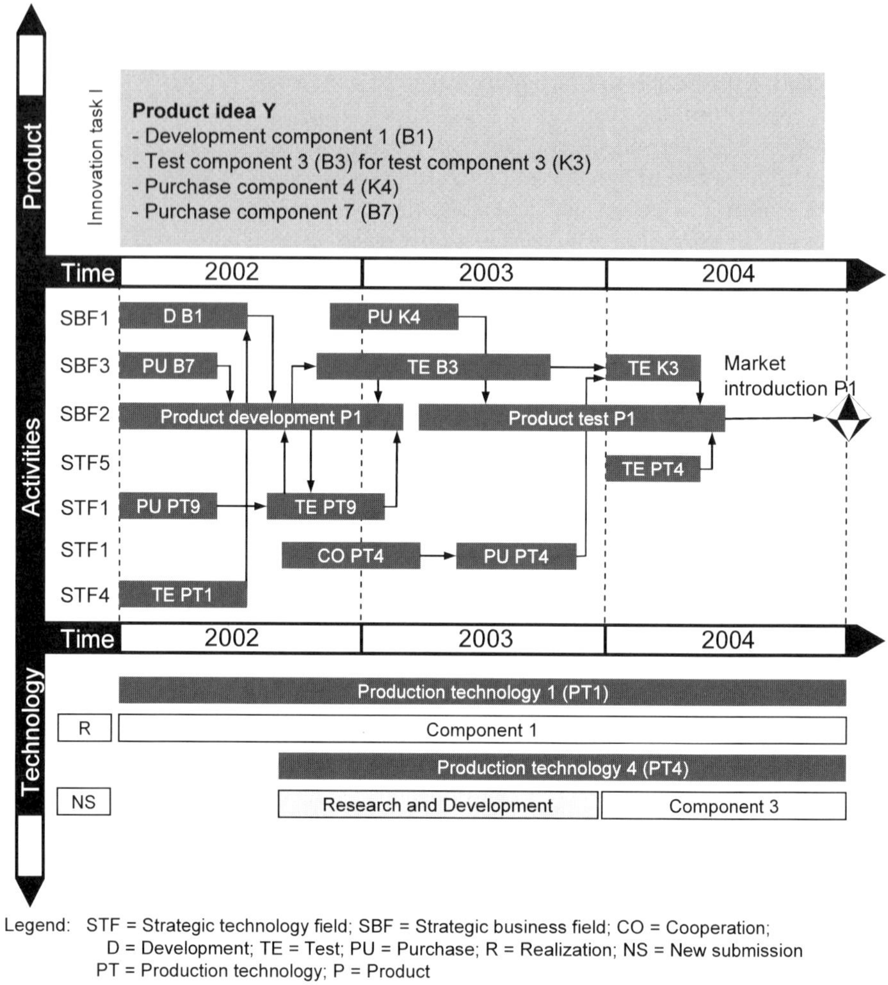

Fig. 4.36 ProjectRoadMap (cf. Walker 2002)

5 Case studies

The following chapter illustrates the practical innovation management in six different case studies.

Case study 1: SCHOTT Glas regularly analyzes its own company situation. The case study describes how these analyses caused SCHOTT to expand its existing competencies in the different business segments. The manufacturer of special glass types SCHOTT transforms from a manufacturer of materials to a provider of system and thus expands its competencies in the usual markets.

Case study 2: Using the "precise look over the customer´s shoulder" – the pro in construction –, the Hilti AG identifies different innovation potentials and strategically implements these potentials in new business segments. On this basis, the company has built up the business segment "Positioning systems".

Case study 3: The third case study shows how the company SUSPA Holding GmbH optimizes its innovation management. Building up on existing (technology) competencies, new products are used to develop yet unknown markets.

Case study 4: MicroMed 2000+ is a study of the Fraunhofer IPT and ILT about the identification of promising application fields for microsystems in medical technology. The aim of the study was to search for potentials in future or growth markets which can be used with own competencies.

The chapter is rounded off by two examples of a more effective realization of the innovation process:

Case study 5: The NEUMAG GmbH & Co. KG is a medium-sized company which offers complex equipments for the production of synthetic fibers as a package solution. It is explained how the product development was optimized by implementing a systematic innovation management and the software-supported assessment of product ideas.

Case study 6: The Dräger Medical AG & Co. KGaA offers machines as well as system solutions and services for the therapy of all relevant fields of the patient process chain. The case study describes how the company implements a worldwide standardized, quick and efficient innovation process.

5.1
SCHOTT Glas
"How a manufacturer of materials became a provider of systems"

UWE H. BÖHLKE, MARKUS GRAWATSCH

In the last few decades, the SCHOTT Group has become a world leader among manufacturers of special glass types.

In this case study, we describe the methods SCHOTT is currently using to examine its market position and deduce measures for securing its future.

The SCHOTT Group

In 1995 – 111 years after the Glastechnisches Laboratorium SCHOTT und Genossen was founded in Jena – the SCHOTT Group can look back on a positive trend in the development of the company. SCHOTT has grown from its origin as a glass working laboratory to become a world leader among manufacturers of special glass types. In the mid-1990s, the company's range of products included material solutions for virtually every world market for special glass, e.g. laboratory glass, TV glass, pharmaceutical glass, glass tubing and optical glass. In the field of ceramic glass for "white goods," SCHOTT has been the world market leader for more than 20 years.

In terms of economic development at SCHOTT, the company has experienced continual growth and in its 111th year will still be operating in the black. That is also the way it should remain in future. In the first year of the new millennium SCHOTT generated turnover worth approximately EUR 2 billion worldwide and employed about 20,000 people internationally in around 100 enterprises.

Really, things couldn't be better!?

Constant further development

Success has been paid for with hard work. Since its founding, SCHOTT has been continually developing and adapting to new situations.

Analyze the position

One of these new situations was the Asian crisis of 1995. Although SCHOTT was not excessively under pressure thanks to its wide range of products, the crisis was a signal to analyze more precisely its own position in the markets it was serving.

The analysis results can be summed up as follows: increasing cost disadvantages in the established production locations accompanied by a wider, but increasingly mature, range of products. Both factors, the restrictive cost situation, and the lack of new products, indicate that a gradual loss of the market position could be expected in future.

When we look at the business portfolio from 1995, the need for action becomes clear. The markets that were being served did not in the main offer further growth options; they were characterized by rising pressure of competition. Partially because of the expiry of patents, it was also to be expected that new competitors would appear on the scene and attack the markets that SCHOTT had accessed. The established business activities lose their productivity over time, due to the natural product life cycle, and activities that have been started to date run the risk of in future not being able to balance out the financial losses.

Business activities follow product life cycle

Changes became necessary – changes for securing as well as expanding the existing position: in the fall of 1995, a company wide productivity campaign was started, which formed the basis for an innovation campaign in the fall of 1998 (Fig. 5.1).

Starting a productivity campaign

Starting an innovation campaign

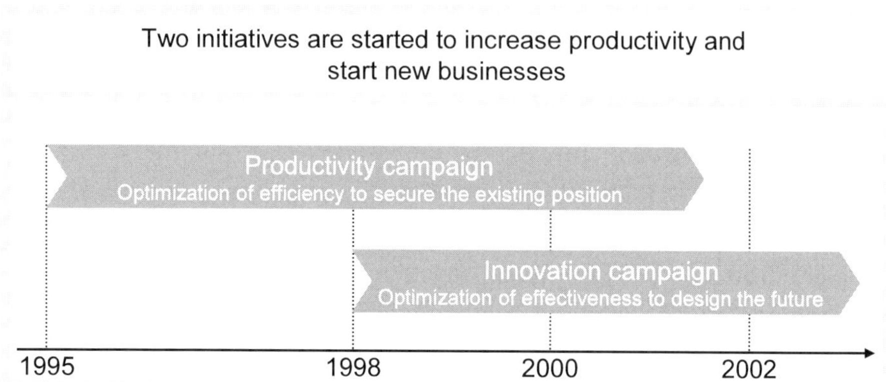

Fig. 5.1 Productivity and innovation campaign

Initiating transition

SCHOTT made preparations for the production of new products and actively began the transition from being purely a supplier of glass to a supplier of systems and components.

For this transition process the following guidelines were defined, among others:

- The plethora of existing business activities was to be countered with a concentration on strategic markets.
- Additional improvement processes were to complement the value added chains that were currently being realized.

- A consistent focus on the markets was to be maintained during the implementation of planned innovations.

Working out the vision

Parallel to the innovation campaign, a new vision, the SCHOTT "Vision 2010" was worked out company-wide to form a mission statement for the further development of the company. In this transition process, a new divisional structuring of the existing business is one of the measures that were carried out. From what was previously a more technology-focused company structure, SCHOTT was developed into a market-focused group. The individual business activities remaining in the portfolio have a minimal volume and, in view of the markets that are to be addressed jointly, are combined in individual Strategic Business Units (SBU). A total of five SBUs have now replaced the eight company divisions that existed before the re-structuring (Fig. 5.2).

Technology/ market focus

Re-structuring the SBUs

Use of inimitability portfolio

The pursuit and further development of the new company structure was carried out using a portfolio developed at SCHOTT, the "inimitability portfolio" (Fig. 5.3).

Analogous to the market portfolio of the Boston Consulting Group, market growth is entered along the vertical axis. The uniqueness or inimitability of the business performance in comparison with competitors is entered on the horizontal axis. This factor refers to the current situation, which will be decisive for the future market position. A detailed distinction is made as to whether SCHOTT is only differentiated from the competition from time to time, or whether the business activities have a long-term inimitability that cannot easily be copied.

This type of depiction offers SCHOTT the advantage that not only the present situation on the markets, but also the explanation for this situation, can be described. Thanks to the better consideration of anticipated business field developments, this form of business field analysis is more practicable for SCHOTT than the classical portfolio approaches.

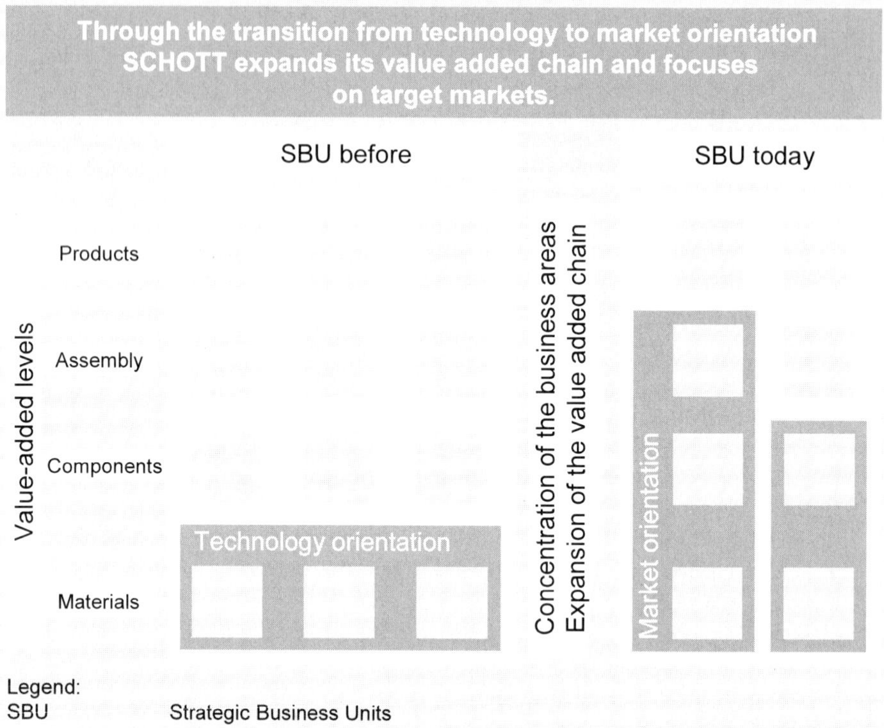

Fig. 5.2 Reduction in number and expansion of the SBUs

The SCHOTT "inimitability portfolio" is subdivided into four quadrants, for each of which unique requirements and strategies for action are developed and assigned to the relevant quadrants. Business activities in the fourth quadrant are thus investigated for their strategic development potential, and in the case of nonexistent USP potential (USP = unique selling point), an exit scenario is worked out and tested.

Financing R&D

In order to support the strategic re-orientation, from 1998 onwards more money will be made available for Research and Development. In the financial year 1997, 3.7% of the turnover will be spent on research and development topics.

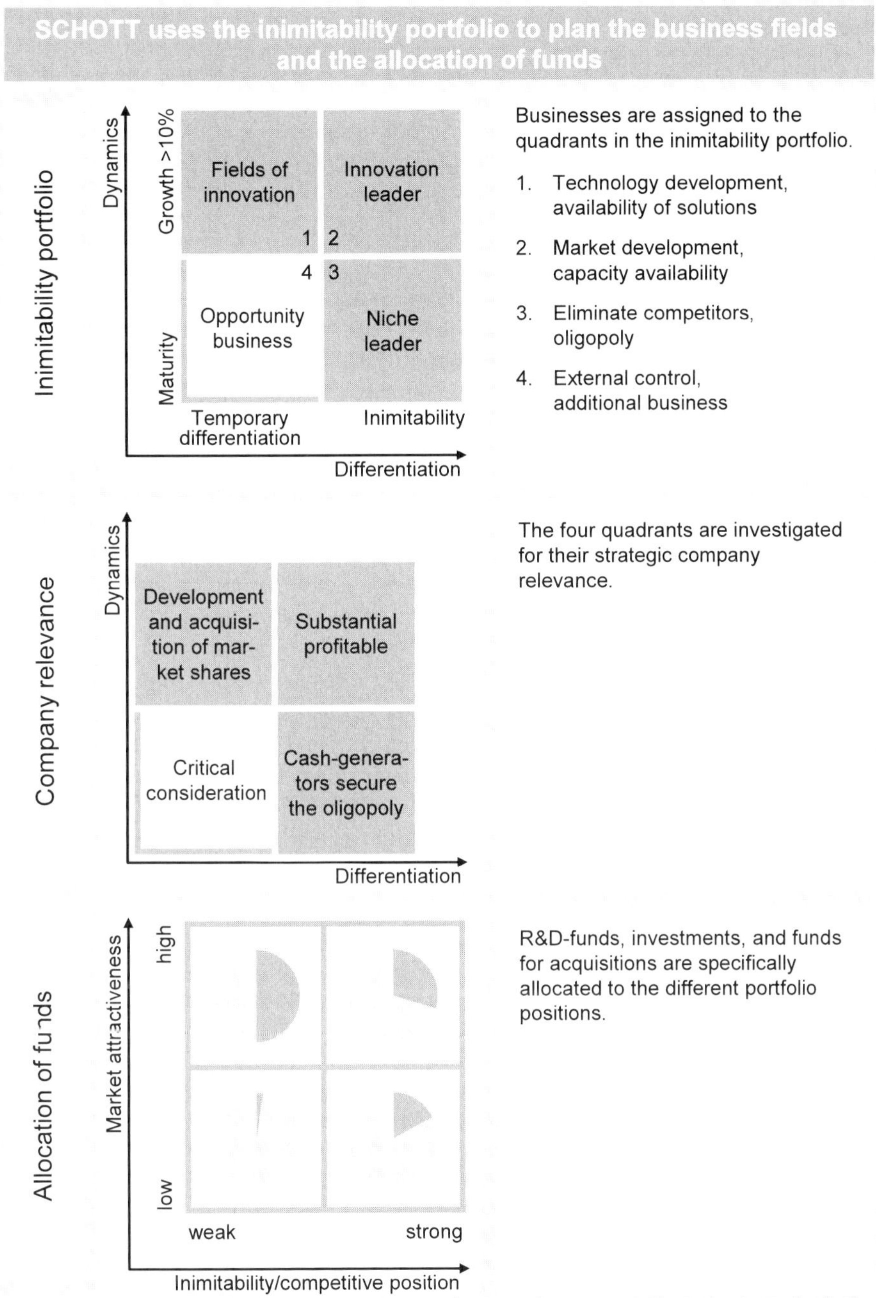

Fig. 5.3 The SCHOTT inimitability portfolio

The budget was increased steadily in the following years and in the year 2001 amounted to more than 6.3% of the turnover, which by that time had also risen. This equals more than EUR 120 million. The allocation or release of investment funding to specific business fields is consistently carried out according to the relevant business position in the SCHOTT inimitability portfolio. For this reason, the allocation of investment funding is over-proportionally high in the innovation quadrant. The assignment in the bottom quadrants is, in comparison with annual depreciation, lower and merely serves to partially compensate for the "loss of substance" (Fig. 5.3).

Motivate employees

The success of the innovation campaign depends primarily on the creativity and willingness of the employees. In order to create the best framework conditions for employees and company, various accompanying measures packages have been launched. Whereas work on the business portfolio has particular reference to the aspects of company vision and strategy, the objective of these measures is to further develop the company culture. For example, one point of this measures package is the consistent introduction of variable salary components for non-tariff employees. A salary variable that is closely coupled to the success of the company reinforces the sense of responsibility of the employee and thus, also, their commitment to the company and its objectives.

Further develop company culture

What results did the innovation campaign have?

Reading the results

The first results of the innovation campaign can already be seen in various indicators: for instance in the increase in the number of patent applications made by SCHOTT (Fig. 5.4).

Patent applications verify success

In 1995, SCHOTT applied for patents for 38 inventions and, compared with other (special) glass manufacturers, was situated at the lower end of the scale. Following the innovation campaign, the number of invention applications had risen rapidly. In 1999, 216 invention applications had been made, and in 2001 this number had risen to 304. SCHOTT had then reached the 36th place for patent applications among German companies.

The inventions continue to be consistently developed by SCHOTT, leading to innovations. Within half a decade, the product range changed dramatically. Increasingly complex components are being supplied, which are located at higher positions in the value added chain.

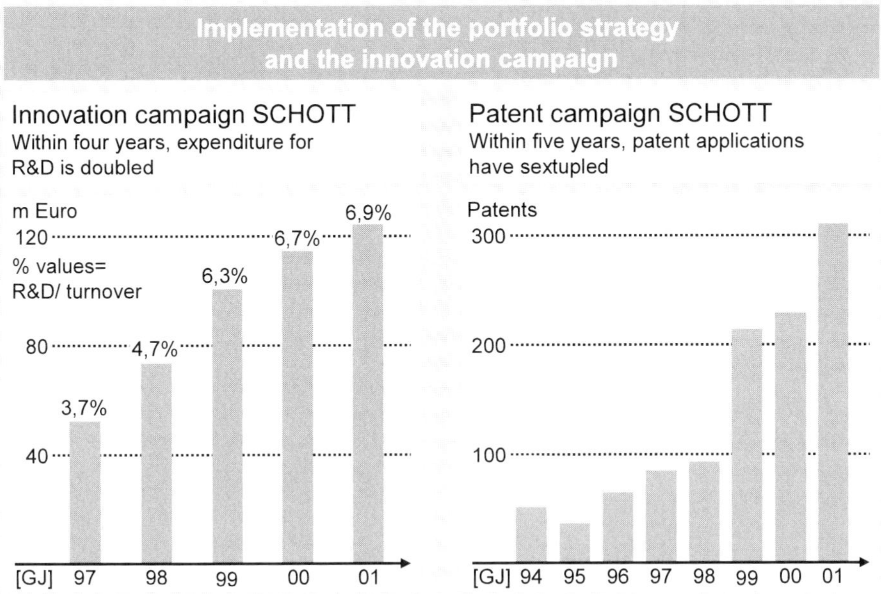

Fig. 5.4 Expenditure for and effects of the innovation campaign

Changes in the product range

One example for this consistency is the development of business in the field of lithography. At SCHOTT Lithotec GmbH, which was founded in 1998, not only expertise in the synthesizing of calcium fluoride crystals is being developed, but further components for lithographic chip manufacture (e.g. photomasks) are also being developed and produced. The importance of this market becomes particularly clear when the following development cascade is considered: when a new generation of computers comes on the market in 2006, the corresponding chips have to be manufactured in 2005. For this, the appropriate manufacturing equipment (wafer steppers) must be ready in 2004, which in turn means that the relevant optical technology must be mastered in the year 2003. The prerequisite for this is, again, that the manufacturing process for the required glass types has reached process maturity in the year 2002.

For SCHOTT, this is a further step out of the century of the electron and into the century of the photon.

5.2
Hilti AG
"Strategic Exploitation of New Business Fields"

WINFRIED J. HUPPMANN, THOMAS BREUER

In 1997, Hilti began marketing laser measuring products for construction industry applications. With the second generation of products, Hilti is now achieving turnover in the three-figure million range. Using the example of the Business Unit "Positioning Systems" we will describe

- how Hilti systematically deduces business potential
- which instruments are used
- what experiences have been gained and
- what constitutes the critical success factors.

Introduction

The business area "Positioning Systems" of Hilti AG achieved a turnover in the three-figure million range in the fiscal year 2002. And the trend is upwards – in spite of the fact that the first Hilti laser positioning tool was only sold in 1997. These systems assist construction workers and craftsmen in the measuring of distances, areas, volumes and circumferences. Later we will describe how Hilti develops this type of innovation, the procedure, the methods used, experiences and critical success factors.

About Hilti AG

Company profile Hilti AG

The Hilti Group employs around 14,000 people and offers the professional user in the construction industry a comprehensive range of drilling and dismantling technology, direct fastening systems, dowel technology, diamond technology and construction chemistry (Fig. 5.5). Included in the range of services are systems with the appropriate tools and consumption elements, consultancy services, application training courses and technical documentation as well as after-sales service.

100,000 direct customer contacts per day

Market cultivation at Hilti is pursued through direct marketing. Two of every three employees are engaged in market cultivation. They shape the company and take care of approximately 100,000 customer contacts daily. Worldwide marketing is organized along market-oriented principles and is represented by the marketing organizations in each land or market.

Internal customer-supplier relationship

Product stewardship lies with the 11 Business Units (BU) that are oriented to the product lines (Fig. 5.5) and, in turn, are centrally organized into 5 business areas. Based on internal customer-supplier relationships, the Business Units place the orders for the manufacture of the required products with Corporate Manufacturing. All production plants of the company are subordinate to Corporate Manufacturing. Together with Corporate Sourcing, Corporate Logistics and Corporate Engineering, this constitutes Hilti's supply chain.

Fig. 5.5 The Hilti product lines

Besides the areas that have already been described, the company also has Central Areas (Fig. 5.6). The Central Areas include those functions of the company for which it is advantageous to have a central configuration. These areas include company treasuring, company personnel, company information systems, company controlling and company development.

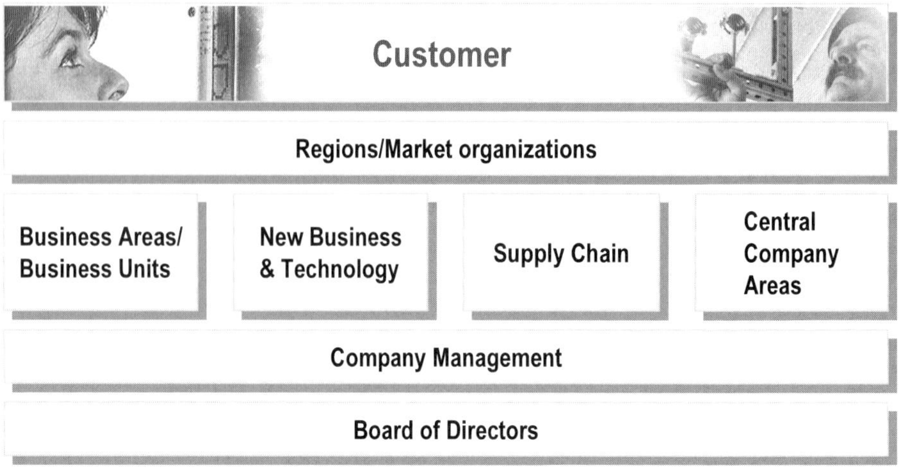

Fig. 5.6 Hilti Company Organization

New Business & Technology: Exploiting synergies and identifying business opportunities

A central role in innovation management is played by the sector New Business & Technology (NB & T); here the specific technical competencies for the regeneration of existing business fields and for the target-oriented development of new business fields are prepared and continually undergo further development. The merger of the areas Company Research, Patent Department and Innovation Management and the Competence Centers in Kaufering, Germany, facilitates the exploitation of synergies in technology management and the systematic identification and exploitation of new business opportunities. Hilti company headquarters are located in Schaan, Liechtenstein.

From mission statement to innovation strategy

Groundbreaking innovations are firmly anchored in company strategy

Constitutive to the innovation capability of the company is Hilti's mission statement "*We want to be the best!*" which is implemented in the strategy "*Champion 3C*" (Fig. 5.7). "3C" stands for the English terms: "customer, competency and concentration." Through clear formulation and intensive communication, this strategy creates a high level of innovation capability: groundbreaking innovations are a core competency at Hilti (Fig. 5.7).

We want to be the best

Customers:
We want to be our customers' best partner.
Their requirements lead our action.

Competency:
We distinguish ourselves by pathbreaking innovations, extensive quality, direct customer contacts and effective marketing.

Concentration:
We concentrate on products and markets in which we can achieve and preserve leading positions.

Fig. 5.7 Excerpt from the Champion 3C Strategy

The customer: starting point for innovation

The individual BUs and the NB & T department plan their product innovations in the sense of the provisions of the "Champion 3C" strategy. The starting point for innovation ideas is the endeavor to meet the requirements of the customer in the best possible way: where can Hilti support the customer even better than before? Which of the customer's work processes can be simplified or made more efficient or pleasant? These innovation ideas are not only connected with workers' and craftsmen's tasks directly on a construction site; the customer also receives support in the acquirement, administration and maintenance of Hilti systems. For example, major customers are supported by Hilti through so-called "fleet management." In this case, Hilti undertakes the complete administration process, so that the customer simply has the required number of systems available for a specific period without any administrative effort.

BU strategies are coordinated throughout the company

In order to coordinate the individual innovation ideas, they are compiled in the innovations strategy of the departments. An important element of the strategy work is agreement between the separate departments with the objective of identifying and exploiting potential synergies. The department NB & T plays a central role here. It has the task of recognizing cross-cutting issues, pursuing them and promoting them as an innovation project in its own departmental strategy.

Underlying "upstream" and "downstream" ideas for positioning systems

In 1992, a typical cross-cutting issue was the inspiration for the present-day BU Positioning Systems: at that time, product stewardship was the responsibility of four so-called divisions, from which the present BUs developed. Three of these divisions had identified activities of the workers, such as measuring, stopping, marking etc., which were carried out "upstream and downstream" of the actual work process, as potential fields for innovation. However, the related potential for innovation had been assessed by the individual divisions as slight. The fact that three of four divisions had identified the topic independently of one another prompted the central Research and Development Department (R&D) of that time to include this field of innovation in its innovation planning. The idea consisted of supporting measuring, stopping and marking, for example of drill holes, with laser measuring systems.

With the objective of more clearly determining the emerging potential, the topic was first taken up in a small research project with a few employees. The research team looked for a method that would permit a quantification of the potential for innovation and discovered the hitherto unknown method of video analysis, which Hilti then used for the first time.

Quantification of innovation potential by video analysis

In video analysis, work procedures are filmed and then analyzed in detail. For the investigation of "upstream and downstream activities," the assembly of cable trays was used as an example. The video analysis revealed that the construction worker's task was divided into 6 phases (Fig. 5.8), each of which took a different amount of time. The work phase of drilling, traditionally supported by Hilti, took up 17% of the work time. Simultaneously, the improvement potential of each work phase was assessed. For the drilling process, an improvement potential of 4% with reference to the total processing time was calculated.

The work phase "measuring and marking" took up 28% of the work time. The tasks of stopping and measuring off using a measuring tape and folding rule were previously done with barely any technical support and, particularly in jobs that were above head height, were relatively complicated. The improvement potential was estimated at 20%. Support for work in this field would bring a high level of customer benefit. The initial conjecture of the high potential for improvement had been confirmed and could now be proven by means of specific value figures. A more detailed description of the field of innovation was included in the innovation planning of the central R&D department for the following year: technical solution possibilities were to be examined and Hilti's commitment in this area was to be nailed down by a business plan.

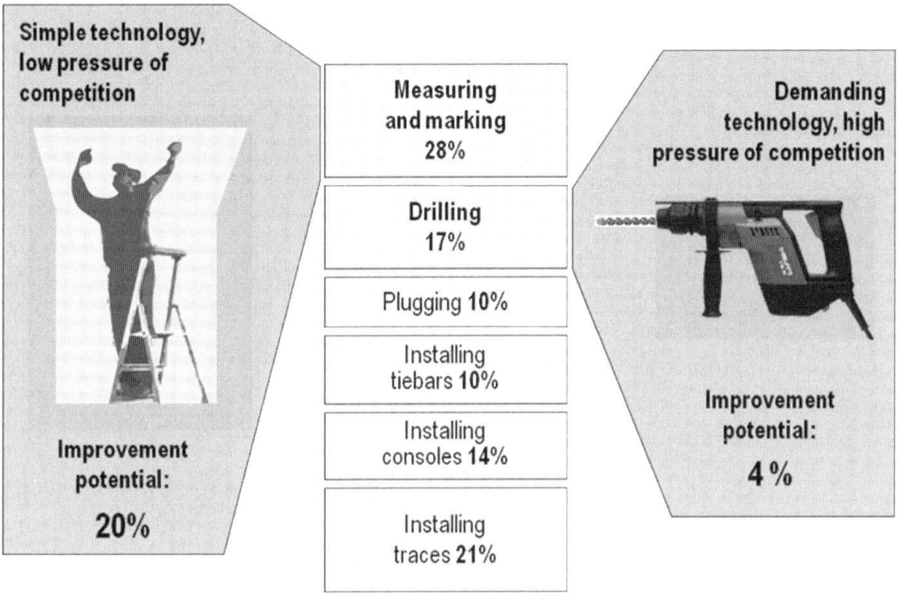

Fig. 5.8 Analysis of work procedures: erection of cable trays

Company-wide coordination of innovation strategies

Before innovation projects involving major expense are started, the individual departments present their innovation strategies at the "Corporate Innovation Workshop." This workshop, which is attended by all company management board members, is one of three workshops held annually at company management level. Here, the innovation strategies of the BUs and the NB & T are presented, discussed and approved.

Identification by integration

Strategy development is an important process for the BUs for which appropriately large amounts of funding are made available. In each of the BUs, 5 to 10 employees work on strategy development and coordination of the strategy with the other BUs. When the strategy work has been completed, about 250 employees have assisted in the strategy development, many points of view from various angles have been pooled and a strong sense of identification with the strategy is ensured.

Portfolio management

Strategy work is supported by portfolio management. This aids the evaluation and prioritization of the innovation projects. The projects are evaluated in 2 portfolios: the Market Attractiveness Portfolio and the Market / Technology Risk Portfolio.

Evaluating chance and risk in the Market Attractiveness Portfolio

In the *Market Attractiveness Portfolio*, the potential profit and the probability of success of an innovation project are juxtaposed, in order to consider the chance and risk of the projects (Fig. 5.9).

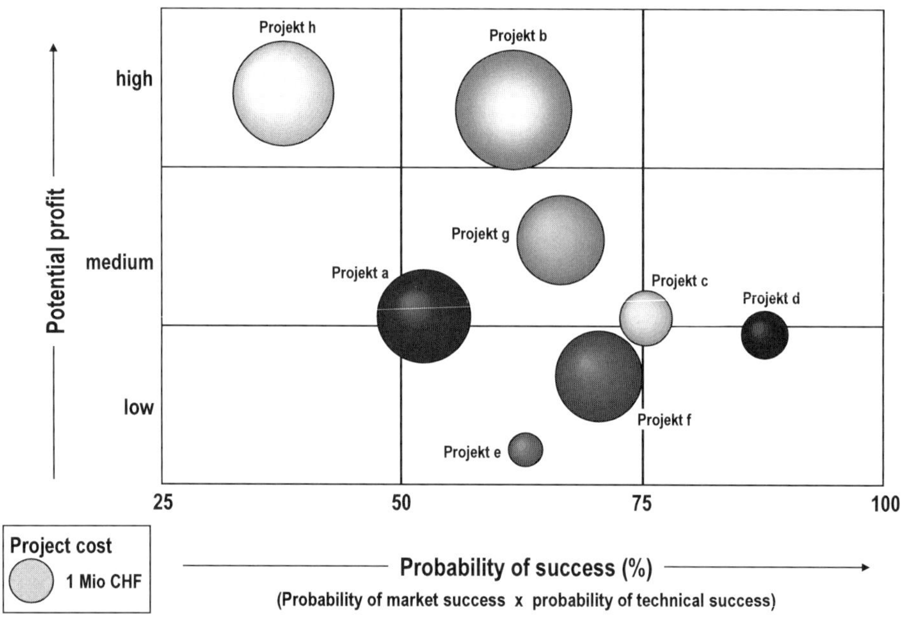

Fig. 5.9 Market Attractiveness Portfolio

Precise definition of assessment parameters

On the horizontal axis, the "probability of success" is accumulated from the "probability of a market success" and the "probability of technical success." The scaling of the axes is qualitative, corresponding to the majority of portfolio applications, (e.g., low, medium or high). However, the individual characteristics are more precisely defined by Hilti for better estimates. Figure 5.10 shows the scaling of the parameter "probability of market success," which bundles the 4 criteria (customer benefit, market potential, internationality, sales experience). The objectivity and comparability of the evaluations are thus increased. This is especially true, when – as is the case at Hilti – many different people are involved in the evaluation of project ideas.

100%	• *Customer benefit*	The new application solution will present an important added value for the customer. It resolves clearly identified problems or substantially improves the existing application.
	• *Market potential*	The target market is growing fast. The intensity of competition is low.
	• *Internationality*	Worldwide sales in all market organizatioins.
	• *Sales experience*	The customer segment is already regularly served by Hilti. The sales department is very experienced in the consultation of these applications.
75%	• *Customer benefit*	The new application solution will present an added value for the customer. It clearly improves the existing application.
	• *Market potential*	The target market is mature or slowly growing. The intensity of competition is medium.
	• *Internationality*	Sales in selected market organizations.
	• *Sales experience*	The customer segment is not part of the regular segments. The sales department is partly experienced but needs additional training.
50%	• *Customer benefit*	The new application solution will present an incremental added value for the customer.
	• *Market potential*	The target market is mature or old. The intensity of competition is medium.
	• *Internationality*	Sales in selected market organizations.
	• *Sales experience*	The customer segments have not been actively worked on so far. The sales department is not experienced in the consultation of this application. Specialists must be trained.

Fig. 5.10 Definition for the evaluation of the parameter "probability of market success"

Balancing of risks through a "healthy" project mix

The second portfolio is the *Market / Technology Risk Portfolio* (Fig. 5.11). Here, the relationship of the innovation projects to previous competencies is examined. It is essential to find a healthy mix of high-risk projects in unknown markets or new technologies and less risky projects in traditional Hilti markets and established technologies.

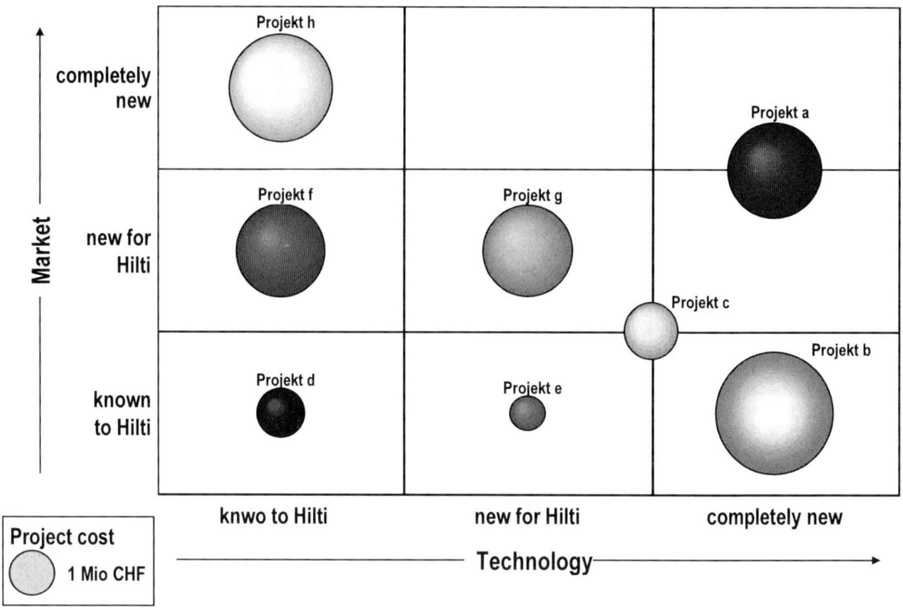

Fig. 5.11 Market / Technology Risk Portfolio

Implementation in TTM projects

The outcome of innovation planning consists of project road-maps for the BUs and the NB & T for the next three years, in which the innovation projects will be scheduled and implemented in so-called "time to money" (TTM) projects. "Time to money" means that the projects run for six months after the product launch. This grouping differs from the more widespread designation time to market, which refers to the period up to the market launch. The project teams utilize the first six months after the product launch to test the market acceptance of the product and to deduce improvement potential for further product generations, or to be able to take immediate action.

The TTM projects are based on stringent mile-stones (Fig. 5.12) and are handled by interdisciplinary teams that are put together specifically for the projects. Employees from all the relevant departments are represented in the teams. The Product Board acts as the controlling authority for the TTM projects; this consists of managers from the company management board and the extended man-agement board who exercise a steering function.

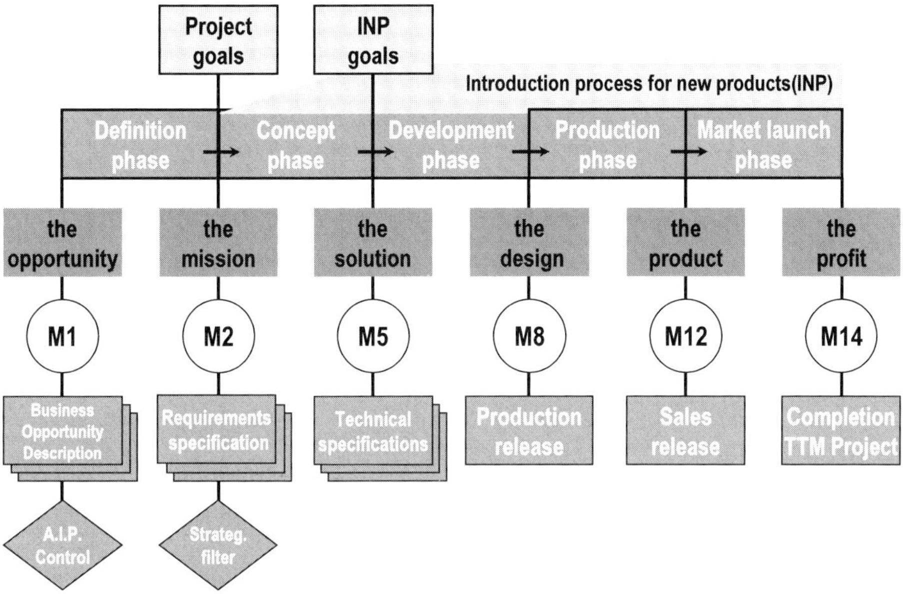

Fig. 5.12 Product development process: Time to Money (TTM)

For the detailed concept of the innovation idea "Laser Measuring Systems," some preliminary work still had to be done before a TTM project could be started. It was decided to carry out first tests with demonstrators in order to test the general suitability of laser measuring systems for use on construction sites, and afterwards a business plan for a new business area was to be drawn up.

Structuring of BU Positioning Systems:
1. Demonstrator
2. Business plan
3. TTM projects

The tests with the demonstrators ran smoothly, and the drawing up of the business plan went ahead. After 8 months of work the business plan was completed. In fall 1995, the management board approved the development of the new business area Positioning Systems. TTM projects were started up.

As time was pressing, the market launch was scheduled for spring 1997. The rapid market entry could not have been achieved without cooperation with the technology suppliers. Cooperation was entered into with a vast range of technology partners throughout the world. An important reason for the rapid market entry was the goal of acquiring experience with the laser systems as quickly as possible. By using direct marketing, Hilti has the advantage of being able to quickly and comprehensively incorporate reactions of the users into the systems it offers, and to include these in the development of subsequent product generations. In this way, Hilti has won a competitive edge over its rivals, who market their systems through the trade and therefore can only have indirect contact with their customers.

Constant customer contact during product development

Lead users give valuable suggestions for product design

Hilti involves its customers in product design at various stages throughout the entire product development process. One of the most important instruments is the cooperation with lead users. *Lead users* are selected customers that are distinguished by their exceptional professionalism. The lead users are selected by means of telephone interviews and invited to workshops held by Hilti. An important criterion for selection is how the customer deals with the products. In the telephone interview with the candidates, for example, the question is asked, "Do you have specific suggestions for how to improve our product?" If the customer has ideas on this matter, he will be invited to a lead user workshop. There he assists in the development of product concepts by contributing his experiences with the problem under consideration and by defining his requirements for a product that has still to be developed. At the same time, Hilti can utilize the solution proposals of resourceful lead users for the solution of the problem. The lead user approach means that Hilti designs its products on the basis of the requirements of the absolute experts and does not settle for just satisfying the "average customer."

Lead users: Products cater to the expert

Product designs using conjoint analysis

In the further development stages, various solution concepts are examined regarding their acceptance by the customer using *conjoint analysis* (chap. 4.8). The intensive exchange of ideas with the customer is only possible because direct marketing provides Hilti with many contact partners and thus also with direct "access" to its customers. Even after the market launch, customer satisfaction for Hilti products continues to be monitored.

Meanwhile Hilti offers a wide range of products based on laser technology for measuring work on construction sites (Fig. 5.13). The existing distance-measuring instruments are constantly being improved using insight gained from lead user workshops and other kinds of customer surveys. At the same time, work continues on new laser positioning products, which will also be of help to the "construction pro" in the future.

Fig. 5.13 Products of the Business Unit Positioning Systems

Newest innovation: electric rock drill for downhole mining

The development of the Business Unit Positioning Systems on the basis of systematic innovation management is not an isolated case at Hilti. One of the most recent projects is the development of the first electric rock drill worldwide for use in downhole mining. Using a study in which Hilti evaluated the market potential for electric rock drills in the mining industry of America, Africa and Australia, the rock drills were developed and were then tested in cooperation with AngloGold, the largest producer of gold in the world. According to current estimates, the new technology has the potential to revolutionize the mining process in downhole mining around the globe.

Summary

- In company strategy, Hilti has set a clear innovation objective: groundbreaking innovations. The organization, processes and company culture all focus on this objective.

- The innovation strategies of the sub-areas are closely coordinated. Cross-cutting issues are taken up by the department New Business & Technology.
- A characteristic feature of the innovation process is the close cooperation and focus on the customer. Intensive "looking over the customer's shoulder" is the source of numerous innovations.
- The key to the customer is direct marketing. This characterizes the company.
- Fundamental methods include portfolio management for the evaluation of innovation projects, working with lead users and the use of video analysis for determining latent and future customer wishes.
- New methods are constantly being integrated into the innovation process.

5.3 SUSPA Holding GmbH "Developing existing markets, discovering new opportunities"

DANIEL E. SPIELBERG

With the systematic expansion of its innovation management, the company SUSPA Holding GmbH is laying a foundation for steady market growth.

Existing markets are being further developed, and on the basis of existing competencies, new markets are being opened up.

SUSPA Holding GmbH

The SUSPA Holding GmbH, with its headquarters in Altdorf, near Nuremberg, is internationally active as a manufacturer of gas springs, hydraulic dampers, vibration dampers, shock absorption systems and height adjustment systems. The company defines itself as a development and system partner of major producers from the office furniture, automotive, durable consumer goods and washing machine industry. Its global marketing network and production locations in Germany, USA, the Czech Republic, India and China ensure close care of the industrial customers. Around 1200 employees worldwide generate an annual sales volume of approximately 174 million euros. The broad customer spectrum is served by market-oriented, globally operating business areas. Through matrix management, the development and production of the product range can take place in decentralized units with a clear focus on product and technology. SUSPA Holding GmbH presents itself to the market as a competent contact partner for all applications in the field of lifting, lowering, tilting and damping.

Research project FASTPRO

In the mid-1990s, a demand for new products had been identified at SUSPA Holding GmbH. Two objectives were taken up for the collaborative research project that was sponsored by the BMBF (German Federal Ministry of Education and Research) and was known as FASTPRO: "Frühaufklärung (early clarification) – Strategien (strategies) – Produktionssysteme (production systems):" firstly, the trial application of various innovation management methods for the development and realization of ideas in new products and, secondly, the development of an ongoing process with which the initially unique innovation efforts were to be institutionalized and implemented organizationally.

Innovation strategy

Innovation strategy: Developing existing markets, discovering new opportunities

The general innovation strategy can be divided into the market-induced development of new technologies for existing customer groups and the competence-based search for new applications for established technologies (Fig. 5.14).

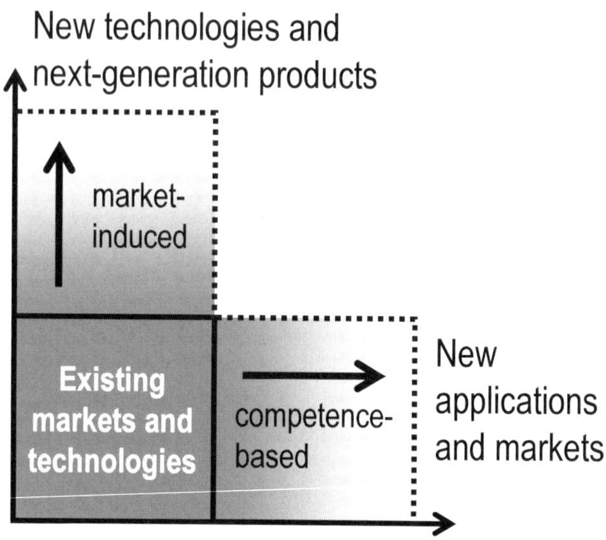

Fig. 5.14 SUSPA is innovative in two major directions

Deriving product ideas from sector trends

With a focus on the three business areas of automotive components, general industrial requirements / white goods (household appliances) and office furniture components, product ideas are created from the relevant sector trends. Depending on the characteristics of the market, this takes place at SUSPA's own initiative, or results from suggestions made by long-standing customers who have come to know and appreciate SUSPA as a system-focused problem-solver. In this type of market-induced innovation, the focus is above all on making the innovation process faster and more effective and on offering the respective customer both, the greatest possible customer proximity as well as the maximum current technical know-how for the specific solution to a problem. An example of this was the development of a so-called crash management system for motor cars, which permits graded impact absorption and which could be integrated into a complete cross-member system. This product will in the middle to long-term future replace the shock absorbers that have been available on the

Market-induced innovation: Optimizing the innovation process, offering customer proximity market up to the present day. Systematic further developments of existing products, such as a lockable gas spring for a chair, with a new type of release mechanism, or a vibration damper for washing machines with an integrated gap sensor also fall into this category of market-induced innovations.

Generating new markets

The second part of the innovation strategy is concerned with the generation of new markets in which the existing technical know-how can be used in new applications leading to the creation of a product. Here, a longer-term observation of general future trends must be considered. Generally speaking, SUSPA is very favorably situated, with its focus on products in the fields of safety and comfort. In a graduated process for the evaluation of ideas, new approaches are selected for implementing established technologies in new applications. Thus, a construction principle for a telescopically extendable footrest was developed for a manufacturer of luxury automobiles. A variant of this principle can also be applied in the field of office furniture: a classic activity of SUSPA is the supply of components for chairs. The new principle of a telescopic height adjustment system was then adopted for desks thus permitting the realization of a sit/stand desk. A further example for a competence-based innovation is the new generation of bicycle dampers, which will be dealt with in more detail below.

Observing future trends

Innovation process

Designing the innovation process

In order to transform the innovation process into actual products, an innovation process was first conceived of by SUSPA. Since the theoretical concept was likewise applied to specific cases, changes and adaptations with practical relevance already emerged in the introductory phase. The concept that exists at present can therefore be considered as tried and tested, but is continually undergoing further development.

Firstly, a rough distinction is made between three sub-processes, which in an ideal situation would take place in a certain order, one after the other, but which, in reality are carried out simultaneously for different projects (Fig. 5.15).

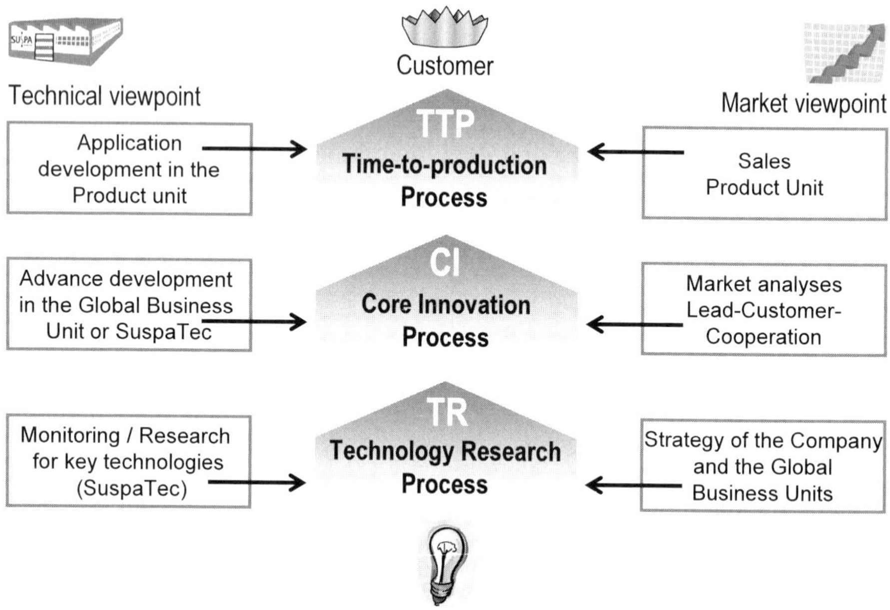

Fig. 5.15 The three main processes leading to innovation at SUSPA

Time-to-Production Process (TTP)

The so-called "Time-to Production Process" (TTP) describes the actual serial development of products. It takes place in accordance with an order from a customer and is based on an existing solution platform, which merely has to be adapted and optimized in some of the details to suit the special requirements. The goal of serial development is the start of serial production. Influenced by the requirements of, for example, the automotive industry, SUSPA has built up strong expertise in the on-schedule development of products that are ready for serial production. The projects are steered using milestones and are assessed by the customer at the various points of decision. The major part of the development outlay as regards costs and resources is used for this stage of the innovation process. Products which reach this stage no longer end up as "flops," as the customer is already relying fully on the realization of the product. The development is carried out in the decentralized product units.

Core Innovation Process (CI)

The "Core Innovation Process" (CI) is the true heart of the innovation process in the more precise sense. Here, the most important strategic decisions are made regarding the continuation or abandonment of a project. The projects are managed in cooperation between the central advance development department and the decentralized product units. Of course, this is only possible in the event that a relevant product unit already exists. The strategic evaluations and decisions rest with the three so-called Global Business Units in conjunction with the management of the company. The objective of CI is the realization of demonstrators and the acquisition of development orders through customers.

Technology Research Process (TR)

The "Technology Research Process" (TR) serves to work out a technological basis, both for promising product technologies, such as actuator or sensor systems as well as the optimization and new acquisition of product technologies, for example, for surface processing. The business areas have the right of proposal, whereas the management board decides on the implementation of the projects. The basic principles are only in rare cases developed through the company's own research. Research projects are frequently carried out in conjunction with the appropriate institutes. For some technological topics, monitoring of current trends and innovations is all that is being done for the time being. The decision on which measures to adopt in a certain area is based on corresponding short surveys regarding the potential of the technologies. The objective of TR is company-wide availability of new technological know-how for the realization of new product ideas. The topics may if necessary be worked on without direct reference to a product.

Gates & Stages Process

TTP and CI are closely connected, as in the ideal situation they should be run in sequence, immediately following one another. In reality, the innovation process is divided into further individual elements that are depicted in figure 5.16. Product ideas pass through this so-called Gates & Stages Process in several stages. This guarantees that, if necessary, certain stages can be omitted or repeated. Particularly the early stages, and especially the CI section, are steered by means of budget planning. This ensures that promising ideas are not condemned to failure through arbitrary setting of deadlines. The project employees have the means of to strongly influence the scheduling through their project budget. An evaluation and decision is always made when a new stage is reached and, thus, when a new budget is to be prepared (e.g. procurement of tools).

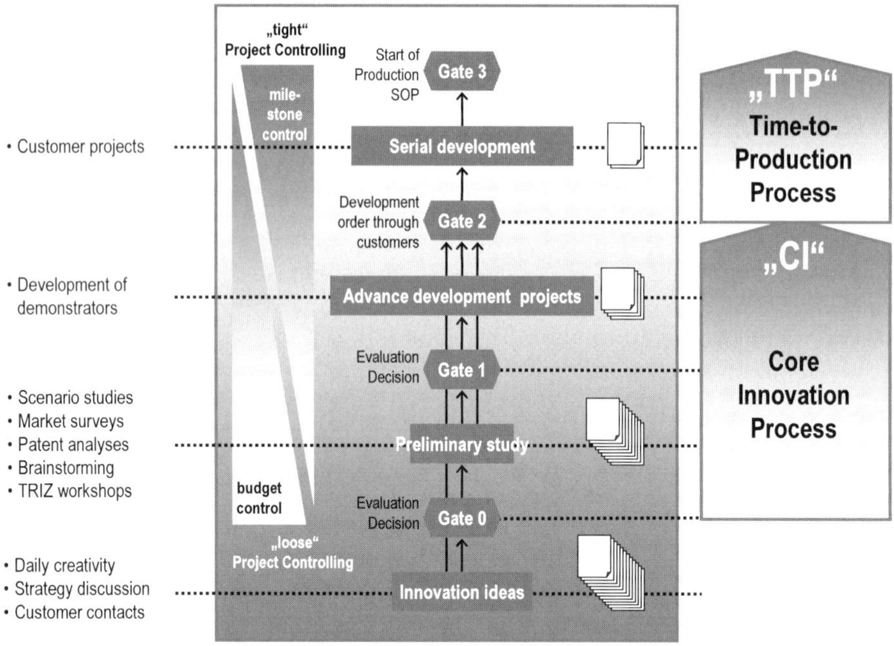

Fig. 5.16 Connecting of advance and serial development in the Gates & Stages Process

Innovation structures

Solving organizational conflict of objectives

For the medium-sized company SUSPA, the bipartite innovation strategy and the tripartite innovation process soon led to an organizational conflict of objectives: on the one hand, it was important to decentralize the development departments and to assign them to market-oriented product units. In this way, a short response time and a high level of customer focus were to be achieved. On the other hand, the necessity arose of making maximum use of synergy effects in the development of new technologies and of bundling the limited resources. For this, the development competence would have had to be centralized.

The dilemma was solved by a so-called virtual organization of development: through the decentralization of the serial development there is actually no possibility of accessing all development competencies at one location. However, in order to nevertheless be in a position to offer the entire range of competencies to potential customers in the early stages of the innovation process, there is now a central advance development, under the auspices of an independent organizational unit: the SUSPATec GmbH (Fig. 5.17). Salaried employees work on the one hand with strategically defined basic projects, for example, on the topics sensor and actuator engineering, or also tribology, ("Technology Research Process," (Fig. 5.15). On the other hand, these employees are involved with developers from the market-oriented company units, in order to jointly develop new products ("Core Innovation Process" and "Time-to-Production Process," Fig. 5.16). The innovation process is run cooperatively, however, with a strong tendency to decentralization in the later stage. Furthermore, the central advance development can be utilized to reduce capacity peaks.

Regulation of budgeting

So that this constellation can also be effectively utilized, and the advance development is not degraded to an extended workbench by the factual constraints of everyday business, special budgeting regulations were made. In its annual planning each business area commits itself to a certain budget for the financing of the advance development. This is initially planned as a general amount that is not tied to specific projects. During the year, the business areas define the project proposals, and the company SUSPATec GmbH in turn draws up an offer for the processing of these projects. The nature of the project is assessed at the Holding level. Thus, technology research projects are financed in equal parts by all business areas, as in this case, technological principles are being worked out that benefit the company as a whole, in the sense of preliminary research. Serial developments, which are only intended to be carried out by SUSPATec as an exception to the rule, are not charged to the previously planned budget. These receive additional financing from the business area. Advance developments, which are the actual purpose of SUSPATec, are financed from the budget that was allocated at the beginning of the year.

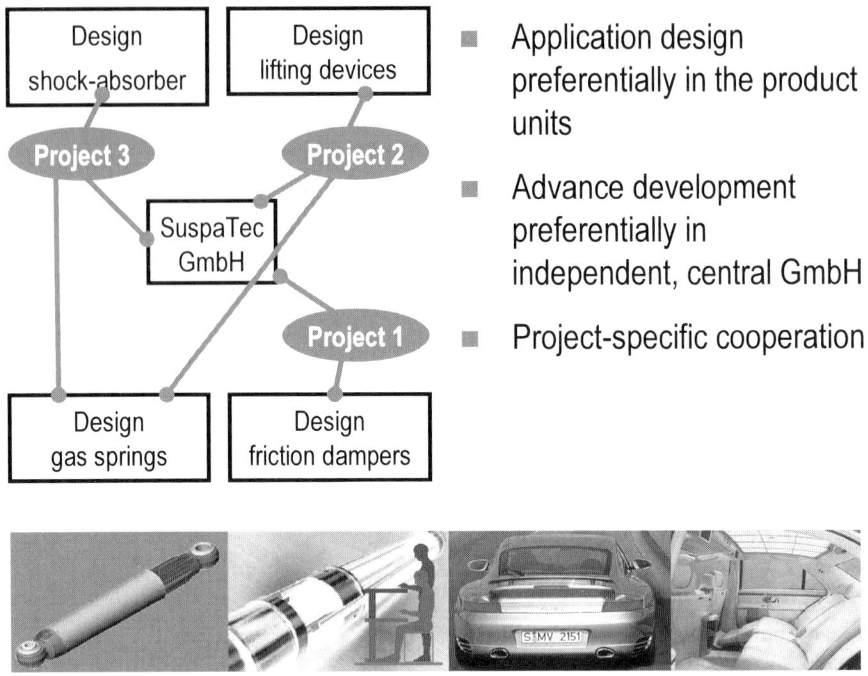

Washing machine damper with sensor | Lifting device for sitting-standing work stations | Lifting device for rear wing (source: Porsche AG) | Adjusting system for lower leg support (source: DaimlerChrysler AG)

Fig. 5.17 The virtual organization of the innovation structure at SUSPA

If a business area does not fully exhaust its advance development budget, the contributions that have not been used up are then used for special projects, which are initiated by the management board and are not tied to one business area. The advance development budget, once it has been planned, can therefore no longer be used for other purposes. At the same time, SUSPATec is only reimbursed with such expenses that are tied to projects that have actually been defined and commissioned. Through this regulation, both SUSPATec and also the business areas have a large and common interest in using the planned budget in the most effective way possible for cooperation projects with advance development character. Figuratively speaking, each business area makes a general innovation contribution in the form of an "innovation tax." Projects that they have defined themselves can, after a positive assessment, be "deducted" from this tax. The objective of all the parties is to completely avoid this "tax" by initiating an adequate number of projects. If one business area aspires to additional, particularly lucrative projects, then a further voluntary commission is, of course, possible.

Case study: the SUSPA bicycle damper

In the year 1997, a workshop on the topic "New Products" was held at SUSPA. Products were being sought, for which the technical implementation could easily be derived from the existing technical expertise. Furthermore, the functions "comfort" and/or "safety" should in some way be concerned. Finally, the precise goal was to open up new customers and growing markets.

Even in the 1950s and 1960s SUSPA had been a supplier of the suitable suspension struts for the small motor cycle industry in France, which was booming at that time. This product was even, in a certain sense, the starting point for all SUSPA activities in the field of springs and dampers.

Formulating the product idea

The step to the formulation of a new product idea in a related segment was, therefore, not so very large: since the early 1990s, the market for mountain bikes had been growing steadily. The latest trend, first developed in the so-called "downhill" field, was to the so-called "fully suspensed" bicycles. These were bicycles which, corresponding to the construction of motorcycles, not only had a spring-mounted front fork, but also a spring-mounted rear swinging fork.

Carrying out secondary market research

The first step was, with the help of secondary market research (publications, visits to trade fairs, generally accessible analyses of the field), to examine the idea with regard to its soundness (Fig. 5.18): admittedly, the rear suspension was at first the smaller segment, but with the spring / damper elements used there it required a component very suitable for this purpose, one that was almost completely independent of the rest of the bicycle. A facilitating factor proved to be the fact that the product idea "bicycle damper" was a component that, like many SUSPA products, was used in a final product for the end user. As a result, it was relatively easy to find the corresponding "hobby" experts within the company and, thus, also to find highly motivated employees for the project team.

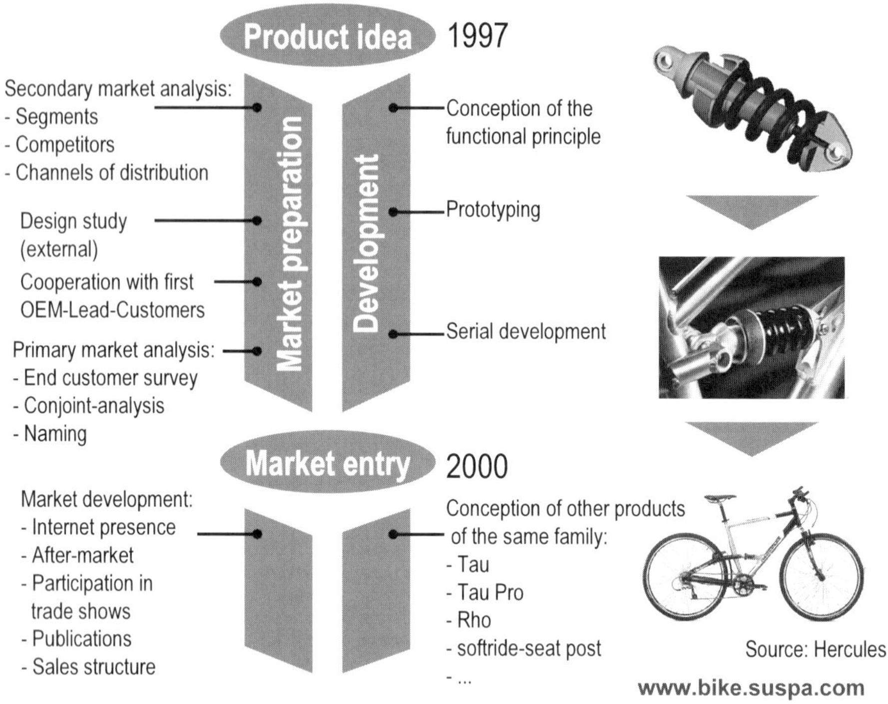

Fig. 5.18 History of the SUSPA bicycle damper

The market for mountain bikes was supposed to follow the trend towards more leisure. But even at that time, alongside the use in the sports branch, the potential comfort function for everyday city bicycles was identified as being particularly significant for the future. This prognosis was to be confirmed in the following years.

Developing a solution concept

After the idea had been initially confirmed from the market viewpoint, a first concept for a solution was developed. The unique twist to the SUSPA solution was the use of a damper cartridge, which could be manufactured on the existing production plant alongside the other serial production. This cartridge is enhanced by components that are specific to bicycles (coil spring, attachment parts) and enables, as in a model kit, the simple derivation of an extremely wide range of variants that can be suited to the customer. In addition, in the event of damage, the pressure cartridge can be replaced very easily and economically.

In the first analysis of the market it was noticed that mountain bikes represent a lifestyle product in which the cost plays a less important role than functionality and design. So, on the strength of the basic technical concept, an external design study was commissioned. The draft that was selected was realized in a demonstrator for customer contacts.

The first contacts with ambitious lead users (meaning low-volume bicycle manufacturers) yielded information about necessary constructive and design-oriented changes to some of the details. The successful presentation at trade fairs and in customer contacts then led to the first development orders from customers.

Carry out a second market analysis

In order to be able to place the product on a broader basis and to avoid a niche existence, a second market analysis was carried out at this point, which was based on primary market research, that is, the surveying of final customers (Fig. 5.19). The systematic development of the study and the corresponding evaluation with the conjoint method yielded essential information for optimizing the product, for the setting of targets for ancillary products of a whole family and, not least, for the expandable naming of the product line. It became apparent that further ancillary products should include adjustable damping, reduced weight or reduced cost. Using these differentiations, clearly identifiable customer and bicycle segments could be covered. This knowledge was then born in mind in the development of the further product line and in the market development.

Use conjoint method

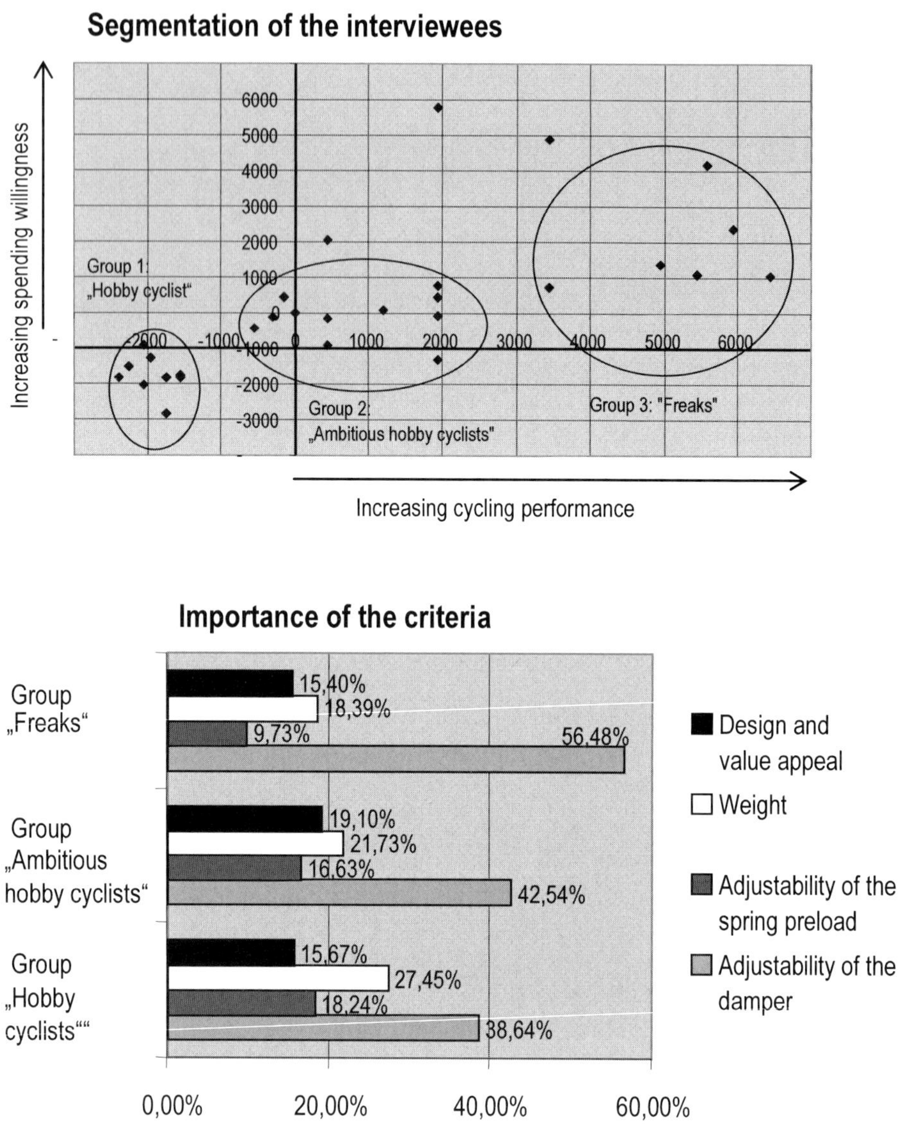

Fig. 5.19 Results of the primary market analysis

Conclusion

Outlay in time and costs

As a conclusion, we can note that in the innovation process, the outlay of time and finances for opening up a new market should by no means be neglected. Alone, the period of time from the idea until the first serial introduction for one customer was around 3 years. At this point in time the product field was still far from achieving economic viability. Not until further development of the market through new orders for serial production, product variants and after-market contacts is the new product gradually able to hold its own. Today SUSPA bicycle dampers are well known to bicycle manufacturers, and new products receive favorable reports in the specialist press. For some manufacturers SUSPA is an exclusive supplier, that is, they cover their entire bicycle range. SUSPA supplies the very small market for high-performance sports equipment only in a subordinate role and concentrates rather on "moderate" leisure bicycles. In this way, SUSPA retains its core competence: reasonably-priced mass production of high-quality technical components.

Procedure for realization

The procedure for the realization of this innovation can be described as systematic and cautious. At the same time, it is characterized by the willingness to take on a long-term entrepreneurial risk. Only in this way can genuine new business be generated with a product at the beginning of its life cycle. The innovation management system of SUSPA Holding GmbH has meanwhile proved its effectiveness, not only with this success story, but also with numerous other new products and has thus made a vital contribution to the renewal of the success positions of SUSPA.

5.4 Survey "MicroMed 2000+" "Discovering applications in growth markets"

CHRISTIAN ROSIER

The case study described here is based on a study carried out by the Fraunhofer Institutes for Production Technology (IPT) and Laser Technology (ILT) in Aachen. During this study promising applications for microsystems in medical technology were identified by involving local medical and technological expertise. The study led to the founding of a competence center for microsystems in medical technology at the Aachen location; the outstanding ranking of this center in production technology is to be extended to new applications.

The Fraunhofer Institutes in Aachen

Fraunhofer Institute for Production Technology IPT

The Fraunhofer Institute for Production Technology (IPT) is concerned with performance requirements from the industrial environment. In order to work on the relevant topics, the institute unites the various disciplines of production technology; these include process technology with its mechanical and control engineering elements, quality and measuring engineering as well as the planning and organizational approach to questions of production technology.

Fraunhofer Institute for Laser Technology ILT

The Fraunhofer Institute for Laser technology (ILT) concentrates its expertise in the field of laser applications and systems. The institute develops and optimizes sources of laser beams and laser components and researches modern laser measuring and testing equipment. Alongside the optimization of process technologies, the Fraunhofer Institute is also engaged in the development of complete laser plants and the related systems and components.

The "obligation" to open up new markets

Innovation strategy of the Fraunhofer Institutes: open up new "business fields"

Research establishments such as the Fraunhofer Institute for Production Technology (IPT) and Institute for Laser Technology (ILT) address themselves to current research topics so as to actively participate in or to influence the shaping of new production technology developments. In order to stay ahead of competing institutions and, in this way, to be a trend-setter, it is important to react to changes in the market and to be aware of new developments at an early stage. The Fraunhofer Institutes IPT and ILT are thus under the "obligation" to continually identify new growth-promising markets and to develop these from a research point of view.

Medical technology as a future field for microsystems technology

Medical technology is considered a growing sunrise industry. The expected demographic trend to an increase in the number of older people as well as the generally growing awareness of health issues will in future lead to an increased demand for (gentle) medical treatment methods and technical aids. Since their use poses a low risk of injury or side effects, minimally invasive treatment methods are particularly suitable as a means of significantly reducing the length of hospital stays for patients and, therefore, of holding down the rapidly growing costs in the health care sector. The trend to miniaturization of medical products resulting from minimally invasive treatment is opening up a major field for future microsystems engineering applications in medical technology.

Carrying out a joint study

Objectives of the study: identification of promising application fields for medical products

The increasing importance of medical technology in society was the inducement for the application-oriented Fraunhofer Institutes to analyze the medical technology market under the aspect of production technology application fields. To facilitate access to this unknown market, the two institutes decided to identify promising application fields for medical products by carrying out a joint study. Since both institutes conduct research in the field of microsystems technology, they focused their attention on applications fields of this type with a connection to microsystems engineering.

Founding a competence center for medical technology

At their location in Aachen, there is a confluence of both medical and technological expertise; for this reason, close cooperation in the field of medical technology was the first thing that was sought after. In the medium to long-term future, the aim was to found a competence center for microsystems in medical technology.

Depiction of results in an innovation roadmap

The results of the study – promising application fields for medical technology and corresponding product ideas – were to be illustrated in an "innovation roadmap." Research and development tasks for product innovation can be derived from this roadmap.

Carrying out the study with the innovation roadmap methodology

Innovation roadmap methodology selected for the procedure

The procedure for the study was selected with reference to the innovation roadmap methodology that was developed at the Fraunhofer Institute for Production Technology (IPT).

Identifying relevant search fields

Search field matrix identifying and structuring search fields

At the beginning of the study, the search fields were first of all defined in order to structure and deal thematically with the comprehensive subject matter of medical technology. With a two-dimensional matrix the individual search fields were defined unequivocally as the point of intersection of the relevant specialist areas and the so-called medical process chain (Fig. 5.20). The specialist areas were listed in the vertical columns of the search field matrix; the (idealized) process of a course of medical treatment (medical process chain) was entered in the horizontal lines. This chain begins with prophylaxis and anamnesis (previous history of an illness) and ends with prosthetics (the insertion of artificial organs into the human organism).

Focus on relevant search fields

With the aim of reducing the number of search fields, extensive research regarding the current problem issues and core themes in the specialist medical areas was undertaken. From the 140 possible search fields, it was possible to identify 28 search fields for which, according to the statements in the literature, a high potential for improvement in medical products was predicted. The selection process will be described below.

Selection method: prepare information data sheets and identify contact partners

First of all an information data sheet was drawn up for each search field, containing a brief description of the search field contents and further search information. The information yielded by the search was filtered and divided into major thematic areas. The more frequently a certain subject matter was discussed in the literature, the more relevant this topic was estimated to be. In this way, the problem issues in medicine were deduced from the frequency of matches for a topic; for each topic at least one contact partner was located through the internet or through a specialist journal. The persons identified were contacted by telephone or in writing and were asked about their view of current medical problem fields from their specialist area. These conversations frequently yielded clues to further problem fields in other medical areas.

For the MicroMed study, 28 relevant search fields were identified with the selection method that we have described; these formed the basis for all further action.

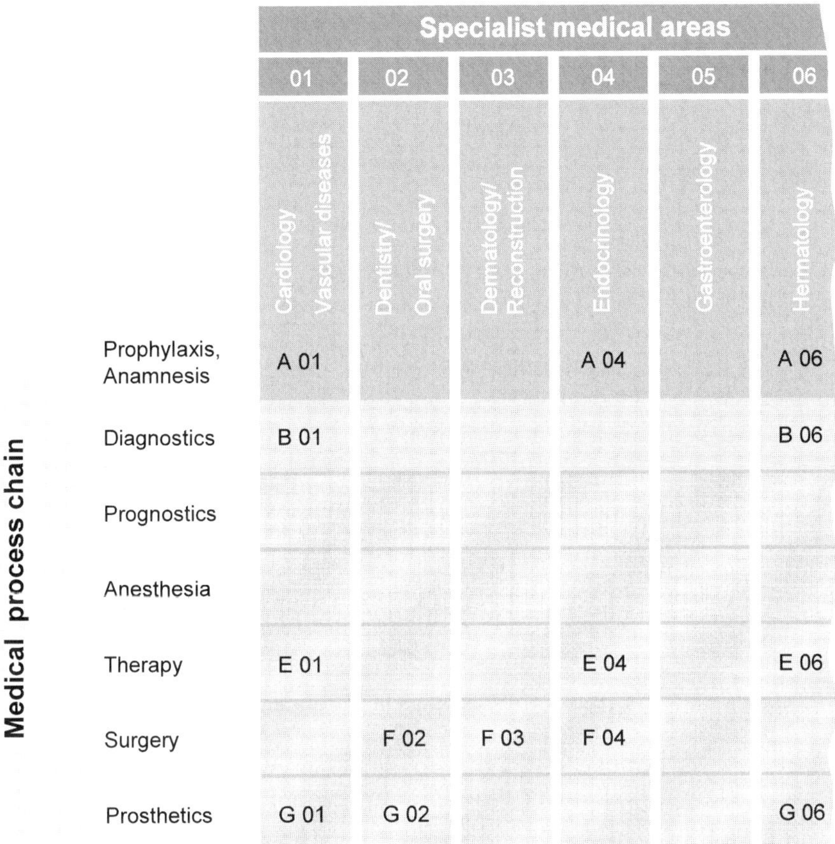

Fig. 5.20 Excerpt from the search field matrix

Recording of full details for the relevant search fields in the information data sheets

Full particulars for the relevant search fields were recorded in the information data sheets; in addition, the activities and tasks of the specialist area (e.g., ENT: ears, nose and throat) were described in the field demarcated by the medical process chain (for example, diagnostics). Furthermore, current treatment methods (i.e. state of the art technology) were summarized and directions of search were determined for potential technical products in the specified discipline. One example: for the search field *B20: HNO diagnostics* the following information was recorded in an information data sheet:

Example contents of an information data sheet

Activities/Tasks

Branch of medicine that is concerned with the diagnostic procedure for illnesses of the ears, nose and throat

Methods

- *State of the art*: endoscopy and micro-biological analysis
- *Directions for search:* miniaturized, minimally invasive endoscopes; continual micro-biological analysis through miniaturized devices that can be worn or implanted in the body (lab-on-a-chip).

Further detail: identifying other experts

In the next step, further experts were identified in the specialist fields that were of particular interest: a search for luminaries in the various fields was carried out in medical literature as well as through the internet; experts who had already been contacted recommended colleagues from other specialist disciplines. The search for and selection of experts was guided by the objective of developing a competence center for microsystems in medical technology at the Aachen location; for this reason, geographical proximity was an important selection factor.

Finding ideas - technologists and medical practitioners jointly generate ideas

Conducting interviews with experts to derive ideas for problem solving

The experts who had been selected were interviewed with the objective of identifying innovation potential for technical products in medical technology. In the conversations, problem fields of the individual specialist areas were systematically discussed; technical ideas for problem solutions – frequently inseparable from product ideas – were derived from the problems that were identified.

Documenting the ideas for problems

The problems mentioned by the medical practitioners were documented in writing and the technical ideas for problems – resulting from the conversation – were noted. The knowledge and results acquired through the interviews could then be used for further interviews with experts, so that the ideas could be evaluated and discussed. This course of action yielded qualified expert opinions on the various approaches and ideas. The ideas that were yielded were recorded with explanations and framework conditions in so-called function data sheets (FDS) (Fig. 5.21).

In order to collect ideas, free interviews were carried out. The experts had sufficient freedom to also describe more complex problems. In addition, the interviews permitted the elimination of problems with communication (e.g. specialist terms) by offering an opportunity for raising queries and by creating a common basis for discussion.

Interview technique: allowing areas of freedom and facilitating consultations

Assessment of ideas

The ideas that were documented – ideas for problems, ideas for solutions or potential for innovation – were assessed in the next phase. The FDS were used for clarifying the potential for innovation from the ideas that had been compiled and for clearly defining the area for consideration; this became the common basis for discussion. The actual assessment of the potential for innovation took place in several workshops in which both medical practitioners and technologists participated. Both elements of the assessment – function data sheets and workshop contents – will be described below.

Function data sheets and workshops for assessing the potential for innovation

Function data sheets (FDS) consisting of two pages were used for documenting the ideas. The first page contained three thematic sections (Fig. 5.21): in the upper section of the page a note was made of how illnesses that are related to the idea are normally treated. The description of the course of the treatment offered an overview of the treatment methods practiced. Next, the problems and challenges to new (not yet developed) products were listed. The third section contained the present technical solutions from which potential product innovations were to be distinguished.

Contents of the function data sheets

The first page of the function data sheet was intended to provide an introduction to the subject matter, as well as describing a context in which the development of the product innovation is to take place.

Function data sheet	Angioplasty with a stent
Course of the treatment 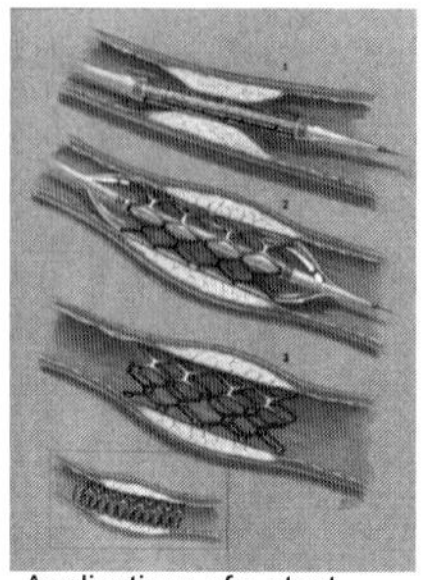 Applications of a stent	▪ When treating an arteriosclerotic narrowing of arteries, the narrowing is enlarged with a balloon catheter. ▪ If the artery is dilated too much, fine longitudinal fissures arise, in which thromboses can develop. ▪ The stent premounted on the balloon catheter is expanded together with the artery and reinforces the latter after the removal of the balloon. When the artery has restabilized, the stent can be removed. ▪ Thromboses can form at the stent which is a foreign object. This is due to flow-dynamic changes (mechanical strain of the blood) as well as to biochemical reactions on the surface of the stent.
Problems / Challenges	▪ Stents need a flow-dynamically optimized surface. ▪ Goal: biochemically optimized surface which does not tend to thrombosis forming. ▪ Biodegradable stent which continuously disintegrates.
Present technical solutions	▪ Metal Stents

Fig. 5.21 Excerpt from a function data sheet

Assessment of the potential for innovation: market and differential potential

On the second page of the function data sheet, the assessment of the potential for innovation was completed; the assessment took place with reference to the market and differential potential. Both overall criteria were more precisely defined by sub-criteria (Fig. 5.22). For every sub-criterion the assessment of the experts that had been reached in the workshops was included and transparently documented in the function data sheet.

Assessment of market potential

The market potential of a product innovation was also assessed according to various criteria, such as the frequency of the illness in question or the impairment in the quality of life of the patients that the illness causes. For example, the assessment of the product idea "movement glove," was made on the basis of how frequently the case of a severed hand occurs, and to what extent the quality of life is influenced by the loss of a hand. The higher the frequency or the greater the impairment, the higher was the assessment of the market potential. As a standard for the assessment a three-fold division into high, medium and low was made.

Market potential	Differentiation potential	
▪ Frequency/morbidity	▪ Risk minimizing	*(0,33)*
▪ Impairment in the quality of life/mortality	▪ Optimization of the result	*(0,27)*
	▪ Time advantage	*(0,20)*
	▪ Cost advantage	*(0,13)*
	▪ Increased comfort	*(0,07)*

Fig. 5.22 Criteria for the assessment of market and differential potential

Assessment of differential potential

The assessment criterion "differentiation potential" combined the criteria with which the unique position of a product innovation was assessed; from the point of view of the patient the assessment took place by comparison of the outstanding features of the new products with (the conventional) alternative possibilities for treatment methods. The assessment criterion "risk minimizing for the patient" was adopted as a sub-criterion of the differentiation potential; the assessment of this criterion dealt in particular with the better treatment result through the application of the product innovation in comparison with traditional alternatives. Further sub-criteria related to the expected advantage in time and money, and the increased comfort through the use of the product innovation. The five assessment criteria for the differentiation potential were weighted in relation to one another in consultation with the experts. "Risk minimizing" was regarded as the most important criterion in this weighting; this criterion included aspects that can have an enormous influence on a person's life.

Finally, the market or differentiation potential from the product was determined using individual assessments of each criterion and the corresponding weighting.

Market potential and differential potential determine market attractiveness

In anticipation, we should mention at this point that for the purpose of entering them in the innovation roadmap, market potential and differentiation potential can be summarized as market attractiveness, which was determined by means of a corresponding portfolio. Market attractiveness expressed in terms of time describes the period of time until a defined (i.e. sufficiently large) demand for a product can be observed on the market. The precise definition of market attractiveness will be explained in more detail later in this case study.

Product life cycle: defining the time at which market maturity of a product idea is reached

In order to be able to assess the timing of market attractiveness of an idea, the positioning of the ideas in the general product life cycle is documented in the function data sheet Fig. 5.23). The corresponding position on the graph was calculated in the workshops; for the timing, the criteria that are proposed by Little for the determining of positions on the life cycle graph (Little 1981) were used. In each case the ideas were reflected against the assessment criteria depicted in figure 5.23 as well as their possible characteristics; at the same time, these characteristics determined the subsequent numerical assessment in which each criterion (insight into R&D of competitors, predictability of development results etc.) was assigned a corresponding numerical value (high = 3, medium = 2 und low = 1). The total value was reached by adding the individual assessments; the amount of the total value was used to position the product ideas in the life cycle graph.

A total of 34 promising ideas for problems

In the search fields that were investigated, 34 ideas that promised success were discovered. The data stored in the function data sheets was used for the integration of the ideas into the roadmap from the point of view of marketing and technology as well as with regard to timing.

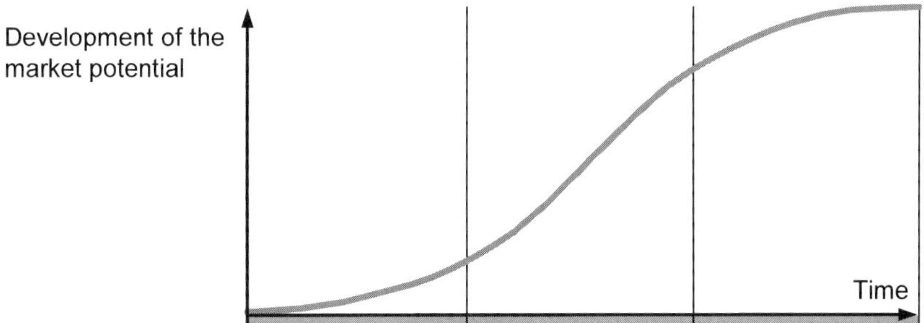

		Pioneer	Follower	Me-too
Insight into R&D of competitors		low	medium	high
Predictability	of development results	low	medium	high
	of market success	low	high	high
	of R&D expenses	low	medium	high
Durability of the competitive edge		high	medium	medium

Fig. 5.23 Guidelines for depicting the timing of product ideas in the life cycle graph (Little 1981)

Drawing up the innovation roadmap

Depiction of results in the innovation roadmap

For depicting the assessment results, the ideas were entered in an innovation roadmap (Fig. 5.24). The areas *market* and *technology* in the innovation roadmap were adapted to the stated intentions for this purpose: on the market side of the roadmap, *market attractiveness* was inserted above the point in time at which a product idea reached market maturity, and, on the technology side, the implementation outlay was entered above the point in time at which technical feasibility was reached.

Market page determining market attractiveness

For determining the market attractiveness – constituted by the market and differentiation potential of an idea – a portfolio was used (Fig. 5.25). It was possible to read off the market attractiveness of a product idea through the positioning in the portfolio; the tendency of a product to be located farther to the right and higher up indicated a correspondingly higher level of market attractiveness.

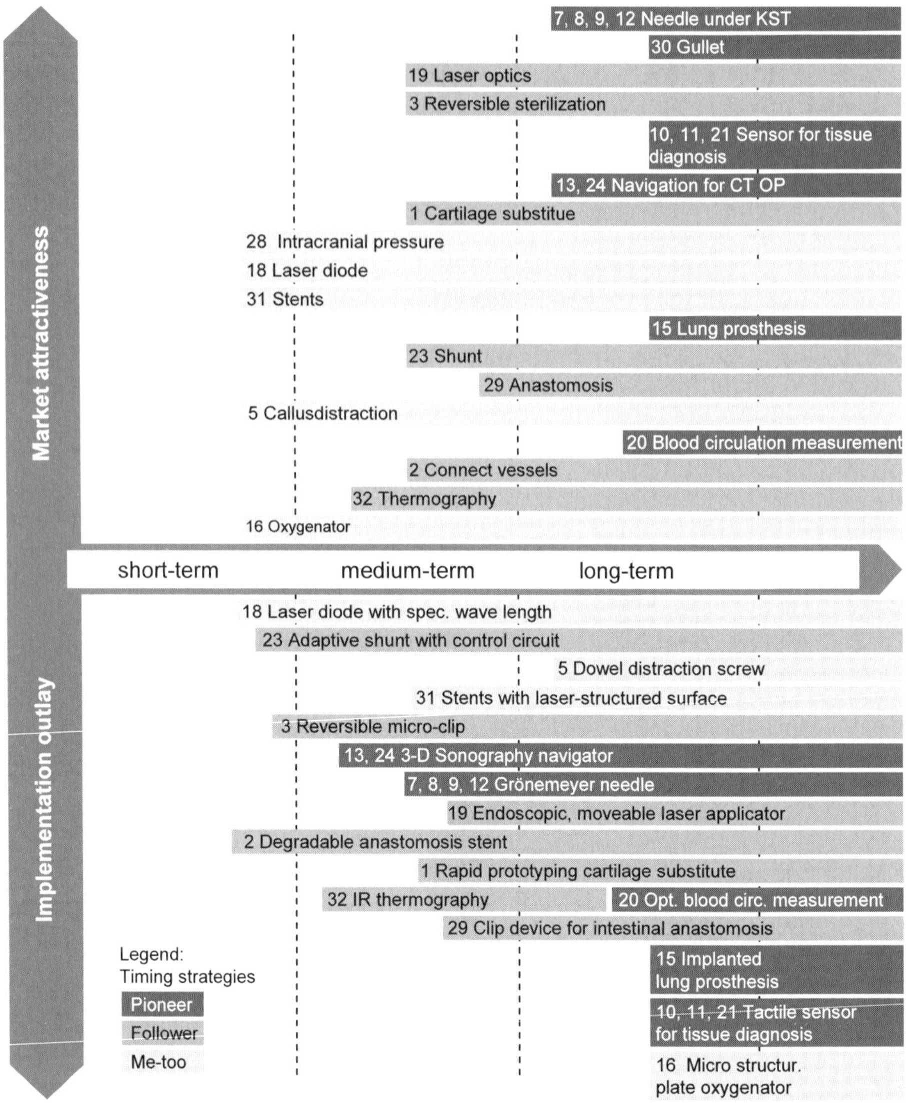

Fig. 5.24 Innovation roadmap for medical product innovations

Market side: determining the timing of market attractiveness

The timing of the market attractiveness of a product idea was determined by means of the product life cycle graph (Fig. 5.26). From the depiction of the life cycle graph it becomes apparent which product ideas were not yet fully developed (time period: pioneer) and which ones had already more or less exhausted their market potential (time period: "me too").

Steps for action were then derived from the positioning of a product idea in the life cycle graph. Ideas of particular interest to the Fraunhofer Institutes were those that could be assigned to the areas "pioneer" and "follower:" these product ideas are (still) relatively undefined, and the further development of these ideas into products with market maturity matches the research strategy of the institutes.

Derivation of steps to be taken from the life cycle graph

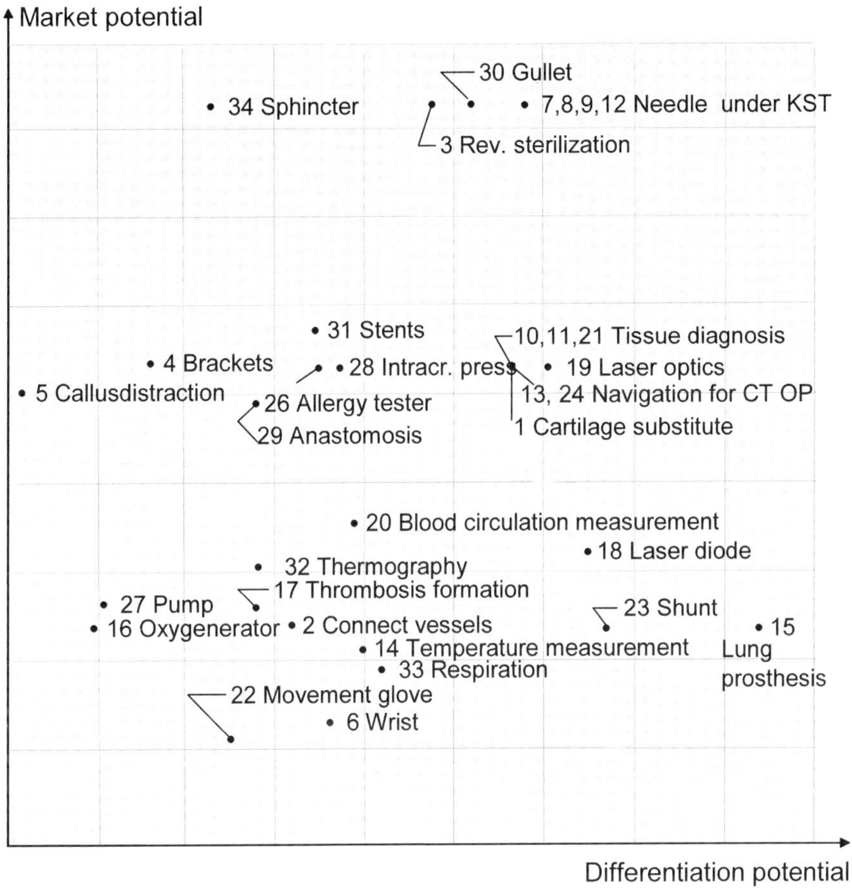

Fig. 5.25 Assessment portfolio for the assessment of market attractiveness

Using the preliminary work (assessment portfolio and product life cycle graph), the product ideas were easily transferred to the innovation roadmap. Increasing market attractiveness was selected as the category criterion for the vertical axis of the market–time area; this was taken over from the assessment portfolio (portfolio for market and differentiation potential).

Insertion into the innovation roadmap

The maturity in terms of time was taken over from the life cycle assignment; the period "pioneer" corresponded to long-term and the period "me-too" to short-term market attractiveness.

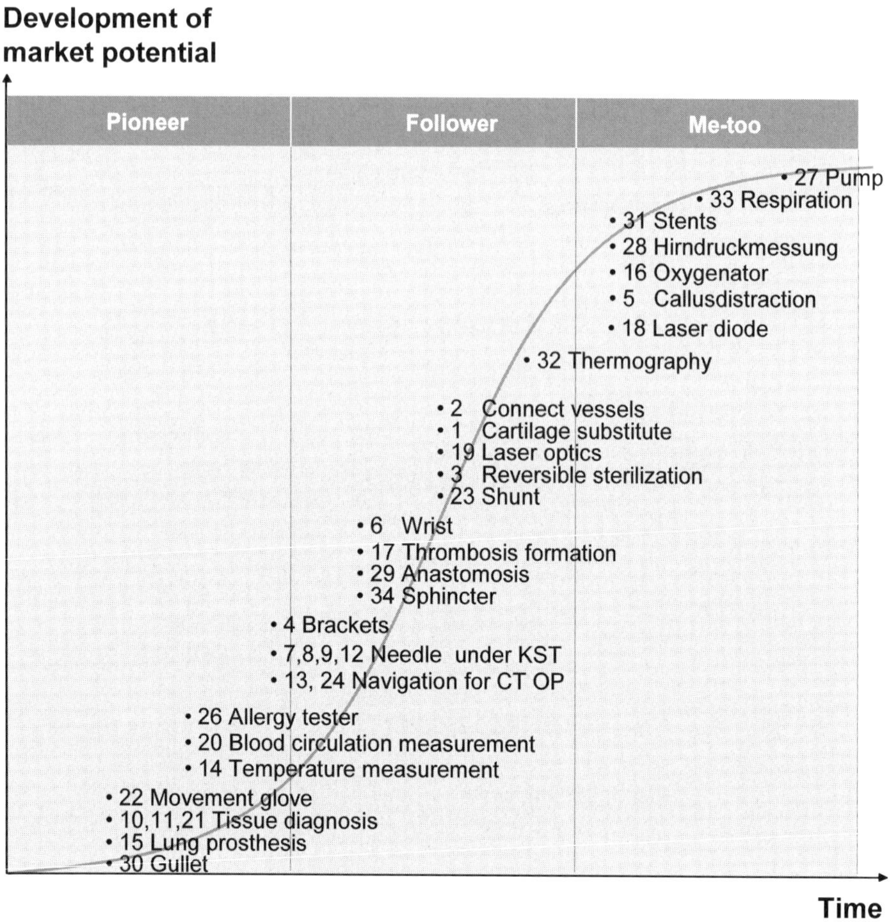

Fig. 5.26 Positioning of selected ideas for problems in the general product life cycle graph

Technology side: determining implementation outlay

The outlay for implementation was assessed from the point of view of the Fraunhofer IPT and ILT. For determining the outlay for implementation, the basic (technical) feasibility of the product ideas was examined and then discussed and assessed with the experts from the two Fraunhofer institutes. Strategic aspects were critical for the selection and timing of the product ideas in the innovation roadmap, for example, the maturity deadlines of R&D performances, that is, in how far the realization of the product idea will contribute to a long-term and sustainable gain in expertise for the institute concerned. The outlay for implementation and maturity deadlines were determined by means of portfolios, in the same way as the market side.

Portfolio 1: Technology potential versus resources

The outlay for implementation (at first without taking the time factor into account) was calculated by comparing existing technology potential with the existing resources (machinery, equipment and specialist personnel). The consideration was guided by the status of the potential available at the time of the consideration, i.e. possible purchasing of additional resources or an eventual expansion of personnel capacity were not accounted for in the positioning of the product ideas in the portfolio.

Portfolio 2: Research consistency versus use of resources

For the timing of the outlay for implementation of the product ideas, the research consistency was compared with the planned use of resources. Research consistency is said to exist when the research required for the implementation of a product idea is congruent with the current main research that is being carried out at the present time. This assessment method was selected in order to utilize synergy effects for current research projects.

Insertion of outlay for implementation into the innovation roadmap

The outlay for implementation that was calculated (by the time at which technical feasibility was achieved) was afterwards entered in the bottom section of the innovation roadmap.

Drawing up a plan of action – prioritizing spheres of activity

Portfolio for prioritizing spheres of activity

For prioritizing and determining the product development projects that were actually to be implemented, a further portfolio was drawn up, which assessed the product ideas with reference to their market attractiveness and the outlay for implementation. This consisted of an innovation roadmap that dispensed with the time axis; the lower section of the former (the axis representing outlay for implementation) was turned upwards (Fig. 5.27).

The special feature of this depiction is that only those product ideas are represented for which the ratio of the timing of market attractiveness to the outlay for implementation expressed in terms of time is approximately 1:1. This means that only product ideas are inserted, for which the demand and the realistic outlay for implementation (on the part of the Fraunhofer IPT and ILT) are in a meaningful ratio.

Derivation of strategies for action

Using the positioning of the product ideas in the (prioritization) portfolio it was possible to derive the following strategies for action:

- *Defer*: the implementation outlay is too high and the market attractiveness too low; a meaningful (and economically viable) realization of the product idea cannot be implemented.
- *Investigate*: According to where the product idea is positioned in the portfolio, an investigation must be made as to whether the market attractiveness justifies the high outlay for implementation or whether the low outlay for implementation can compensate for the low market attractiveness.
- *Implement*: both outlay for implementation and the market attractiveness support a realization of the product idea.

The advantage of the portfolio depiction that was used is that, in comparison with the roadmap, it offers a better overview of all product ideas which the company (in this case the Fraunhofer IPT and ILT) can realize in a meaningful ratio of market demand and technical effort for implementation. This (project) filter can reduce a host of ideas to a manageable number.

Suitable product ideas

In the MicroMed study, the product ideas 18 (laser diode) and 19 (endoscopic moveable laser applicator) were the most promising product innovation ideas.

Results of the MicroMed study

Cooperation between technologists and medical practitioners: generating numerous technical ideas for problems and solutions

In the results of the study it became clear that specific questioning of medical experts meant that a large number of previously disregarded technical performance requirements could be determined. On the basis of the problems described by the medical practitioners, and the resulting technical and scientific abstraction, clear proposals for solutions could be developed, which were defined in more detail in specific performance requirements.

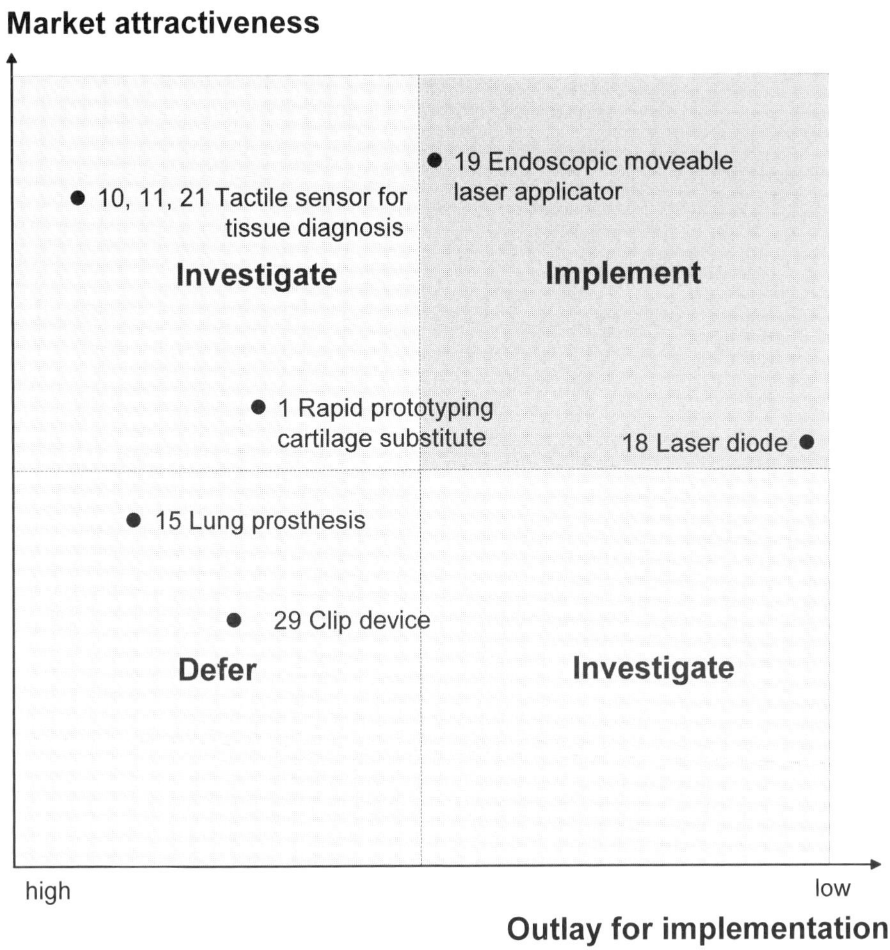

Fig. 5.27 Prioritization of selected product ideas

During the interviews it became apparent that the medical practitioners had generally not recognized the technical potential. The technologists at the Fraunhofer institutes, on the other hand, were already developing initial concepts for technical solutions while still participating in the consultations with the experts. The interdisciplinary creative workshops proved to be a useful tool for transferring medical problems directly into technical ideas for problems.

Innovation road-map: deriving errors in spheres of activity

The assessments of the market side and the technological side which were first of all undertaken separately, depending on the performance requirement, and the combination of these in the innovation roadmap facilitated a comparative assessment of the methods. This meant that strategic recommendations for action could be worked out for continuing the work. The prioritized methods were handed over to the specialist department of the Fraunhofer ILT and the Fraunhofer IPT where they were further developed.

Recommended spheres of activity

In order to exploit synergy effects of the existing major research topics at the Fraunhofer IPT and ILT, a focus on the following spheres of activity was recommended:

- optical diagnostic techniques
- pressure sensors
- navigation systems
- minimally invasive instruments for tomography-assisted surgery
- applications for biodegradable work materials

All in all, the cooperation between the institutes and the interdisciplinary method had a very positive effect on the quality of the performance requirements or the approaches to solutions that were yielded from this work. It was therefore agreed that the work begun with this study should be continued. To achieve systematic processing of the field of work "medical technology," the various areas of expertise of the partners involved and the geographical proximity to the medical experts should be exploited.

5.5 Neumag GmbH & Co. KG "Systematic innovation management as the basis for effectiveness in product development"

Carsten Voigtländer, Thomas Bauernhansl,
Jens Schröder

For more than 50 years, the company Neumag has been an icon of reliability, experience and high quality in the development and production of systems for the manufacturing of synthetic fibers.

In this case study we will describe how Neumag ensures the effectiveness of its production by systematic innovation management.

Neumag - the company

The company Neumag is an independent business unit of the world's largest textile machinery manufacturer, the Swiss Saurer Group. Neumag is a medium-sized company that supplies complex systems as a total solution for the manufacture of chemical fibers. Using these production systems for synthetic fibers, their customers all over the world produce, for example, yarn for carpets and upholstery or textile fibers. Neumag is the market and technology leader in the fields of BCF (bulked continuous filament) and staple (spun) fibers, with testimonials from around the world. As a trailblazer for many new technologies and processes, Neumag's in-house technical school develops highly innovative machinery with the label "Made in Germany," that is perfectly designed to match customer requirements.

Location: Neumünster

With around 500 employees, the complete process chain is covered at the company's location in Neumünster: from development and design, marketing, project planning and manufacturing, up to the pre-assembly stage. The commissioning of the systems, as well as maintenance and service, is carried out by Neumag employees at the customer's site.

Starting Point

Customers around the world as facility operators

Neumag's situation is characterized by a relatively limited number of international competitors and a very large number of customers – also spread around the globe. Neumag's customers (= facility operators) are players in a fiercely contested world market for synthetic fibers or synthetic fiber products, which is consequently characterized by high pricing pressure. Furthermore, many of Neumag's customers are located in Asia, as in this region of the world energy and personnel costs are comparatively low.

Competitive environment triggers performance spiral

The competitive situation is shaped by constant further development of the systems – continually driven by rivalry among competitors – with regard to increasing their productivity and their potential for saving costs, that is, a mixture of product and process innovation. At the relevant trade fairs, competitors regularly demonstrate systems that, by means of higher speeds, feature a larger output than that of the rival systems. Cannibalization effects are the result, as with larger output and constant market prices, the potential market for these systems is in continual decline. Even Neumag is not exempt from the effects of this spiral.

Innovation leaps as a basis for sustainable company success

Sustainable advantages over competitors cannot, however, be achieved by increasing performance alone. It is important to be able to present unique selling points, that is, to develop systems or system components as well as new processes or technologies, which lead to significant cost advantages for the customer, accompanied by a performance advantage, or even a new (final) product or new features for the final product. Besides these technical unique selling points, innovative services and maintenance are gaining in importance and can lead to a more astute perception on the part of the customers.

These unique selling points allow correspondingly higher prices for systems – at least until the competition has caught up with the technology – and also increase company turnover and profits. Because of the innovative systems technology, the company also derives medium and long-term benefit from its established reputation and outstanding market image.

Central importance of product development

In order to maintain, or rather, to expand the very good competitive position, it is necessary to work continually at implementing innovation leaps, and to do so within suitable time frames. Thus, the area of product development plays a central role in the securing and expansion of competitiveness. Certainly, product development exists in an area of tension that makes concentration on the "real" and important tasks more difficult (Fig. 5.28). Besides the external influences that we have mentioned, internal restrictions also exist at Neumag, which the development department can only evade with difficulty, if at all. The competencies and capacities available in the development division are limited – as is also the case in other medium-sized companies. In addition, daily business

Limited capacity in the development division

already occupies a major part of these capacities with developments for customer orders, customer tests and trouble shooting. It is therefore necessary to concentrate the existing capacity of a development division and to focus on the projects and activities which will ensure the continued success of the company.

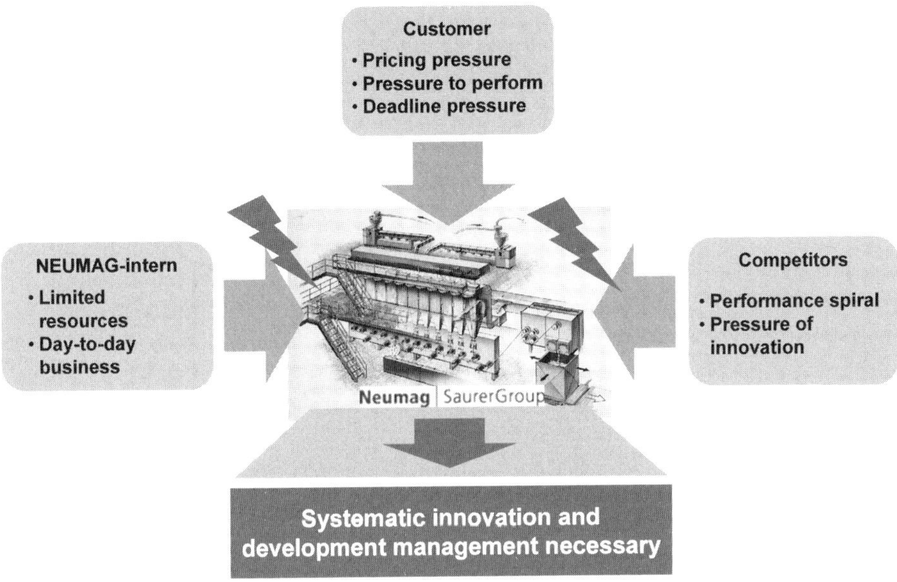

Fig. 5.28 Tension in the area of product development at Neumag

Focusing of capacities is essential

As in many other branches, customer benefit is the most important issue for Neumag. Customized systems are therefore developed and constructed in close cooperation with the customers. Consequently a major part of the know-how is concentrated in the development department; which thus makes a significant contribution to the competitiveness of the company.

Far-reaching consequences of errors in product development

Because comprehensive warranties for product quality, performance and energy consumption are generally accepted for the systems, development of fully matured components is very important. This truth therefore applies especially to Neumag: errors that are made at the development stage can have far-reaching consequences. If customers do not receive fully developed products, and, if, as a result, problems occur during operation of the systems or even lead to system down-times, immediate action must be taken. The result is high claims and warranty costs, resulting in particular from the high personnel costs and travel expenses for service employees. Furthermore, an immeasurable loss of image and reputation will be caused.

Stating the problem

Earlier: no prioritization of projects

In recent years, Neumag experienced an increase in problems in its development division. A study that was carried out in cooperation with the *Laboratorium für Werkzeugmaschinen und Betriebslehre* (*WZL)* – "Laboratory for Machine Tools and Production Engineering" of the RWTH (*Rheinisch-Westfälische Technische Hochschule*) Aachen University demonstrated that 30% of the entire development capacity had to be invested in troubleshooting (Fig. 5.29).

Major causes of problems that were identified were the lack of prioritization and insufficient focusing of the development activities. The resulting bottle-necks in development capacity – intensified by sketchy planning and inadequate information flows – led to a situation in which new developments, that in some areas were not fully developed, were delivered to customers. This led to further inquiries and complaints and, in some cases, meant that subsequent improvements had to be made, which in the final analysis only caused further deterioration in the capacity situation. Because of rising warranty costs and the danger of the loss of image, there was an acute need for action to be taken.

Earlier: inefficient R&D meetings

Even with the help of the annual R&D meeting, in which all relevant departments took part (management board, development, design, marketing, service etc.) this vicious circle could not be broken. These meetings, which, as a rule, were very intensive, were, on the contrary, typified by recriminations and disunity with regard to the contents and priority of future projects. Interpersonal discord further intensified this disunity so that a project landscape that could be agreed to and supported by all parties was, as a rule, never achieved, or was not implemented.

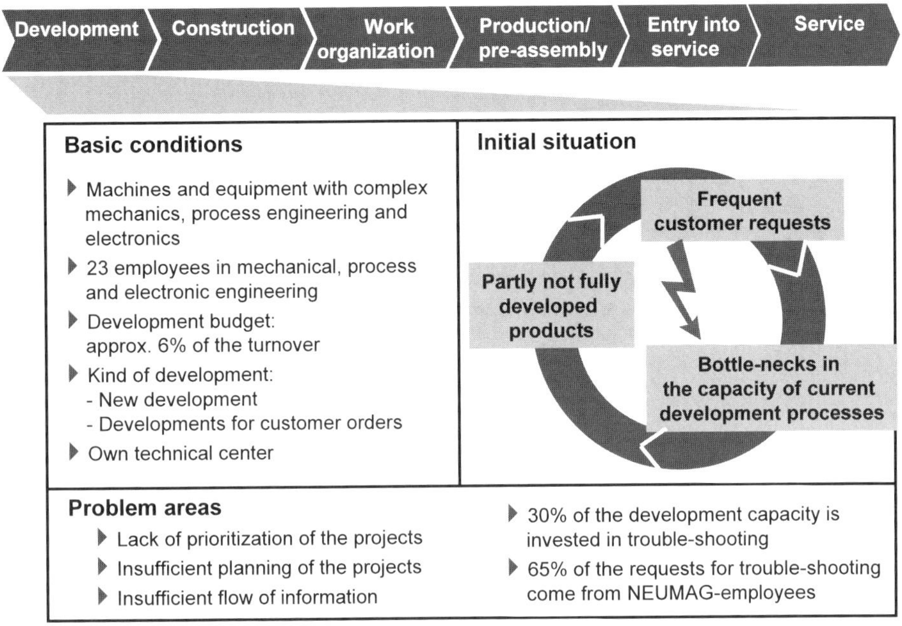

Fig. 5.29 Development division of Neumag before the reorganization

Transferability of the problems

The situation that has been described can be said to be true of many development departments in medium-sized companies. Many companies are not sufficiently aware of the importance of the development process and frequently do not have a consistent innovation and management structure for multiple projects. Furthermore, they do not have binding criteria for the setting up and assessment of a project landscape, in order to efficiently exploit existing development capacities.

Analysis of problems and approach to a solution

Reorganization of the Neumag innovation process

Against the background of the presentation of the problem described above, it was important to reorganize the innovation process at Neumag and to integrate it into a systematic and company-wide development management structure. This task was tackled jointly with the Laboratory for Production Engineering (*WZL*) at the *RWTH* Aachen University.

In the early phases of idea and demonstrator development, 80% of the total development costs were already specified (Fig. 5.30). On the other hand, 90% of the costs are only incurred at the stage of development which sees a product become ready for serial production. Therefore it is important, especially in the early stages of product development, to set the appropriate course. Higher outlay and effort at this early stage will be repaid during the life cycle of the product.

Effectiveness and efficiency in product development

The areas of activity for optimizing product development can be divided into measures for increasing effectiveness and measures for increasing efficiency. To increase effectiveness ("doing the right things"), measures must be taken that are aimed at concentrating the activities in product development on the essential tasks. On the other hand, measures for increasing efficiency ("doing things right") is aimed at completing the tasks with the lowest possible effort and expense.

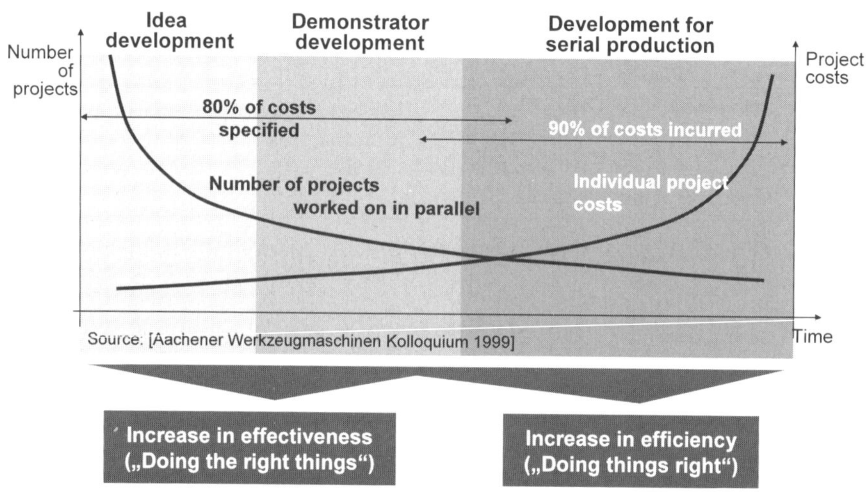

Fig. 5.30 Areas of activity for optimizing product development (AWK 1999)

Lever: effectiveness

The reorganization of the innovation process at Neumag required a comprehensive analysis of completed projects, an analysis of activities and documenting of the current situation. As already depicted in figure 5.29, insufficient prioritization and inadequate planning of the projects were identified as major problems, i.e. the effectiveness of the product development was not guaranteed. Here it was important to apply leverage and to deduce measures and implement them rapidly.

Approach to a solution: the PORTAL method

PORTAL method

The objective of the PORTAL method that was developed in conjunction with the *WZL* (Laboratory for Machine Tools and Production Engineering) is to systematically support the innovation process in order to concentrate the available, limited capacities of the development division on the "right" projects (effectiveness). This is a user-value-based method in which development projects are prioritized step by step (Fig. 5.31).

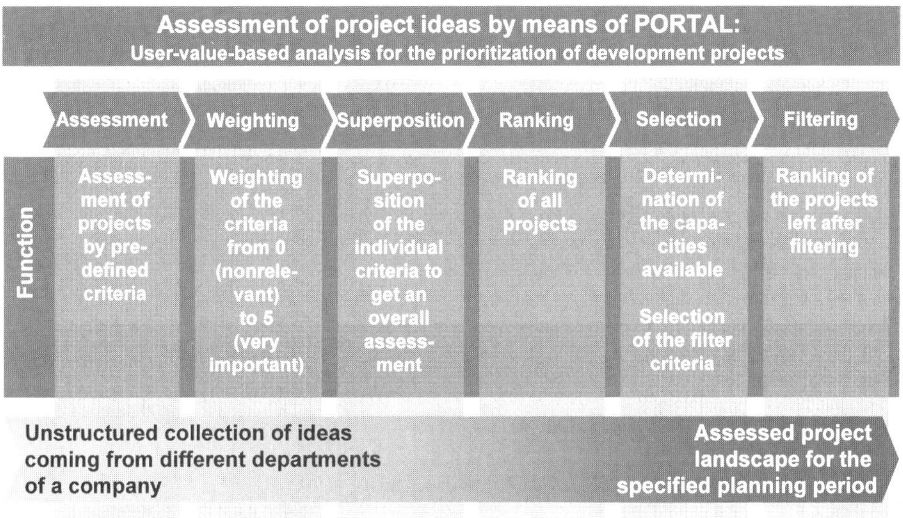

Fig. 5.31 PORTAL method

Documenting project ideas

In the first step, the project ideas are documented and structured. The documentation takes place on the basis of a formalized, computerized questionnaire, tailored to the specific activities of each department. This has the advantage that project or product ideas are first collated, structured and assessed within the various departments, without the necessity of taking into consideration the interests of other departments. At the same time, the pre-defined questionnaire ensures that this assessment is compatible with the views of the other departments. The completed questionnaires are collated and compared centrally by the development manager.

Assessment of project ideas

In the second step, the individual projects or product ideas are assessed. The list of comparisons is again made available to all departments involved. Using pre-defined criteria with pre-defined characteristic features, each department makes its own assessment of the individual projects or product ideas. Criteria for the assessment, such as size of the market, strategic importance and urgency are defined in terms specific to Neumag (Fig. 5.32).

Pre-defined assessment criteria

In addition, the development costs and the capacity that will be required must be estimated. Each project or product idea is assessed by means of these criteria.

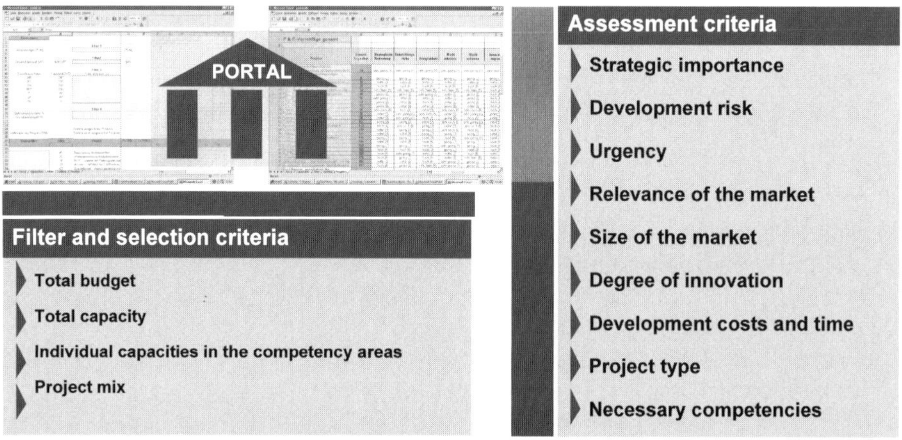

Fig. 5.32 Neumag PORTAL: criteria for project assessment

Comparison of the individual assessments

The individual assessments from each department are presented in a joint R&D meeting, discussed and then amalgamated to form a joint assessment. In spite of the partially divergent views of the different departments, the PORTAL method demonstrated that the assessments differed only slightly and that the comparison took merely a fraction of the time that had previously been required for the earlier, unstructured R&D meetings.

Weighting of the assessment criteria

At the annual R&D meeting, the individual criteria that are specific to Neumag are weighted. This is done by the method of comparing pairs. The weighting is checked annually and, if required, is adjusted to any modification in company strategy.

Normalization of the assessment

In the course of a normalization step the individual criteria are superimposed, that is, a normalized overall assessment is calculated for each single project. The assessment procedure itself was implemented in a software tool – based on MS-

Excel – so that the evaluation takes place automatically. By means of these numerical values, an initial ranking of the projects or product ideas can be determined.

Preliminary project ranking

This ranking at first represents the "ideal" project landscape, which cannot normally be fully implemented due to limited resources (budget, capacity etc.) or because it does not conform to Neumag's specific development strategy. A filtering stage was therefore introduced that takes account of these constraints.

The criteria that were defined for filtering the projects were the *available total budget*, the available *capacity in the development division* and standards for a *specialist project mix*.

Filtering the projects

The ranking of the individual projects is filtered accordingly. The capacities available for development activities are documented in the review of capacities – if required, separately, according to specialist disciplines – and thus the total available capacity as well as the available specialist capacity is determined. The capacity requirement in terms of the projects is cumulated according to the project rankings. The project list is "cut off" at the point at which the necessary capacity requirement exceeds the available capacity – if needed, the latter can also be expanded by adding external capacities.

Result: assessment of project landscape

As a result of the PORTAL method, the remaining projects are on hand in the form of a project landscape. These projects are of great strategic importance for the company; they do not meet any relevant exclusion criteria and are compatible with the available resources (capacity, budget) for the period under consideration.

The selected projects are assembled to form a project landscape and released to the participating divisions, according to capacity. This project landscape meets the requirements for high effectiveness of product development to a high degree.

Experiences with implementation

Increasing effectiveness with PORTAL

With the assistance of the PORTAL method, as described above, both existing and new projects are systematically accepted and transferred into a project landscape. The method thus represents an essential tool for increasing the effectiveness of the development division at Neumag.

High degree of acceptance in the company

Through the introduction and implementation of the PORTAL method, it was possible to focus the existing capacities and strengths of the development division on the "right" projects at the "right" time. The involvement of all the relevant decision-makers and the transparent decision-making process led to a high level of acceptance within the company.

Potential for savings

Thanks to the systematic procedure, a savings potential of around EUR 20,000 per annum – merely with regard to the selection of projected projects – could be demonstrated.

The "human" resource is still the most important factor in the innovation process

In spite of the systematic procedure required by the PORTAL method, the annual R&D meeting, that is, the inter-departmental agreement and discussion, remains an important tool for obtaining company-wide acceptance for decisions. Only by involving all relevant decision-makers, is it possible to guarantee that the project landscape can be promptly and consistently implemented, and that failures (not only successful ventures) are also born jointly by all those involved.

Intranet-based development manual

The software tool for the PORTAL method was integrated into Neumag's intranet-based development manual (Eversheim et al. 2001b) and this represents an important module in the reorganized innovation and development process at Neumag (Fig. 5.33).

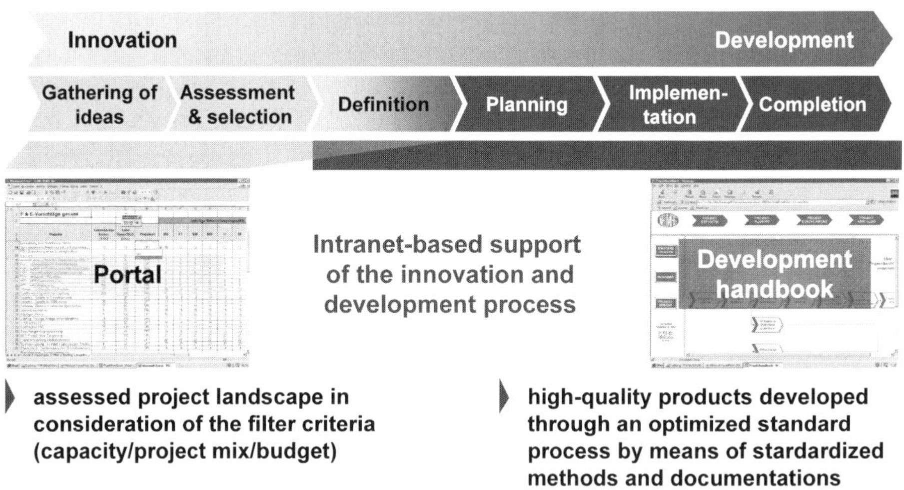

Fig. 5.33 The reorganized innovation and development process at Neumag GmbH

5.6 Dräger Medical AG & Co. KGaA "Reorganization of the business process Innovation"

RUDOLF-HENNING LOHSE, MICHAEL HILGERS

Innovation is more than merely Research & Development: it requires ideas management; market observation; roadmaps for markets, technologies and products; complexity management by concentrating on core competencies and through modularization and the involvement of suppliers at an early stage.

This case study will demonstrate how Dräger Medical set about creating and introducing a standardized, rapid and efficient innovation process worldwide.

Introduction

Company profile Drägerwerk AG

The Dräger Group is one of the world's leading companies for medical technology, and safety and aeronautical engineering (2001: 9535 employees, turnover: EUR 1,137 million; (EBIT: earnings before interest and taxes): EUR 49.5m). The products and services protect, monitor and support essential functions of human beings. The company Drägerwerk AG aims to be among the top three suppliers in each of its international markets and to generate a return of more than 20% (EBIT) by the year 2005 (Dräger 2002).

Company profile Dräger Medical

Dräger Medical AG & Co KGaA (referred to below as Dräger Medical) provides single products as well as integrated system solutions and services for cost-efficient therapy for all relevant areas of the patient process chain: from the emergency workplacc to ancsthctics and thc opcrating thcatcr as wcll as the areas of intensive, neonatal and home care (2001: 4837 employees, turnover EUR 805m (EBIT: earnings before interest and taxes). Dräger Medical has 45 subsidiary companies throughout the world; development and production sites are located in Germany, Holland, the USA and China (Dräger 2002).

Business Excellence Program

Because of internal weaknesses that were becoming apparent, the strategy committee of the Dräger Group decided in 1998 to introduce a Business Excellence Program. The following vision was formulated:

- all processes are focused on the customer
- all employees are motivated
- excellent business results

Process management: vision...

The vision requires a working method which is committed to the process of continual improvement of daily work and which improves processes with the aim of raising quality and lowering outlay.

...and reality

In the year 1998, Dräger medical was still far removed from this vision. The business processes were as a whole not sufficiently productive or secure. Some attempts had already been made to improve a wide range of malfunctions in processes, but due to barriers in the processes, organizational structures and company culture, these had not been eliminated successfully. This will be illustrated using the example of the business process "innovation."

Current status: innovation process

Current status: business process "innovation"

The development period for a new core product generally lasted four years. Conception phases were often not long enough to be able to minimize technical risks. New technologies were made "ready for serial production" while partly still in the product development stage. This was one of the reasons why frequent delays in sales launches were normal. The length of delay in a development project was on average one year. Development was not focused on core competencies. This led to the situation in which "make or buy" decisions were often taken late and – as a result – suppliers were only involved at a later stage (frequently not until after the start of series production). The "further development" was carried out parallel to the serial production; accordingly, the organization was forced to process numerous technical alterations. This led to high stock levels and the scrapping of numerous items as well as long delivery times. Areas of responsibility were not always clear: efficient methods, tools and sequences were implemented differently in each development location. This was a clear indication that the innovation process at Dräger medical was inadequately governed and not efficient enough.

Crisis in the year 2000

It was foreseeable that without profound change an EBIT target of 20% was unattainable. In the year 2000, a program for increasing productivity was resolved. The accruals for this program were the reason for negative earnings at Dräger Medical, and also at Drägerwerk AG, for the first time in the year 2000. By focusing on the main business, streamlining the organization, defining business processes and reinforcing the innovative strength, the turnaround was achieved in the following year, 2001.

Introduction of process management

In September 2000, it was clear to Dräger Medical that a significant improvement could not be achieved without a thorough restructuring of all business processes. This knowledge, and the associated consequences, ensured that today Dräger Medical is on the road to becoming a process organization that operates consistently around the globe.

Process organization – the third dimension besides business units and markets

Beside the business unit managers (global responsibility for products) and the regional managers (regional market responsibility for sales and service), there are strategic process owners (global responsibility for efficiency and security of a business process). The management board is responsible for the cross-company general framework. Strategy, course of action and cross-process topics are coordinated and agreed on in the network of strategic process owners. Every process owner is assisted by a committee that is made up of the most important sub-process owners. The committee is responsible for the reshaping of the process and for its continual improvement.

Process reorganization – finding the right starting point

The first step (early February 2001) was the differentiation of the business processes by the management team (Fig. 5.34). Whereas at the beginning of the discussion there were still four business processes: "generate products," "management of installed base," "generate orders" and "fulfill orders," it soon became clear that product generation would have to be broken down into further stages. The business process "innovation" was introduced from the idea stage to the start of serial production, and the business process "life cycle management" from the start of serial production to the end of the product's life cycle.

In April 2001, the reorganization of the business process "fulfill orders" was begun. The reorganization has the objective of introducing fast and punctual direct delivery to end customers throughout the world. In June 2001, the process project "generate orders" was started in order to correctly understand and process the market. On the basis of the conceptions that had been developed for "fulfill orders" and "generate orders," the reshaping of the business process "innovation" could begin in February 2002. By the end of the year 2003, the business processes for "life cycle management" and "management of installed base" should also have been drawn up and the rollout should have begun.

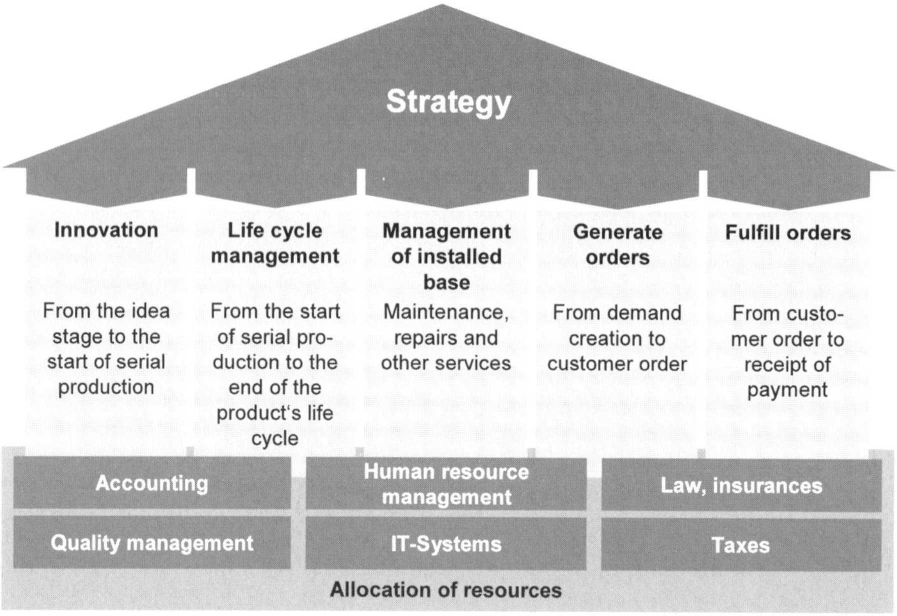

Fig. 5.34 The process house at Dräger Medical

An integrated approach leads to success

In the past, repeated attempts were made to improve processes in the long term. For example, the gate concept of the development process was revised in order to shorten the development time. The real cause, the inadequate quality of requirement specifications – which could be traced back to market surveys that were incomplete – could not be corrected in this way. Therefore, after many external and internal discussions it became clear: without an integrated approach and the willingness to question everything, nothing could be done.

Initiative program "*go* BEST !"

For this reason, Dräger Medical started the initiative program "*go* BEST !" in June 2001, with the intention of creating an integrated framework for all cross-company improvement initiatives (Fig. 5.35).

Fig. 5.35 Initiatives of the "*go* BEST !" program

Standard procedure for process projects

Dräger Medical decided in favor of a standard procedure in the first process project. It foresees that at the conception phase the future process will first of all be divided into four steps. These steps require the following action:

1. developing a vision for the future process; fixing overall planning objectives; defining project organization and other framework conditions
2. analyzing and evaluating the current process
3. planning the target process
4. planning implementation

Afterwards, the implementation stage begins. It includes the detailed concept as well as a pilot implementation; this is followed by the global rollout.

Central organization for process-redesign projects

In September 2000, the employees in the division Quality & Processes were gathered together in order to prepare and coordinate the process centrally. The employees prepared the process projects and in part also took over the management of the projects. The preparations yielded project plans and schedules, an itemization of the required resources, detailed planning of a kick-off workshop and the selection of management consultancy firms.

Use of external consultants

Depending on the process project, Dräger Medical frequently preferred to obtain external support. Here Dräger Medical primarily made use of management consultant firms, in order to quickly achieve good results through cooperation with their employees. The following roles were assigned to the management consultancy firms:

- moderator for inter-cultural project teams
- experts for knowledge of processes and methods
- minutes-keeper for ensuring the consolidation of the project results

For the conception of the future innovation process, Dräger Medical selected a management consultancy firm from the field of knowledge and innovation management (referred to below as consultant). Their tasks included demonstration of best-practice examples, preparation and follow-up of workshops, conducting the said workshops and the documentation and consolidation of the project results.

Documentation of business processes

In order to achieve a standardized documentation of the business processes, Dräger Medical has, in addition to the process stages (business process, work process, sub-process, process step), defined templates (work process representation, sub-process sheet, process sequence plan) that are to be used. Standard Microsoft® programs are used for the documentation of the process sequence plans in the company language, English. For the administration and the worldwide representation in Dräger's intranet in various national languages, an IT solution based on Microsoft® is being investigated.

The reshaping of the innovation process is described in detail below.

Start: reshaping

Reshaping of the innovation process

Dräger Medical is aiming to reach a position of innovation leadership and cost leadership in its markets. For this reason the efficiency of the development process is a vitally important topic, as 80–90 percent of the cost structure during the full life cycle of a product is already decided in the early stages of the innovation process.

Kick-off workshop – taking up the organization

Early in February 2002, a meeting in Lübeck was attended by 20 representatives from all six business units, the sales division and other process projects. In a kick-off workshop with representatives from the USA and Holland, an initial analysis of the weak points of the business process "innovation" was made, before a joint vision: "Nine, nine, six." was formulated. This formula describes a global innovation process at world-class level for all development locations of Dräger Medical (Fig. 5.36).

Reducing development time by half

"Nine-nine-six" requires integrated thinking in the conception phase and at the same time the shortening of the "time-to-market" process. The following lead times are to be achieved for a new generation of equipment

- nine months for a secured concept
- nine months for development
- six months for validation

In future, Dräger Medical aims to develop a new generation of equipment within a maximum period of 24 months.

Concentrating on core competency

Besides shortening development times by half, the second objective was a radical reduction in the complexity of the procurement chain. This reduction begins with concentrating on the company's core competency. The question is, "What must Dräger Medical be able to do better than its competitors in order to hold its own in the markets in the long term?" and not, "What can Dräger Medical already do well?" Until the present time, Dräger Medical has developed many single parts itself and administered them technically and logistically. In future, Dräger Medical plans to focus on the integration of modules, hardware and software architecture and the development of core competency modules. Everything else will gradually be transferred to suitable suppliers.

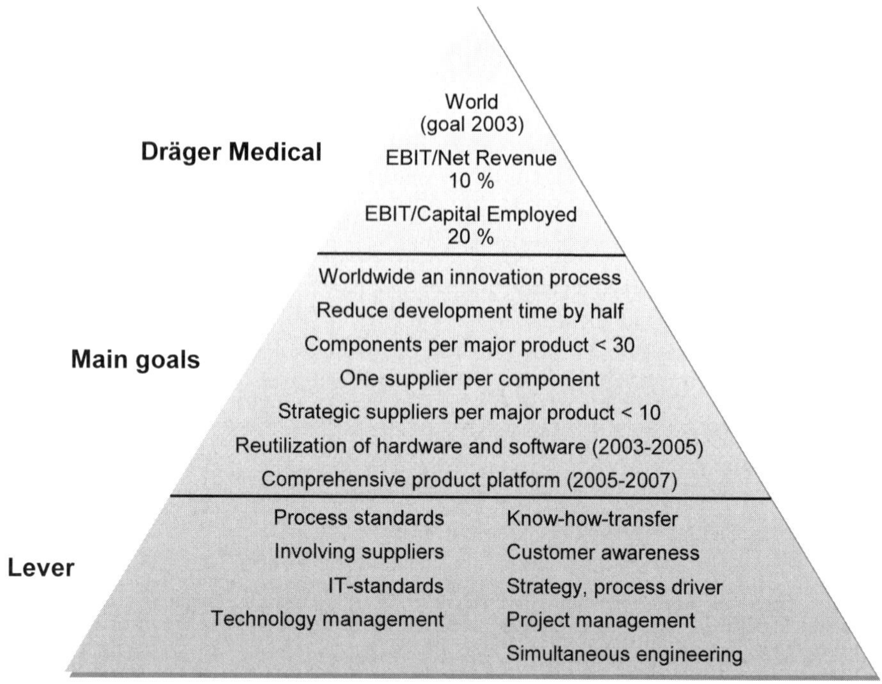

Fig. 5.36 Targets and levers for the reshaping of the innovation process

Early integration of suppliers

In future, the "make or buy" decisions will be made at the start of the conception phase, and from this point onwards, development, manufacture and life cycle management will be carried out by strategic partners, so as to obtain the maximum benefit from the core competencies of these partners. Factors for the selection of strategic suppliers are (in this order, weighting in parentheses):

- quality (35%)
- logistics (25%),
- costs (25%)
- technological potential (15%)

At the location in Lübeck alone, the purchasing department looks after 1350 suppliers. A computerized system is currently being installed in order to assess suppliers systematically and to involve strategic and preferred suppliers in development projects.

Product platform – Lego-like modular structure

In order to achieve ongoing cost leadership, Dräger Medical has resolved to introduce a product platform as its third major objective. This platform will include not only the hardware but also the software – the latter nowadays makes up approximately 50% of development costs, and the trend is upwards. By the year 2007, Dräger Medical hopes to have filled the platform: that is, all modules of the platform should exist by then. Ideally, a major product should in future consist of not more than 30 modules. Each module is to be developed, manufactured and maintained during its life cycle by a strategic supplier. This means, for example, that the supplier will be responsible for the provision of spare parts after series production has ended or for providing an alteration service.

Identifying levers

Furthermore, during the kick-off workshops levers were identified which must be applied in order to significantly improve the innovation process (Fig. 5.36). The course of action and the rough timetable were also agreed upon. In addition, two consultancy firms were invited to provide support in the discussion "What is 'world class' or 'best practice'?"

Based on the examples of the consultancy firms and taking into account the improvement levers, the future innovation process was roughly divided into four work processes:

- define market, product and technology
- research technology
- develop product
- manage and support innovation

Successful buy-in is essential for the project start

Whereas at the beginning of the two-day workshop the 20 participants were inclined to view matters in terms of their business unit and location, at the end, all the participants were convinced that a globally standardized innovation process is both necessary and attainable. After the kick-off, the project organization for the conception phase was planned in detail, so that the resources of the global organization can be implemented as efficiently as possible.

Conception of the future innovation process

Starting from the results of the kick-off workshop, the concept for the future innovation process was worked out in detail in four subsequent one or two day workshops.

Separation of technology and product development

First of all, the four work processes that had been determined in the kick-off stage were critically scrutinized. In doing so, it became apparent that the work process "research technology" had not yet been sufficiently clearly defined. Technology management had been identified as an important improvement lever, with which in future technological development was to be separated from the actual product development. So that in future only technologies that were ready for serial production could flow into the work process "develop product" (DP[32]), it was noted after the first workshop that in future the aspects laboratory prototype and prototype must be taken into consideration as well as the aspect of foundational research. To do this, the work process "develop a prototype" (DC[33]) was defined.

The five work processes

As a result of the action described above, the future innovation process at Dräger medical now consists of the following five work processes (Fig. 5.37):

- Define market, product and technology (DMPT[34]),
- Research technology (RT[35]),
- Develop prototype (DC)
- Develop product (DP),
- Manage and support innovation (MS[36]).

Based on the strategy, market-led roadmaps[37] for care areas, technologies, products and important components are developed in the work process DMPT. This long term projection – of up to ten years – for technologies and products is, in the view of Dräger Medical, an essential success factor for cutting development time by half. The process DMPT is used to steer the work processes RT, DC and DP by means of clearly defined project charters. In RT, feasibility studies for important technologies are made, either in theory (degree of maturity 1) or in practice (degree of maturity 2). In DC – using the results of RT – technologies are evaluated with the aid of a prototype (degree of maturity 3).

In DP, a product with its components is developed in three phases (conception, development, validation). At this point the results of DC are also included. The process is complete when series production is running reliably. The last work process, MS, forms a framework for the other four work processes. Here the following aspects are principally dealt with:

---

32 DP: Develop Product.

33 DC: Develop prototype-like Component.

34 DMPT: Define Market, Product and Technology.

35 RT: Research Technology.

36 MS: Manage and Support Innovation

37 "Roadmaps" in this context refers to plans depicting events that are related logically and chronologically with regard to market trends, competitors, products and technologies.

- ideas management
- coordinating and defining IT support
- basic principles and courses of action for work processes and sub-processes, tools and project organization
- measurement of efficiency of the innovation process and the continual improvement thereof

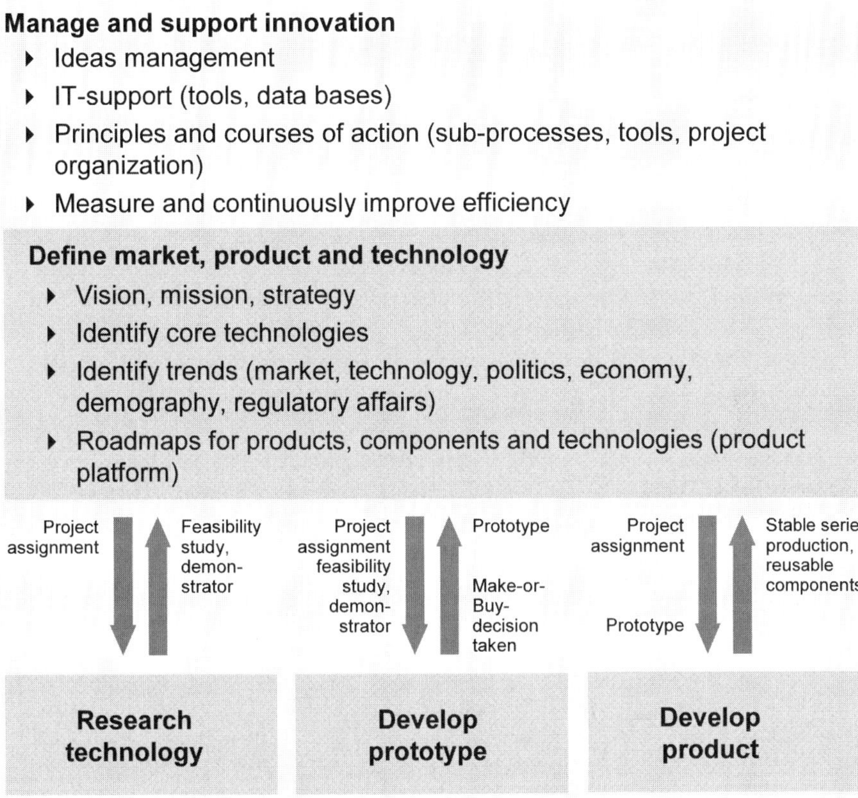

Fig. 5.37 The work processes of the future innovation process

Below, some of the most important points of the work processes DMPT and DP will be explained briefly.

DMPT – drawing up of meaningful roadmaps

Identify trends

In future, trend information about markets, technologies and other fields will be collated systematically and regularly. The market data will be made available by the Dräger marketing companies.[38] After the evaluation and prioritization of the data – based on a trend and competitor roadmap – then roadmaps for products, platforms modules and technologies will be drawn up, first of all for each business unit (Fig. 5.38).

Across-the-board coordination of roadmaps

Afterwards across-the-board coordination of care areas and the consolidation of the roadmap for products, platform modules and technologies will take place. Then gaps can be identified, e.g. necessary technological developments (Fig. 5.39) or product developments ("face lift" or "next generation," Fig. 5.40). For this purpose possible projects will be specified and initiated in alignment with the business process "life cycle management.

Allocate resources according to time horizon

Depending on the time horizon, the project resources will be assigned to the solving of a variety of questions (Groenveld 1997):

- one to two years: How can we achieve it?
- three to five years: What can we achieve?
- five to ten years: What is possible?

DP—the critical path must be identified

High process quality reduces costs and time outlay

As a producer of medical products, Dräger Medical is required to fulfill a wide range of norms. The work process "develop product" is therefore at the heart of the innovation process. Only a high standard in these work processes will achieve lower costs and shorter development times as well as high product quality (VDA 1998).

The work process "develop product" was divided into three stages: "conception," "implementation" and "validation" (Fig. 5.40). A total of nine sub-processes were identified, each of which includes a further 10 to 25 process steps (a total of around 180). The main emphasis of the reshaping lies in the conception phase.

---

38 The determining of market data will in future take place in the business process "generate orders" with the work process "analyze the market." This again demonstrates the necessity of an integrated approach. For without the simultaneous reshaping of the business processes "generate orders" and "innovations" it would not be possible to design essential across-the-board interfaces

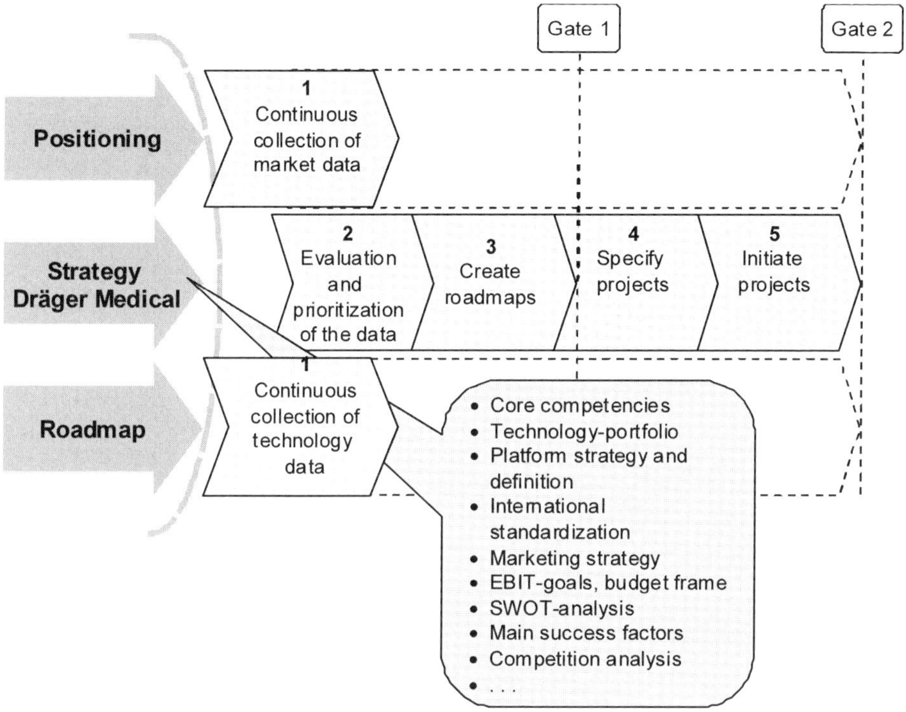

Fig. 5.38 Define sub-processes within the work process "Market, product and technology"

The sub-processes must run consistently through all three phases; if this is not the case, the user will become lost in the complexity of the process steps. Developing this structure and coordinating it across three development locations was one of the main activities in the conception of the future innovation process.

Objective of the conception phase

The objective of the conception phase is to plan, evaluate and select alternatives, which in the best way possible fulfill the demands of the requirement specification. First of all the requirements of the new product are determined, in order to specifically define the product alternatives and manufacturing alternatives. Project planning is also part of this process. The requirement specification sums up the results of the conception phase.

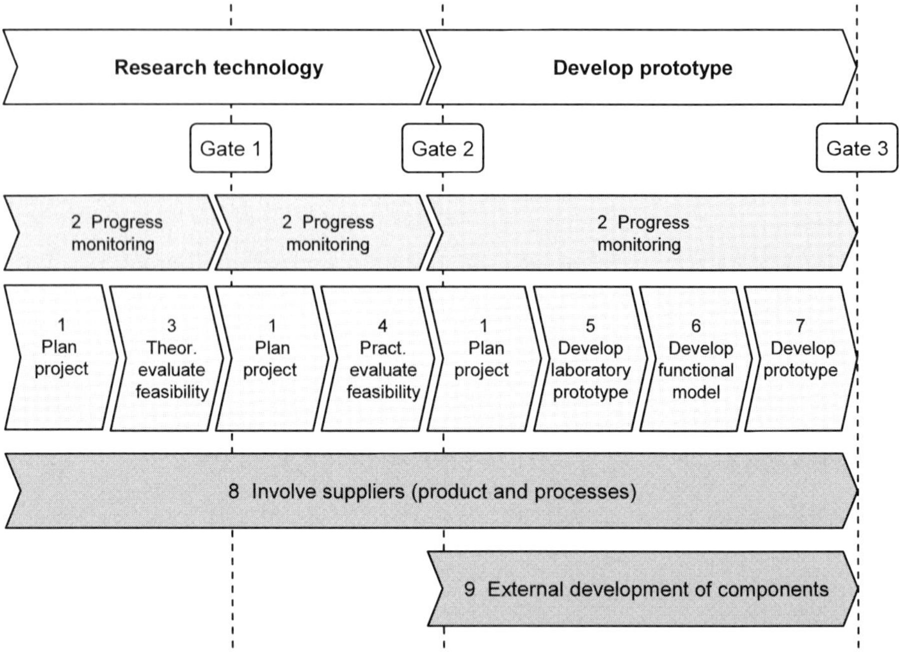

Fig. 5.39 Synergy of the sub-processes of "research technology" and "develop prototype"

Early "make or buy" decision in the conception phase

Complex tool-related parts frequently determine the development time at Dräger medical (e.g. complex housings made of synthetic materials). For this reason, "make or buy" decisions must already be made at the beginning of the conception phase. The involvement of suppliers at an early stage is one of Dräger Medical's major goals. This permits the company to concentrate on its core competencies. To achieve this, Dräger Medical must on the one hand purposely cultivate suppliers that are capable of rapidly producing serial parts through laboratory prototypes and prototypes (Fig. 5.39).

On the other hand, the developers at Dräger Medical need to understand that only the early involvement of suppliers will shorten the learning curve.

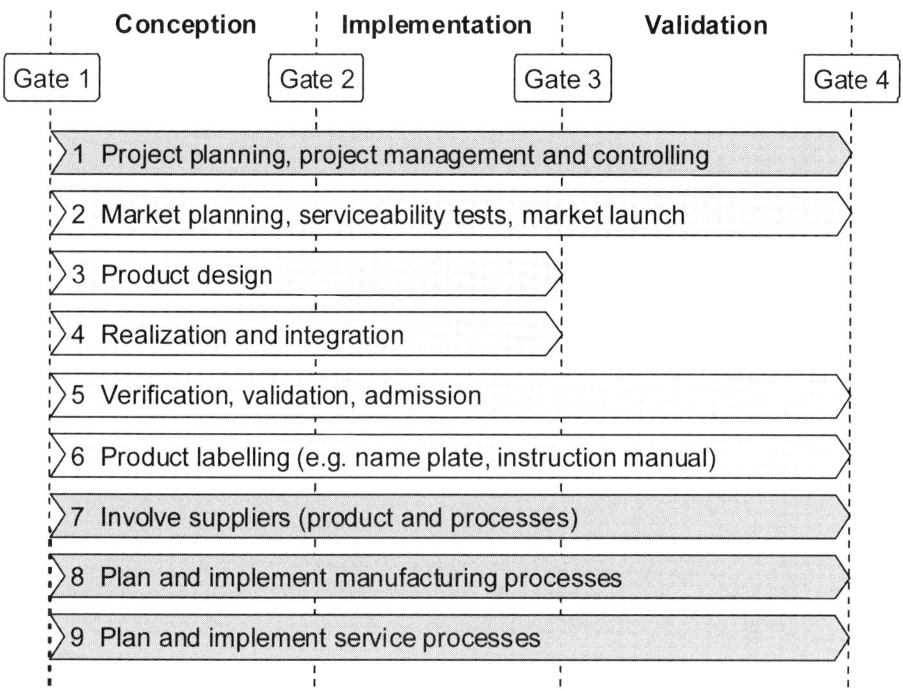

Fig. 5.40 Synergy of the sub-processes of "develop product"

A clear "gate" concept

So-called "gates" are used to separate the various phases. Each phase is checked by a project team of independent employees, using checklists for risks and completeness. If a result is not satisfactory, it must be reworked before the next phase begins.

Implementation and validation phase

In the implementation phase, the product and the manufacturing and service processes are further refined. The validation of the product and of the manufacturing and service processes concludes the development project.

Project organization in the conception phase

Planning the conception phase

During the four-month conception phase (Fig. 5.41), 35 employees of the global organization were involved consistently in all project steps for drawing up the conception and implementation plan. Here an 80/20 approach was rigorously pursued in order, on the one hand, to describe in adequate detail the future globally standardized innovation process, and, on the other, to remain sufficiently abstract to not become lost in the detailed points in the conception.

Project organization

For the organization of the project, a steering committee, a core team and four project teams – one for each work process

– were set up. A project manager from the field of "Quality & Processes" was responsible for the entire conception phase; for each work process a project manager from that specialist field was appointed. One employee from the consultancy firm was assigned to each project leader.

Good preparation ensures success

The standard course of action as described above was used for the conception phase. In addition a special workshop was held for the platform strategy and the selection of a consultancy firm for the development and launch of a product platform. In order to be able to pay sufficient attention to the coordination needs of the teams and to efficiently utilize "know-how carriers," Dräger Medical had planned workshops lasting one to two days (Fig. 5.41).

Ease the burden through external advisors

A total of 420 man-days were expended for the conception phase "innovation" (Fig. 5.41), with 300 man-days for Dräger Medical, including project management, and 120 man-days for the consultancy firm. From Dräger's point of view this course of action has proved to be worthwhile. The participants were not burdened with preparations or follow-up; all agreements could be achieved simultaneously, the use of a consultant released the organization from time-consuming but necessary documentation and consolidation activities.

The implementation of the future innovation process is described in detail below.

Implementation of the future innovation process

Make it happen!

All in all, three implementation phases have been defined (Fig. 5.42). Besides the preparation of the first roadmaps and an inter-care-area product platform, the focus in the first phase was on the implementation of the early involvement of suppliers, as well as the implementation of the work process "develop product." In the second phase, an IT-based parts standardization will be introduced, and the design and implementation of Competence Centers will be planned and tested with a pilot project. Furthermore, standardized management and controlling will be introduced world-wide. This also includes the introduction of a workflow solution for the release of master data.

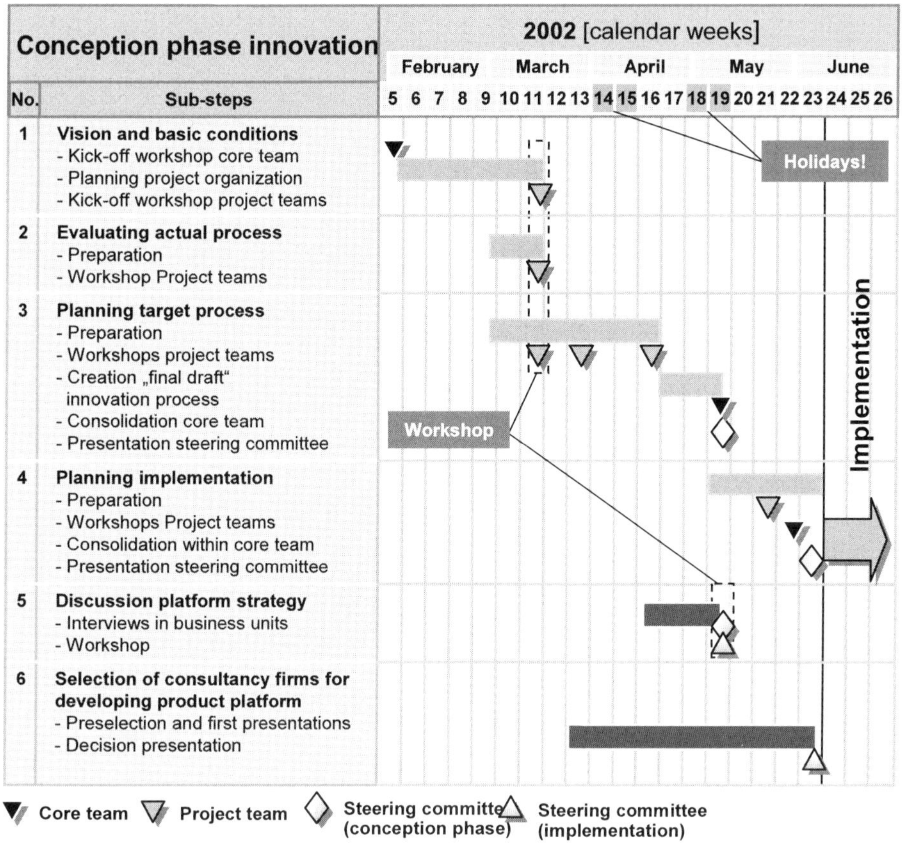

Fig. 5.41 Schedule for the conception phase "innovation"

From the present standpoint, 4100 man-days will be necessary to introduce the future innovation process. As a result, a delay of approximately 3 to 5 months in other projects must be accepted. After brief discussion, however, it was clear on all levels: "It's now or never!" This acceptance in the organization is indispensable for the ongoing implementation of the new innovation process.

A high level of acceptance in the organization is indispensable!

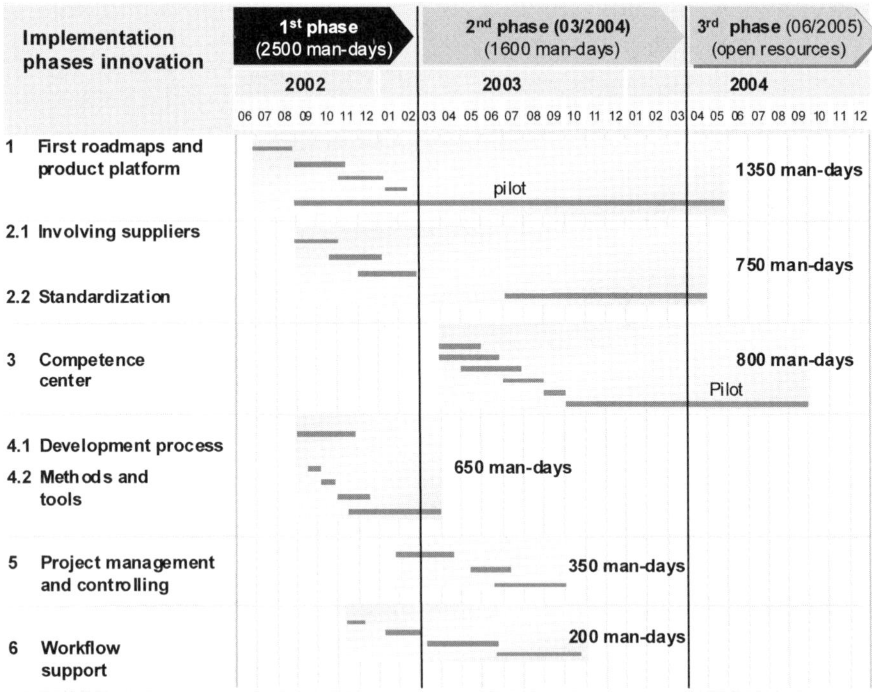

Fig. 5.42 Schedule for the implementation of the future innovation process

Looking to the future

Securing long-term competitiveness with new innovation process

With its new innovation process Dräger Medical is creating a basis for long-term competitiveness. With three project processes being implemented, the organization has now reached the limits of its capacity. The implementation for the innovation process has just begun; the main work thus still remains to be done. Nonetheless, Dräger Medical is certain that with this course of action a workable and practical concept has been created. Dräger Medical's procedures and experiences during the conception of a new innovation process may perhaps be of help to the reader, when starting a project of this kind himself.

Bibliography

Abell DF (1978)
Strategic Windows
In: Journal of Marketing, Juli 1978, p. 21-26

Aeberhard K (1996)
Strategische Analyse: Empfehlungen zum Vorgehen und zu sinnvollen Methodenkombinationen
Dissertation, Verlag Lang, Bern, 1996

Agamus (1998)
Stars der Innovation – Die Agamus-Consult Innovations-Studie
Agamus-Consult, Starnberg, 1998

Akao Y (1992)
QFD – Quality Function Deployment.
Liesegang G (Editor)
Verlag Moderne Industrie. Landsberg/ Lech, 1992

Albers S, Eggers S (1991)
Organisatorische Gestaltung von Produktinnovations-Prozessen Führt der Wechsel des Organisationsgrades zu Innovationserfolg
In: zfbf, Volume 43, Issue 1, 1991, p. 44-64

Altschuller GS (1973)
Erfinden – (k)ein Problem?: Anleitung für Neuerer und Erfinder
Tribüne Verlag, Berlin, 1973

Altschuller GS (1984)
Erfinden – Wege zur Lösung technischer Probleme
VEB Verlag Technik, Berlin, 1984

Altschuller GS (1998)
40 Principles: TRIZ Keys to Technical Innovation.
Edition 1., TRIZ Tools, Volume 1
Technical Innovation Center, Worcester, 1998

Ansoff IH (1975)
Managing Strategic Surprise by Response to Weak Signals
In: California Management Review, Volume 18, Issue 2, 1975, p. 21-33

AWK Aachener Werkzeugmaschinen-Kolloquium. Eversheim W, Klocke F, Pfeifer T, Weck M (Editor) (1999)
Wettbewerbsfaktor Produktionstechnik – Aachener Perspektiven
Shaker Verlag, Aachen, 1999

AWK Aachener Werkzeugmaschinen-Kolloquium. Eversheim W, Klocke F, Pfeifer T, Schuh G, Weck M (Editor) (2002)
Wettbewerbsfaktor Produktionstechnik – Aachener Perspektiven
Shaker Verlag, Aachen, 2002

Backhaus K, Erichson B, Plinke W, Weiber R (1996)
Multivariate Analysemethoden – eine anwendungsorientierte Einführung
Springer-Verlag, Berlin et al., 1996

Berekoven L, Eckert W, Ellenrieder P (1996)
Marktforschung.
Edition 7, Gabler, Wiesbaden, 1996

Binder V, Kantowsky J (1996)
Technologiepotentiale – Neuausrichtung der Gestaltungsfelder des strategischen Technologiemanagements
Dissertation, Hochschule St. Gallen, Dt. Universitäts-Verlag, Wiesbaden, 1996

Bleicher K (1999)
Das Konzept integriertes Management
Edition 5, Campus Verlag, Frankfurt/ M., New York, 1999

Bonissone P (1982)
A fuzzy sets based linguistic approach: theory and applications.
In: Gupta M (Editor) Fuzzy information and decision processes. North-Holland Publishing Company, Amsterdam et al., 1982

Boutellier R (Editor), Gassmann O, von Zedtwitz M (1999)
Managing global innovation: uncovering the secrets of future competitiveness
Springer-Verlag, Berlin et al., 1999

Brandenburg F (2002)
Methodik zur Planung technologischer Produktinnovationen
Dissertation, RWTH Aachen, Shaker Verlag, Aachen, 2002

Brandenburg F, Spielberg D (1998)
Implementing New Ideas into R&D-Strategies - Innovation Management for the Automotive Industry
31st ISATA, Advanced Manufacturing in the Automotive Industry, Düsseldorf, Juni 1998

Brankamp K (1971)
Planung und Entwicklung neuer Produkte
de Gruyter, Berlin, 1971

Brassard M, Ritter D (1994)
Memory Jogger II
Methuen/ USA: GOAL/ QPC, 1994

Brauchlin E (1995)
Problemlösungs- und Entscheidungmethodik
Edition 4, Haupt, Berlin, 1995

Breiing A, Knosala R (1997)
Bewerten technischer Systeme – Theoretische und methodische Grundlagen bewertungstechnischer Entscheidungshilfen
Springer-Verlag, Berlin et al., 1997

Brockhoff K (1985)
Die Produktinnovationsrate als Instrument der strategischen Unternehmensplanung
In: Zeitschrift für Betriebswirtschaft, Volume 55, Issue 5, 1985, p. 451-476

Brockhoff K (1999)
Forschung und Entwicklung – Planung und Kontrolle
R. Oldenbourg Verlag, Munich et al., 1999

Brose P (1982)
Planung, Bewertung und Kontrolle technologischer Innovationen
Erich Schmidt Verlag, Berlin, 1982

Buchholz W (1996)
Time-to-Market-Management: Zeitorientierte Gestaltung von Produktinnovationsprozessen
Dissertation, Universität Gießen, Verlag W. Kohlhammer, Stuttgart, 1996

Buck A, Herrmann C, Lubkowitz D (1998)
Handbuch Trendmanagement – Innovation und Ästhetik als Grundlage unternehmerischer Erfolge
FAZ-Verlag, Frankfurt/ Main, 1998

Bugdahl W (1990)
Methoden der Entscheidungsfindung
Edition 1, Vogel Verlag, Würzburg, 1990

Bullinger HJ (1994)
Einführung in das Technologiemanagement
B. G. Teubner Verlag, Stuttgart, 1994

Chmielewicz K (1970)
Forschungskonzeption der Wirtschaftswissenschaften – Zur Problematik einer entscheidungstheoretischen und normativen Wirtschaftslehre
Poeschel Verlag, Stuttgart, 1970

Clark KB, Fujimoto T (1991)
Automobilentwicklung mit System – Strategie, Organisation und Management in Europa, Japan und USA
Campus Verlag, Frankfurt/ Main et al., 1991

O´Conner J, Seymour J (2000)
Neurolinguistisches Programmieren
German translation: Dolke G, Edition 10, Kirchzarten: VAK, 2000

Cooper RG (1993)
Winning at new products: accelerating the process from idea to launch
Edition 2, Addison-Wesley Publishing Company, Reading, Massachusetts, 1993

Corsten H, Reiß M (Editor) (1995)
Handbuch Unternehmensführung: Konzepte – Instrumente – Schnittstellen
Gabler Verlag, Wiesbaden, 1995

Corsten H, Reiß M (Editor) (1999)
Betriebswirtschaftslehre
Edition 3, Oldenbourg Verlag, Munich et al., 1999

Dang N, Lenz B (1992)
Das Technologieprofil - Ein Instrument für die strategische Technologieplanung
In: io Management, No 61, 1992, p. 36-38

Delphi (1998)
Delphi ´98 - Studie zur globalen Entwicklung von Wissenschaft und Technik
Studie des Fraunhofer ISI im Auftrag des BMBF, Karlsruhe, 1998

Deschamps JP, Nayak RP, Little AD (1996)
Produktführerschaft – Wachstum und Gewinn durch offensive Produktstrategien
Campus Verlag, Frankfurt/ Main et al., 1996

Deutsches Institut für Normung e.V. (DIN) (2006)
DIN EN 60812:2006-11 - Analysis techniques for system reliability - Procedure for failure mode and effects analysis (FMEA) (IEC 60812:2006); German version EN 60812:2006, 2006

Dräger (2002)
Geschäftsbericht der Drägerwerk AG 2001

Droege (1999)
Barrieren und Erfolgsfaktoren der Umsetzung von Innovationen – Weltweite Studie Innovationsmanagement (Volume II)
Studie der Droege & Comp. AG und des Bundesverbandes der deutschen Industrie e. V. (BDI), Düsseldorf, Cologne, 1999

Dyckhoff H, Ahn H (1998)
Integrierte Alternativengenerierung und -bewertung
In: DBW, Volume 58, Issue 1, 1998, p. 49-63

Ehrat M (1997)
Kompetenzorientierte, analysegestützte Technologiestrategieerarbeitung: Methodisches Konzept zur Unterstützung der Technologiestrategieentwicklung mit externen Datenbanken
Dissertation, Hochschule St. Gallen, 1997

Eggers S (1993)
Existenz und Erfolg eines wechselnden Organisationsgrades in Innovationsprozessen
Verlag Josef Eul, Bergisch Gladbach, Cologne, 1993

Ehrlenspiel K (1995)
Integrierte Produktentwicklung – Methoden für Prozessorganisation, Produkterstellung und Konstruktion
Hanser Verlag, Munich, 1995

Eisenführ F, Weber M (1999)
Rationales Entscheiden
Edition 3, Springer-Verlag, Berlin et al., 1999

Eßmann V (1995)
Planung potentialgerechter Produkte – Ein Beitrag zur Produktkonversion
Dissertation, Universität Dortmund, Dt. Universitäts-Verlag, Wiesbaden, 1995

Eversheim W, Schuh G (Editor) (1996)
Die Betriebshütte – Produktion und Management
Springer-Verlag, Hamburg, 1996

Eversheim W, Schmidt R, Saretz B (1994)
Systematische Ableitung von Produktmerkmalen aus Marktbedürfnissen
In: io Management, Volume 63, No 1, 1994, p. 66-70

Eversheim W, Klocke F, Brandenburg F, Fallböhmer M (2000)
Integrated Manufacturing technology Planning
33rd CIRP International Seminar on Manufacturing Systems. Stockholm, 2000

Eversheim W, Breuer T, Grawatsch M (2001a)
Combining the Scenario Technique with OFD and TRIZ to a Product Innovation Methodology.
Conference, TRIZ future, Bath, UK, November 2001, p. 273-281

Eversheim W, Breuer T, Grawatsch M (2002)
Erfinden nach Plan mit TRIZ
2nd Innovationswerkstatt Strategische Produktplanung, Nürnberg, Januar 2002

Eversheim W, Voigtländer C, Bauernhansl T, Schröder J (2001b)
Intranetbasiertes Entwicklungsmanagement – der schnelle Weg zu Neuprodukten
In: Konstruktion, Mai 2001

Fallböhmer M, Brandenburg F, Trommer G (2000)
Generation and Evaluation of Manufacturing Alternatives in Early Design Phases
International Workshop on Multi-Criteria Evaluation MCE, Neukirchen, 2000

Fraunhofer (1998)
Erfolgsfaktoren von Innovationen: Prozesse, Methoden und Systeme
Ergebnisse einer gemeinsamen Studie von Fraunhofer-Instituten, 1998

Friese W (1975)
Ein System zur koordinierten Produktplanung in Unternehmen der Investitionsgüterindustrie
Dissertation, RWTH Aachen, 1975

Gabler (1997)
Gabler-Wirtschafts-Lexikon
Edition 14, Gabler, Wiesbaden, 1997

Gassmann O (1996)
Internationales Innovationsmanagement
Vahlen Verlag, Munich, 1996

Gassmann O (1997)
Internationales F-&-E-Management
R. Oldenbourg Verlag, Munich, 1997

Gassmann O, Kobe C, Voit E (2001)
Quantensprünge in der Entwicklung erfolgreich managen
Springer-Verlag, Berlin et al., 2001

Gausemeier J, Ebbesmeyer P, Kallmeyer F (2001)
Produktinnovation - strategische Planung und Entwicklung der Produkte von morgen
Hanser Varlag, Munich, 2001

Gausemeier J, Fink A, Schlake O (1996)
Szenario-Management – Planen und Führen mit Szenarien
Edition 2, Carl Hanser Verlag, Munich, Wien, 1996

Geisinger D (1999)
Ein Konzept zur marktorientierten Produktentwicklung: Ein Beitrag zur Steigerung der Erfolgsquoten neuer Produkte
Dissertation, Universität Karlsruhe, 1999

Gerhards A (2002)
Methodik zur Interaktion von F&E und Marketing in den frühen Phasen des Innovationsprozesses
Dissertation, RWTH Aachen, Shaker Verlag, 2002

Gerpott TJ (1991)
Globales F&E-Management. Bausteine eines Gesamtkonzeptes zur Gestaltung eines weltweiten F&E-Standortsystems
In: Booz, Allen & Hamilon (Editor) Integriertes Technologie- und Innovationsmanagement: Konzepte zur Stärkung der Wettbewerbskraft von High-Tech-Unternehmen, Schmidt, Berlin, 1991

Geschka H (1986)
Innovationsmanagement
In: o. V. (1983) Management-Enzyklopädie, Das Managementwissen unserer Zeit, Verlag Moderne Industrie, Landsberg/ Lech, 1986

Graebsch M, Lindemann U, Weiß S (2007)
Lean development in Deutschland : eine Studie über Begriffe, Verschwendung und Wirkung
Verlag Dr. Hut, Munich, 2007

Green PE, Tull DS (1982)
Methoden und Techniken der Marktforschung
Poeschel, Stuttgart, 1982

Groenveld P (1997)
Roadmapping Integrates Business and Technology.
In: Research Technology Management, Volume 40, No 5, Sept.-Oct. 1997, p. 48-55

Grob HL (1999)
Einführung in die Investitionsrechnung
Vahlen Verlag, Munich, 1999

Haberfellner R, Becker M, Büchel A, von Massow H, Nagel P (1999)
Systems Engineering – Methodik und Praxis
Daenzer, W.F.; Huber, F. (Editor), Edition 10, Verlag Industrielle Organisation, Zürich, 1999

Hansmann K (1983)
Kurzlehrbuch Prognoseverfahren
Gabler Verlag, Wiesbaden, 1983

Hartung S (1994)
Methoden des Qualitätsmanagements für die Produktplanung und –entwicklung
Dissertation, RWTH Aachen, Shaker Verlag, Aachen, 1994

Hauschildt J (1996)
Innovationsförderliche Führung, Organisation und Unternehmenskultur
In: Eversheim (1996), p. 4-21 to 4-26

Hauschildt J (1997)
Innovationsmanagement
Vahlen, Munich, 1997

Herb R (2000)
TRIZ – der systematische Weg zur Innovation: Werkzeuge, Praxisbeispiele, Schritt-für-Schritt-Anleitungen
Verlag Moderne Industrie, 2000

Herstatt C (1998)
Dialog mit Kunden und Lead-User-Management
In: Barske H, Gerybadze A, Hünninghausen L, Sommerlatte T (Editor): Das innovative Unternehmen, Gabler Verlag, Wiesbaden, 1998

Herzhoff S (1991)
Innovations-Management: Gestaltung von Prozessen und Systemen zur Entwicklung und Verbesserung der Innovationsfähigkeit von Unternehmungen
Dissertation, Universität Siegen, Verlag Josef Eul, Bergisch Gladbach et al., 1991

Higgins JM, Wiese GG (1996)
Innovationsmanagement
Springer-Verlag, Berlin, 1996

Hill B (1989)
Biologische Konstruktionen, Vorbilder für die Technik
Maschinenbautechnik 38, 1989

Hill B (1997)
Innovationsquelle Natur, Naturorientierte Innovationsstrategie für Entwickler, Konstrukteure und Designer
Shaker Verlag, Aachen, 1997

Hill B (1999)
Naturorientierte Lösungsfindung, Entwickeln und Konstruieren nach biologischen Vorbildern
expert-Verlag, Renningen-Malmsheim, 1999

Hinterhuber H (1996)
Strategische Unternehmungsführung,
Bd.1: Strategisches Denken, Edition 6, de Gruyter, Berlin, 1996

Horvath P (1994)
Controlling
Edition 5, Vahlen, Munich, 1994

Horx M, Wippermann P (1996)
Was ist Trendforschung?
Econ Verlag, Düsseldorf, 1996

Huxold, S (1990)
Marketingforschung und strategische Planung von Produktinnovationen: ein Früherkennungsansatz
Erich Schmidt Verlag, Berlin, 1990

Invention Machine (1998)
TechOptimizer
Professional Edition, Version 3.0, Softwareanwendung, Boston, 1998

IWB Software (1997)
Ideation International Inc. Southfield, 1997

Jacob H, Voigt KI (1997)
Investitionsrechnung
Gabler Verlag, Wiesbaden, 1997

Jürgens V (1998)
Ressourcenorientierte Leistungsgestaltung
Dissertation, Universität St. Gallen, 1998

Kahn H, Wiener AJ (1967)
The Year 2000
Macmillan, London, 1967

Kahneman D, Slovic P, Tversky A (Eds.) (1986)
Judgement under uncertainty: Heuristics and biases
Cambridge University Press, Cambridge et al., 1986

Kamiske GF, Brauer JP (1993)
Qualitätsmanagement von A bis Z – Erläuterung moderner Begriffe des Qualitätsmanagements
Hanser Verlag, Munich, 1993

Kano N (1995)
Upsizing the organisation by attractive quality creation
In: Kanji GK (Editor): Total Quality Management. Proceedings of the first world congress, London, 1995

Kehrmann H (1972)
Die Entwicklung von Produktstrategien
Dissertation, RWTH Aachen, 1972

King B (1994)
Doppelt so schnell wie die Konkurrenz
St. Gallen: gfmt, 1994

Klaus G, Liebscher H (1976)
Wörterbuch der Kybernetik
Deutscher Verlag der Wissenschaften, Berlin, 1976

Kleinschmidt EJ, Cooper RG, Geschka H (1996)
Erfolgsfaktor Markt – Kundenorientierte Produktinnovation
Springer-Verlag, Berlin et al., 1996

Klocke F, Eversheim W, Fallböhmer M, Brandenburg F (1999)
Einsatzplanung von Fertigungstechnologien – Hilfsmittel für die Technologieplanung in frühen Phasen der Produktentwicklung
In: Zeitschrift für wirtschaftlichen Fabrikbetrieb ZWF. Volume 94, Issue 4, Carl Hanser Verlag, Munich, 1999, p. 186

Koller R (1994)
Konstruktionslehre für den Maschinenbau. Grundlagen zur Neu- und Weiterentwicklung technischer Produkte mit Beispielen
Springer-Verlag, Berlin, 1994

Koppelmann U (1993)
Produktmarketing – Entscheidungsgrundlage für Produktmanager
Edition 4, Springer-Verlag, Berlin, 1993

Kotler P, Bliemel F (1999)
Marketing-Management : Analyse, Planung, Umsetzung und Steuerung,
Edition 9, Schäffer-Poeschl, Stuttgart, 1999

Krubasik EG (1982)
Technologie - strategische Waffe
Wirtschaftswoche, Volume 36, No 25, 1982, p. 30-46

Kruschwitz L (1993)
Investitionsrechnung
Walter de Gruyter Verlag, Berlin, 1993

Kursawe F, Schwefel HP (1998)
Künstliche Evolution als Modell für natürliche Intelligenz
In: Von Gleich A (Editor) Bionik, Ökologische Technik nach dem Vorbild der Natur?
B.G. Teubner, Stuttgart, 1998

Lenk E (1994)
Zur Problematik der technischen Bewertung
Carl Hanser Verlag, Munich et al., 1994

Linde HJ, Hill B (1993)
Erfolgreich erfinden: widerspruchsorientierte Innovationsstrategie für Entwickler und Konstrukteure
Hoppenstedt Technik Tabellen Verlag, 1993

Litke HD (1993)
Projektmanagement: Methoden, Techniken, Verhaltensweisen
Edition 2, Hanser Verlag, Munich, 1993

Little AD (1981)
The Strategic Management of Technology
European Management Forum, Davos, 1981

Livotov P (2002)
TRIZ Computer-Aided Innovation, Technologien für Innovationen
TriSolver Edition, 2002

Mann D (2001)
Ideality And 'Self'
Conference, TRIZ future, Bath, UK, November 2001

Manns JR (1992)
Produktinnovation als Ergebnis der Koordination von F&E und Marketing: Ursachen, Auswirkungen und Lösungsmöglichkeiten, dargestellt am Beispiel eines Unternehmens der mobilen Kommunikation
Verlag Wissenschaft & Praxis, Ludwigsburg et al., 1992

Martino JP (1995)
Research and Development Project Selection
John Wiley & Sons, New York et al., 1995

Mascitelli R (2004)
The Lean Design Guidebook
Edition 1, Technology Perspectives, Northridge, 2004

Mattheck C (1998)
Der Baum hat es schon immer gewusst – Design in der Natur und nach der Natur.
In: Von Gleich A (Editor) (1998) Bionik, Ökologische Technik nach dem Vorbild der Natur? B.G. Teubner, Stuttgart, 1998

Meffert H (1998)
Marketing – Grundlagen marktorientierter Unternehmensführung
Edition 8, Gabler Verlag, Wiesbaden, 1998

Meyer J (1995)
Benchmarking
Schäffer-Poeschel, Stuttgart, 1995

Michel K (1987)
Technologie im strategischen Management – Ein Portfolio-Ansatz zur integrierten Technologie- und Marktplanung
Erich Schmidt Verlag, Berlin, 1987

Micic P (2000)
Der ZukunftsManager: Wie Sie Marktchancen vor Ihren Mitbewerbern erkennen und nutzen
Hauffe, Freiburg (Breisgau), 2000

MMP Management Partner GmbH (1999),
Macht Not erfinderisch? – Innovationen als Motor für Wachstum und Unternehmenserfolg, Umfrage,
Press release, www.Management-Partner.de, 01.11.1999

Mintzberg H (1988)
Opening up the Definition of Strategy
In: Quinn, J. B.; Mintzberg, H.; James, R. M. (Editor): The Strategy Process – Concepts, Contexts and Cases; Englewood Cliffs, 1988, p. 13-20

Mißler-Behr M (1993)
Methoden der Szenarioanalyse
Dissertation, Universität Wiesbaden, DUV (Deutscher Universitäts-Verlag), Wiesbaden, 1993

Möhrle M, Isenmann R (2002)
Technologie Roadmapping Zukunftsstrategien für Technologieunternehmen
Springer-Verlag, Berlin et al., 2002

Müller J (1990)
Arbeitsmethoden der Technikwissenschaften – Systematk, Heuristik, Kreativität
Springer-Verlag, Berlin et al., 1990

Nachtigall W (1986a)
Technische Biologie und Bionik.
In: Von Gleich A (Editor) Bionik, Ökologische Technik nach dem Vorbild der Natur? B.G. Teubner, Stuttgart, 1986

Nachtigall W (1986b)
Biostrategie, Eine Überlebenschance für unsere Zivilisation
dtv Deutscher Taschenbuch Verlag, Munich, 1986

Nachtigall W (1998)
Bionik, Grundlagen und Beispiele für Ingenieure und Naturwissenschaftler
Springer-Verlag, Berlin, 1998

Naisbitt, J (1991)
Megatrends 2000. Ten new directions for the 1990´s
New York, 1991

Opaschowski HW (1997)
Deutschland 2010: Wie wir morgen leben – Voraussagen der Wissenschaft zur Zukunft unserer Gesellschaft
Edition 1, British-American Tobacco, Hamburg, 1997

Ossadnik W (1999)
Planung und Entscheidung
In: Corsten (1999), p. 127-207

Pahl G, Beitz W (1993)
Konstruktionslehre - Methoden und Anwendungen
Edition 3, Springer-Verlag, Berlin et al., 1993

Pannenbäcker T (2001)
Methodisches Erfinden in Unternehmen
Gabler Verlag, Wiesbaden, 2001

Patzak G (1982)
Systemtechnik – Planung komplexer innovativer Systeme: Grundlagen, Methoden, Techniken
Springer-Verlag, Berlin et al., 1982

Pelzer W (1999)
Methodik zur Identifizierung und Nutzung strategischer Technologiepotentiale
Dissertation, RWTH Aachen, Shaker Verlag, Aachen, 1999

Peples W (1999)
Innovationsmanagement, Praktische Betriebswirtschaft
Cornelsen Giradet, Berlin, 1999

Perillieux R (1987)
Der Zeitfaktor im strategischen Technologiemanagement
Erich Schmidt Verlag, Berlin, 1987

Perillieux R (1995)
Technologietiming
In: Zahn (1995), p. 267-284

Petroski H (1992)
The Evolution of Useful Things
Alfred A Knopf, New York, 1992

Petrov V (2001)
The Laws of System Evolution
Tagung, TRIZ future, Bath, UK, November 2001

Pfeifer T (1996)
Qualitätsmanagement – Strategien, Methoden, Techniken
Edition 2, Hanser Verlag, Munich, 1996

Pfeiffer W, Weiß E (1995)
Methoden zur Analyse und Bewertung technologischer Alternativen
In: Zahn E (Editor) Handbuch Technologiemanagement. Schäffer-Poeschel Verlag, Stuttgart, 1995

Pfeiffer W, Metze G, Schneider W, Amler R (1991)
Technologie-Portfolio: zum Management strategischer Zukunftsgeschäftsfelder
Edition 6, Vandenhoeck & Ruprecht, Göttingen, 1991

Pleschak F, Sabisch H (1996)
Innovationsmanagement
Schäffer-Poeschel Verlag, Stuttgart, 1996

Popcorn, F (1995)
The Popcorn Report
Heyne, Munich 1995

Popper KR (1994)
Logik der Forschung
Edition 10, Mohr Verlag, Tübingen, 1994

Porter ME (1986)
Wettbewerbsvorteile: Spitzenleistungen erreichen und behaupten
Campus, Frankfurt, 1986

Porter ME (1996)
What is Strategy?
Harvard Business Review, Volume 74, Issue 6, Nov./Dec. 1996, p. 61-78

Porter ME (1997)
Wettbewerbsstrategie (Competitive Strategy) – Methoden zur Analyse von Branchen und Konkurrenten
German translation: Brandt V und Schwoerer TC, Edition 9, Campus Verlag, Frankfurt/ Main et al., 1997

Prahalad CK, Hamel G (1991)
Nur Kernkompetenzen sichern das Überleben
In: Harvard Business Manager, Volume 13, No 2, 1991, p. 66-78

Probst GJB, Gomez P (1989)
Vernetztes Denken: Unternehmen ganzheitlich führen
Gabler Verlag, Wiesbaden, 1989

Probst GJB, Gomez P (1991)
Vernetztes Denken: ganzheitliches Führen in der Praxis
Gabler Verlag, Wiesbaden, 1991

Pümpin C (1992)
Strategische Erfolgspositionen – Methodik der dynamischen strategischen Unternehmensführung
Verlag Paul Haupt, Bern, 1992

Pümpin C, Prange J (1991)
Management der Unternehmensentwicklung – Phasengerechte Führung und der Umgang mit Krisen
Campus Verlag, Frankfurt/ Main et al., 1991

Reinhart G, Lindemann U, Heinzl J (1996)
Qualitätsmanagement – Ein Kurs für Studium und Praxis
Springer-Verlag, Berlin, 1996

Robens H (1986)
Modell- und Methodengestützte Entscheidungshilfe zur Planung von Produktportfoliostrategien
Campus, Frankfurt/ Main, 1986

Saad KN, Roussel PA, Tiby C (1991)
Management der F&E-Strategie
Gabler Verlag, Wiesbaden, 1991

Sabisch H (1991)
Produktinnovationen
Poeschel Verlag, Stuttgart, 1991

Schelker T (1976)
Problemlösungsmethoden im Innovationsprozess – Ergebnisse einer empirischen Untersuchung
Verlag Paul Haupt, Bern, 1976

Schierenbeck H (1993)
Grundzüge der Betriebswirtschaftslehre
Munich, 1993

Schlicksupp H (1992)
Innovation, Kreativität und Ideenfindung
Edition 4, Vogel Verlag, Würzburg, 1992

Schmidt G (1996)
Organisationsmethodik und -technik
In: Eversheim (1996), p. 3-34 to 3-42

Schmidt R (1996)
Marktorientierte Konzeptfindung für langlebige Gebrauchsgüter
Gabler, Wiesbaden, 1996

Schmitz WJ (1996)
Methodik zur strategischen Planung von Fertigungstechnologien: Ein Beitrag zur Identifizierung und Nutzung von Innovationspotenzialen
Dissertation, RWTH Aachen, Shaker Verlag, 1996

Schramm R, Ludwig J, Töpfer B (1997)
Patentanalyse und Patentstrategie
in: PATINFO 97. Innovationsstimulierung durch Patentinformation, 19. Kolloquium über Patentinformation, Ilmenau, 5.-6.6.1997 Ilmenau: TU Ilmenau, 1997, p. 47-104

Schröder HH (1995)
F&E-Management
In: Corsten (1995), p. 599-614

Schröder A (1998)
Vorentwicklung – Schlüsselfaktor für die Innovationskraft des Unternehmens
Tagungsband Euroforum-Konferenz, Munich, 1998

Schubert B (1991)
Entwicklung von Konzepten für Produktinnovationen mittels Conjoint-Analyse
Poeschel, Stuttgart, 1991

Schuh G, Schwenk U (2001)
Produktkomplexität managen
Carl Hanser Verlag, Munich, Wien, 2001

Schuh G, Wiendahl HP (Editor) (1997)
Komplexität und Agilität
Springer-Verlag, Berlin et al., 1997

Schultz-Wild L, Lutz B (1997)
Industrie vor dem Quantensprung – Eine Zukunft für die Produktion in Deutschland
Springer-Verlag, Berlin et al., 1997

Servatius HG (1985)
Methodik des strategischen Technologie-Managements
Erich Schmidt Verlag, Berlin, 1985

SFB o.V (1998)
Modelle und Methoden zur Integrierten Produkt- und Prozessgestaltung
Arbeits- und Ergebnisbericht 1996-1998 of SFB 361, RWTH Aachen

Sommerlatte T (1997)
Bedürfnisse entdecken
Campus Verlag, Frankfurt/ M., New York, 1997

Specht G, Behrens S, Kahmann J (2000)
Roadmapping – ein Instrument des Technologie managements und der Strategischen Planung,
In: Industriemanagement, Issue 5, 2000, p. 42-46

Spielberg DE (2002)
Methodik zur Konzeptfindung basierend auf technischen Kompetenzen
Dissertation, RWTH Aachen, Shaker Verlag, Aachen 2002

Staudt E (1996)
Innovationsstrategien
In: Eversheim (1996), p. 4-6 to 4-14

Staudt E, Hafkesbrink J, Barthel R (1991)
Neue Techniken im Spannungsfeld alter Systeme – Entscheidungshilfen bei der Einführung von CIM
In: Milling P (Editor) Systemmanagement und Managementsysteme. Duncker & Humblot, Berlin, 1991

Stippel N (1999)
Innovationscontrolling
Vahlen, Munich, 1999

Stummer C (1998)
Projektauswahl im betrieblichen F&E-Management – Ein interaktives Entscheidungsunterstützungssystem
Gabler Verlag, Deutscher Universitäts-Verlag, Wiesbaden, 1998

Terniko J, Zusman A, Zlotin B (1998)
TRIZ – Der Weg zum komkurrenzlosen Erfolgsprodukt
In: Herb R (Editor), Verlag Moderne Industrie, 1998

Teufelsdorfer H, Conrad A (1998)
Kreatives Entwickeln und innovatives Problemlösen mit TRIZ/ TIPS: Einführung in die Methodik und ihre Verknüpfung mit QFD
Publicis-MCD Verlag, Erlangen, 1998

Thom N (1980)
Grundlagen des betrieblichen Innovationsmanagements
Edition 2, Peter Hanstein Verlag, Königstein/ Ts., 1980

TriSolver Group (2002)
TriSolver Ideengenerator & Manager
Professional Edition, Softwareanwendung, Hannover, 2002

Trux W, Müller G, Kirsch W (1985)
Das Management strategischer Programme
Munich, 1985

Tschirky H, Birkenmeier B, Brodbeck H (1996)
Die Handshake-Analysis: eine neue Methode des Technologie- und Innovationsmanagements
In: io Management, Volume 65, No 11, 1996, p. 19-22

Tversky A, Kahneman D (1986)
Judgement under uncertainty: Heuristics and biases
In: Kahnemann (1986), p. 3-20

Ulrich P, Fluri E (1992)
Management: Eine konzentrierte Einführung
Edition 6, Verlag Paul Haupt, Bern 1992

Utterback JM (1994)
Mastering the Dynamics of Innovation – How Companies Can Seize Oppotunities in the Face of Technological Change
Harvard Business School, Boston, 1994

VDA (1998)
Qualitätsmanagement in der Automobilindustrie
Volume 4, Issue 3: Sicherung der Qualität vor Serieneinsatz – Projektplanung, Verband der Automobilindustrie e.V. (VDA), Frankfurt/ Main, April 1998

VDI (1983)
Systematische Produktplanung
VDI-Gesellschaft Konstruktion und Entwicklung (Editor), Edition 2, VDI-Verlag, Düsseldorf, 1983

VDI (1993)
VDI-Richtlinien 2221, Methodik zum Entwickeln und Konstruieren technischer Systeme und Produkte
VDI-Gesellschaft Entwicklung Konstruktion Vertrieb, 1993

VDI (1995)
Wertanalyse: Idee – Methode – System
Zentrum Wertanalyse der VDI-GSP (Editor), Edition 5, VDI-Verlag, Düsseldorf, 1995

Vinkenmeyer R (1999)
Roadmapping als Instrument für strategisches Innovationsmanagement
In: technologie & management,
Volume 48, Issue 3, 1999, p. 18-22

Voegele A (1999)
Das große Handbuch Konstruktions- und Entwicklungsmanagement
Landsberg/ Lech: Verlag Moderne Industrie, 1999

Von Gleich A (1998)
Bionik, Ökologische Technik nach dem Vorbild der Natur?
B.G. Teubner, Stuttgart, 1998

Von Hippel E (1988)
The Sources of Innovation
Oxford University Press, New York, 1988

Von Nitzsch R (1998)
Strategisches Management: Planung und Kontrolle
Vorlesungsskript, RWTH Aachen, 1998

Wagner MH, Thieler W (2001)
Wegweiser für den Erfinder – Von der Aufgabe über die Idee zum Patent
Edition 2, Springer-Verlag, Berlin et al., 2001

Walker R (2002)
Informationssystem für das Technologiemanagement
Dissertation, RWTH Aachen, 2002

Walter W (1997)
Erfolgversprechende Muster für betriebliche Ideenfindungsprozesse.
Dissertation, Universität Karlsruhe, 1997

Warnecke HJ (1997)
Komplexität und Agilität – Gedanken zur Zukunft produzierender Unternehmen
In: Schuh (1997), p. 1-8

Weber K (1993)
Mehrkriterielle Entscheidungen
R. Oldenbourg Verlag, Munich et al., 1993

Webster JL, Reif WE, Bracker JS (1989)
The Manager´s Guide to Strategic Planning Tools an Techniques
In: Planning Review, Volume 17, No 6, November/December 1989, p. 4-13

Weis HC (1997)
Marktforschung
Kiehl, Ludwigshafen, 1997

Weiss PA (1992)
Die Kompetenz von Systemanbietern
Springer-Verlag, Berlin et al., 1992

Wengler MM (1996)
Methodik für die Qualitätsplanung und –verbesserung in der Keramikindustrie – Ein Beitrag zum Qualitätsmanagement bei der Planung neuer und der Optimierung bestehender Prozesse
Dissertation, RWTH Aachen, VDI Verlag, Düsseldorf, 1996

Werners B (1993)
Unterstützung der strategischen Technologienplanung durch wissensbasierte Systeme
Habilitation, RWTH Aachen, 1993

Westkämper E (1987)
Strategische Investitionsplanung mit Hilfe eines Technologiekalenders
In: Wildemann, H (1987)

Wildemann H (1987)
Strategische Investitionsplanung: Methoden zur Bewertung neuer Produktionstechnologien
Gabler, Wiesbaden, 1987

Wöhe G (2000)
Einführung in die allgemeine Betriebswirtschaftslehre
Edition 20, in collaboration with Ulrich Döring, Verlag Franz Vahlen, Munich, 2000

Wolfrum B (1992)
Grundgedanke, Formen und Aussagewert von Technologieportfolios (I, II). In: WISU, No 4, 1992, p. 312-320 and No 5, p. 403-407

Wolfrum B (1994)
Strategisches Technologiemanagement
Gabler Verlag, Wiesbaden, 1994

Wolfsteiner WD (1995)
Das Management der Kernfähigkeiten – Ein ressourcenorientierter Strategie- und Strukturansatz
Dissertation, Hochschule St. Gallen, 1995

Wüstenberg D (1998)
Kreativität bei der Konstruktion von Maschinen
In: Von Gleich A (Editor) Bionik, Ökologische Technik nach dem Vorbild der Natur? B.G. Teubner, Stuttgart, 1998

Zadeh LA (1965)
Fuzzy Sets
In: Information and Control, No 8, 1965, p. 338-353

Zahn E (Editor) (1995)
Handbuch Technologiemanagement
Schäffer-Poeschel Verlag, Stuttgart, 1995

Zahn E, Weidler A (1992)
Integriertes Innovationsmanagement: Die Zukunft wird im Kopf gewonnen
In: Gablers Magazin, No 10, 1992, p. 17-23

Zehnder T (1997)
Kompetenzbasierte Technologieplanung – Analyse und Bewertung technologischer Fähigkeiten im Unternehmen
Dissertation, Hochschule St. Gallen, Gabler Verlag, Wiesbaden, 1997

Zimmermann HJ (1993)
Fuzzy Technologien – Prinzipien, Werkzeuge, Potentiale
VDI-Verlag, Düsseldorf, 1993

Appendix

Appendix A
Method Data Sheets (Gerhards 2002)

Anne Gerhards

● well suited
◐ suited

Methods and tools	1. Goal setting	2. Future analysis	3. Idea generation	4. Idea evaluation	5. Idea detailing	6. Concept evaluation	7. Implementation planning
	Phases of the innovation process						
20 Cost-Levers Method	◐						◐
Abstraction			●				
Affinity diagram						●	
Analogical consideration		●	●		●		
Analysis of orders and offers	●	●					
Balanced Scorecard	●						
Benchmarking	●	●	●	●	●	●	●
Bionic			●		●		
Brainstorming		●	●		●		
Business area profile							●
Chance-Risk portfolio				●		●	
Competitive analysis	●	●					
Component tree					●		●
Conjoint analysis						●	
Core competency analysis	●						
Cost-benefit analysis				●		●	
Cost of poor quality calculation	◐					●	
Customer process monitoring		●	●				
Delphi method, The	●	●	●	◐	●	◐	
Design best-practice guideline			●		◐		
Design review	◐	◐		●		●	
Economic analysis						●	
Efficiency analysis				●		●	
Experience curve analysis	●	●				●	
Field of force analysis		●		●		●	●
FMEA				●		●	●
Force-Fit method		●	●		●		
Function analysis			●		●		
Function modeling			●				
Guidance questions			●				
Ideality			●				
Innovation checklist				●		●	
Innovation road map							●
Ishikawa diagram			●		●		
Kano diagram	●	●					
KO criteria catalog				●		●	
Lead-User concept					●		
Lean development	●						●
Lean QFD	●	●		●		◐	

● well suited
◐ suited

Methods and tools	Phases of the innovation process						
	1. Goal setting	2. Future analysis	3. Idea generation	4. Idea evaluation	5. Idea detailing	6. Concept evaluation	7. Implementation planning
Matrix of priorities			◐	●		●	
Metaplan technique				●		◐	
Method 635		◐	●		●		
Method for concept selection				●		●	
Mind mapping			●		●		
Module optimization checklist	●	●	●			●	
Morphological matrix			●		●		
Must/should/could prioritization	◐	●		●		●	●
Neighbor field integration		●	●		●		
Patent analysis	●	●					
Pairwise comparison			●		●		●
Platform plan	●	●					●
Plan for information procurement	●	●	●	●	●	●	●
Polarity profile		●	●		●		
Portfolio analysis				●		●	
Problem solution tree			●		●		
Product potential evaluation				●		●	
QFD			●	◐	●	◐	
Quick-Look value engineering event				●		●	
R&D portfolio				●		●	
Requirement list					●	●	
Risk analysis				●			●
Scenario management		●					
Selective list		●	●		●		●
SIS-Method		●	●		●		
Strategy definition of the innovative strategy	●						●
Success factor portfolio	●						
SWOT Analysis	●	●					
Synectics		◐	●		●		
Target costing					●	●	
Technology calendar						●	
Technology tree	●	●	●				
TILMAG-Method		●	●		●		
Trend extrapolation		●					
TRIZ-Contradiction matrix			●		●		
Value analysis				◐		●	
Weighted point evaluation				●		●	

Twenty-Cost-Levers Method

Goal/ Result	Reducing the costs with consideration of frequent dependences 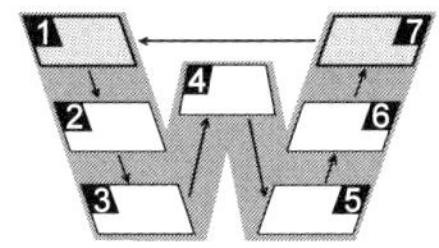
Informa-tion input	• Project ideas
Content/ Procedure	The Twenty-cost-levers method serves cost reduction. Here, the particular "levers" are weighted for each turning knob and are evaluated in relation to a reference concept.

Pros	Cons
• Consideration of several, also indirect, influencing variables	• Not suitable for application everywhere

Literature/ Reference	• Mascitelli 2004
Notes:	

Abstraction

Goal/ Result	Concentration on essential / functional aspects 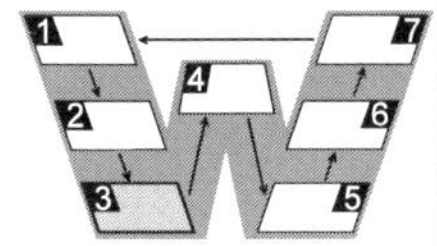
Informa-tion input	• Setting of tasks, problem
Content/ Procedure	Abstraction serves the focus on important facts. Here, the examined problems are first reduced to quintessential points and details are neglected. This way a concentration on the functional aspects is reached.

Pros	Cons
• Creates new solution methods • Creation of analogies to other disciplines	• Not particularly integrative

Literature/ Reference	• Pahl, Beitz 1993 Ehrlenspiel 2003
Notes:	

Affinity Diagram

S. Shiba

Goal/ Result	Structuring and classification of a large number of ideas (e.g. from brainstorming meeting) ⇒ Establishment of priorities 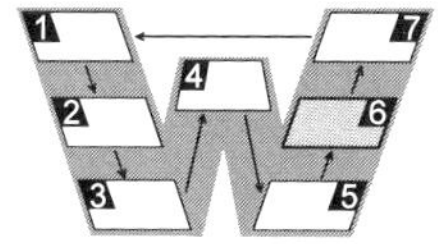
Information input	• Solution ideas • Detailed solutions
Content/ Procedure	All ideas are written on separate sheets of paper. These notes are attached to a wall. An attempt is made to find generic terms for different ideas. This procedure is continued until all ideas are divided into groups and the groups into further superior groupings. A structural tree develops, in which the ideas and their relations among one another are illustrated.

Pros	Cons
• Classification of ideas • 1st selection of ideas	• Sole main solutions can be misjudged in their importance, since they are possibly subordinated to another group

Literature/ Reference	• Brassard 1994
Notes:	

Analogical Consideration	S. Shiba
Goal/ Result	Solution of technical problems with the examination of examples, outside of the subject area. ⇒ Future projections, ideas and a suggestion for a solution 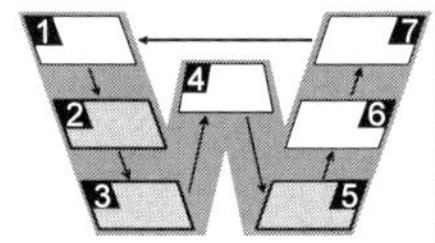
Information input	• Trends • Innovation tasks
Content/ Procedure	The analogical consideration is an creativity technique, which is organized in groups. Specified are first of all, the volitional characteristics and functions of the technical approach area. Afterwards models are needed, which represent similar characteristics or functions. The system which offers those characteristics or functions is examined. The fungibility of the effectiveness is examined afterwards.

Pros	Cons
• If those examples are discovered the solution is progressed already very far.	• Certain restriction during the trouble-shooting (problem solution), since only examples or something similar is examined.

Literature/ Reference	• Haberfellner et al. 1999 • Bionic, TRIZ
Notes:	

Analysis of Orders and Offers

Goal/ Result

Clear picture of the connections between the enterprise and the customers

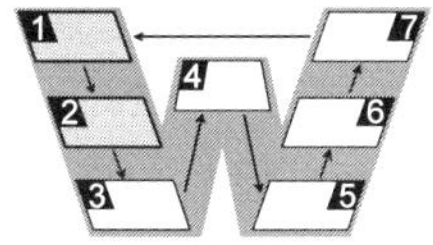

Information input

- Offers, orders
- Sales statistics (if available)

Content/ Procedure

The analysis of orders and offers serves for pointing out connections in the customer interaction of an enterprise. Here first of all the periods under consideration are specified, as well as the characteristics of the orders and offers which are to be considered. This is followed by the procurement of the data and their systematic evaluation. On that basis the order and offer ratio and the proportional distribution of the sold/offered options can be calculated. Now the results, measured on the basis of the implementation rate of the options, can be interpreted and customer requirements and customer value can be defined more exactly.

Pros	Cons
• Intuitive feasibility • Simple integration of additional parameters	• Only referring to past

Literature/ Reference

Notes:

Balanced Scorecard

Goal/ Result

Balanced goal setting with the four perspectives finances, customer, processes, learning/ developing

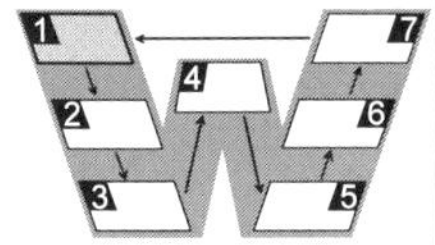

Information input

- Strategic goals
- Performance indicators (e.g. net yield increases, also qualitative ratios such as image increase etc.)

Content/ Procedure

The Balanced Score Card is a tool for goal setting. First, a cause-and-effect chain of strategic objectives is built up. Subsequently, the strategic objectives are combined with indicators and operational objectives and from that the measures to be implemented are defined.

Pros	Cons
• Strategy is operationalized, represented and communicable • Complexity reduction	• Danger of defining too many and too complex goals • Manipulation of characteristic numbers

Literature/ Reference

Notes:

Prognosis

Analysis

Creativity

Problem solving

Evaluation

Benchmarking	S. Shiba
Goal/ Result	Determination of the aspired specification of methods, practices, products ⇒ Information system for the measurement of the achievement of objectives. 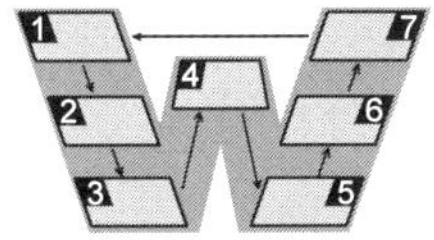
Information input	• Performance requirements • Problems
Content/ Procedure	Benchmarking is an continuous and systematic process for the determination of outstanding methods and activities, which allow best performance. During the procedure, entrepreneurs are analyzed and the specification of particular measurement categories are compared with the measurement categories of the own company. With the comparison method of the activities with those, which are controlled by other enterprises excellently, market-oriented and own realistic target settings must be determined for the own company. Also the way for success must be defined.

Pros	Cons
• Progression of the competitiveness • Knowledge about competitive behaviour	• Insecure definition of the class winner • Difficult analysis of the competitors business process

Literature/ Reference	• Meyer 1995, Eversheim, Schuh 1996, Kamiske, Brauer 1993
Notes:	

Bionic

J.E. Steel

Goal/ Result

Transmission of biological structures, mechanisms and systems on technical solutions
⇒ Ideas and suggestions

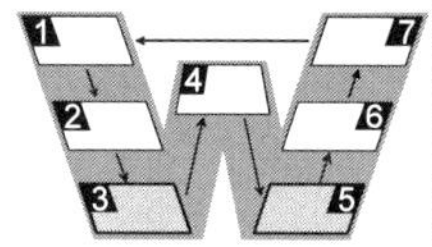

Information input

- Problems

Content/ Procedure

In the ideal case an interdisciplinary team of technical and biological range is arranged. For the processed functions equivalent functions are searched in the nature. It is particular analyzed how functions are solved in the nature. Next step is to transfer the solution to technical problem definitions.

Pros	Cons
• Identification of unusual and new solutions	• Difficulties in the composition of interdisciplinary teams • Knowledge in natural science is required

Literature/ Reference

- Brauchlin 1995, Schlicksupp 1992, Ehrlenspiel 1995
- TRIZ-Method

Notes:

Brainstorming

A. Osborn

Goal/ Result

Brainstorming and solution determination with creativity advancement, respectively.
⇒ Innovation potential, ideas and proposed solutions

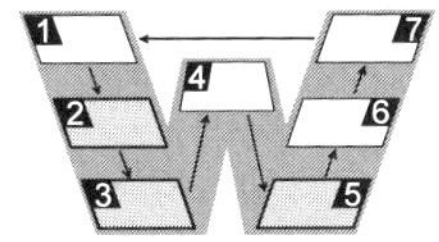

Information input

- Innovation tasks
- Trends

Content/ Procedure

Brainstorming is a group creativity technique that was designed to generate a large number of ideas for the solution of a problem. Brainstorming is a group technique which includes 6 to 12 persons which cooperate in an interdisciplinary way. During the first phase – the creativity phase – each person of a team expresses all ideas, which he has for the defined keyword. The suggestions may not be criticized by the other members of the team. All ideas should receive attention, whereby the quality of the ideas should be the centre of attention. In the second phase all suggestions are structured and evaluated together by a team.

Pros	Cons
• Finding innovative suggestions with leaving conventional patterns of thinking • Creativity promotion of the team members	• Complex procedure by the evaluation of all suggestions • Strong personalities are dominating the group

Literature/ Reference

- Kamiske, Brauer 1993, Eversheim, Schuh 1996, Haberfellner et al.: 1999, Schlicksupp: 1992
- Brainwriting, Discussion 66, Synectics

Notes:

Business Area Profile

Goal/ Result	Documentation of the relevant information on market, market performance, competition, competencies, economic goals and necessary measures 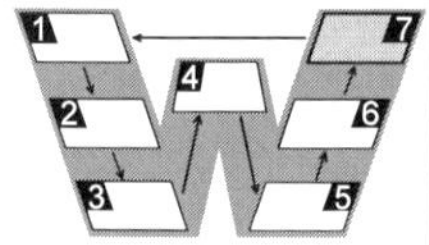
Information input	• Numbers, data, facts - both internal and external
Content/ Procedure	The business area profile serves the complete collection of relevant information. To this end the data on market and market performance is first completed. Subsequently, the success factors and characteristic data of the competition are measured. On the basis of the evaluation of the own possibilities and competencies goals and measures are defined as a last step.
Literature	• Gausemeier, Ebbesmeyer, Kallmeyer 2001
Notes:	

Pros	Cons
• Structured data collection • Comparability of the data with different ideas	• Assessing the required data is not very easy

Prognosis
Analysis
Creativity
Problem solving
Evaluation

Chance-Risk Portfolio

Goal/ Result	Comparison of chances and risks of individual ideas
Informa-tion input	• Within the company: Ideas, strengths, weaknesses • External: Chances, risks
Content/ Procedure	The chance-risk portfolio serves the comparison of idea-specific chances and risks. Here the most important chances and risks are designated to each idea, documented and then evaluated. After that the ideas are entered in the portfolio according to the results.

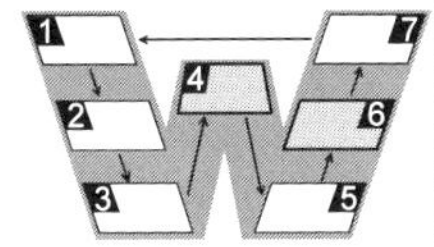

Pros	Cons
• Very intuitive variant to set chances and risks in correlation	• Classification very subjective • Low comparability of the respective chances and risks

Literature/ Reference	• Gausemeier, Fink 1999
Notes:	

Prognosis
Analysis
Creativity
Problem solving
Evaluation

Competitive Analysis

Goal/ Result

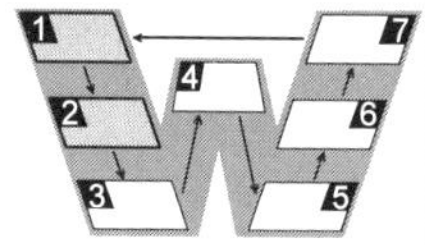

Systematic examination of the competitive environment
⇒ Clear picture of the current competition situation and the importance of the various specific competitive forces

Information input

- Information on the five competitive forces (Porter)

Content/ Procedure

The competitive analysis serves systematical examination of the company's competitive environment. The initial aim is the analysis of the decisive factors of each competitive force. These factors include particularly the threat posed by substitute products and new entrants, as well as the bargaining power of suppliers and customers. After that, the determined factors are weighted.

Pros	Cons
• Structured consideration of all relevant competitive factors in a specific market	• Markets with high competitive dynamics (Schumpeter competition) are not easily ascertainable • No consideration of possible cooperation

Literature/ Reference

- Porter 1996
- EFA- Assessment Tool

Notes:

Component Tree

Goal/ Result

Representation of component structure of technical systems
⇒ General structure of components

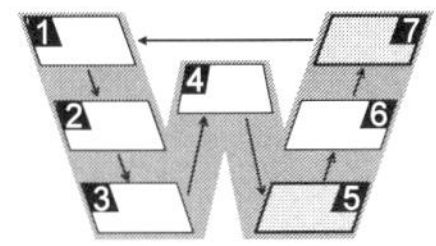

Information input

- Innovation tasks
- Detail solutions

Content/ Procedure

A component tree is responsible for the representation of the system structure. Outgoing from the total system in an iterative proceeding each component of the system is tested for its function and its expenditure input and output. After the analysis and description of further requirements (achievement goals, site conditions) is the component illustrated in the component tree. If necessary the component is divided into further partial components and the analysis is iterated until the necessary grade of detail of the component tree is reached.

Pros	Cons
• Clear representation of the components of a system • Several levels of detail for different purposes	• Risk of confusing information with to many components

Literature/ Reference

- Pfeifer 1996
- Fault tree

Notes:

Conjoint Analysis

Goal/ Result	Determination of the total utility of a product ⇒ Evaluation and prioritization of product characteristics 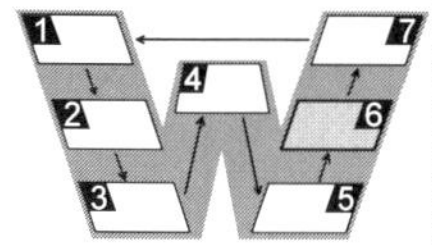
Informa-tion input	• Product characteristics and their distinctness
Content/ Procedure	The conjoint analysis serves to estimate the acceptance of a product and its functions by the customer. During the process it is assumed that the total utility of a product consists additively of the utility of the individual product components. Initially, all important product characteristics and their possible distinctness are determined. Several concepts are realized by combination of different grades of distinctness of the different characteristics. These concepts are presented for evaluation to the potential customer, who arranges the concepts according to their preferences. The number of product characteristics must be planned carefully, so that mathematically the underlying evaluation of individual characteristic distinctness can be determined from the indicated order of the concepts.
	Pros: • Prioritization of individual product characteristics • support by evaluation programs **Cons**: • High expenditure with complex products • Duration up to weeks and months
Literature/ Reference	• Eversheim, Schuh 1996 • Value analysis
Notes:	

Core Competency Analysis

Goal/ Result	Identification of the core competencies ⇒ Clear picture of the core competencies
Informa-tion input	• Structured overview of own company's skills and those of the main competitor.
Content/ Procedure	The core competency analysis is used to identify skills within a company and check their suitability as core competencies. As a first step, the specific skills needed for the company's product development are identified. The skills with customer value are classified as competencies. The competencies are then evaluated regarding how well they are mastered and their significance for customer value. Those competencies which are well mastered and have a high significance are classified as core competencies.

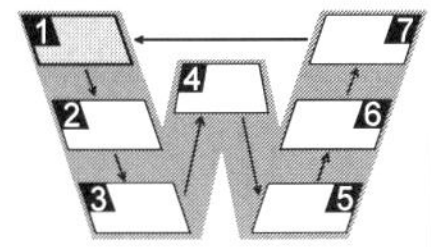

Pros	Cons
• Overview of economically relevant skills within the company	• Hard to measure

Literature/ Reference	• EFA-Assessment Tool
Notes:	

Cost-Benefit Analysis

Goal/ Result

Evaluation of solutions concerning cost-potential
⇒ Cost-benefit calculation and selection of solutions

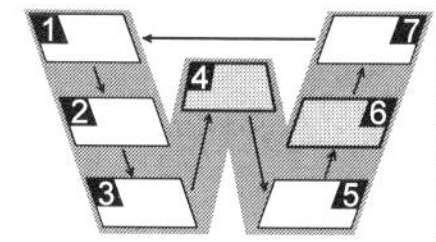

Information input

- Alternatives
- Analogical solutions with well known costs

Content/ Procedure

The cost-benefit analysis (CBA) is a process of weighting the total expected costs against the total expected benefits of one or more actions in order to choose the best or most profitable option. Everything is made within a team. Different methods can be used like the cost estimation method. Beside direct costs personnel costs and/or similar costs are considered also. Acquired costs and benefit are compared. The solution which shows the best relationship is selected.

Pros	Cons
• Direct flowing of the financial aspect into the solution • Early consideration of the cost aspect of solution	• Qualitatively good solutions with an unfavorable relationship are not considered • Inaccurate results due to difficult predictable market tendencies

Literature/ Reference

- Haberfellner 1999, Horvath 1994
- Cost-performance analysis

Notes:

Cost of Poor Quality Calculation

Goal/ Result

Consideration whether to invest in improvement of a process or in checking and repairing

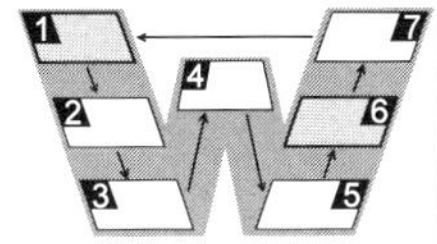

Information input

- Data and estimates on various cost factors

Content/ Procedure

The cost of poor quality calculation is a decision-making aid. First the cost factors of the regarded process are determined. Subsequently, Cx is determined. An investment in a process improvement is recommendable, if Cx < 0.

Pros	Cons
• Simply accomplishable decision-making aid	• Difficult to accomplish with complicated and extensive processes

Literature

- Mascitelli 2004

Notes:

Customer Process Monitoring

Goal/ Result

- Particularly identification with regard to latent customer needs

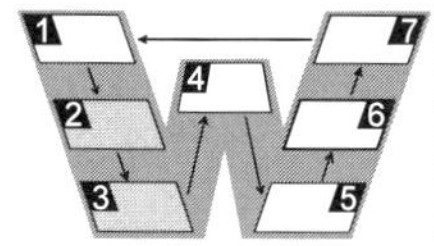

Informa-tion input

- Exact definition of the target group

Content/ Procedure

Customer Process Monitoring (CPM): The main aim of CPM is the coverage of customer needs. During the method a specific process is analyzed in order to discover improvement opportunities for the product application. Transferred to the change analysis the operator has the possibility to analyze the customer behavior under influence of the defined future acceptance to recognize problems which are created for the customer in handling the future situation.

Pros	Cons
• Possibility to discover new latent customer needs • Creation of customer understanding	• To some extent very complex • A good contact to the customer is necessary

Literature/ Reference

- Schröder 1998

Notes:

Delphi Method

H.L. Geschka

Goal/ Result	Obtaining forecasts ⇒ Market forecasts and technology forecasts
Informa-tion input	• Problems • Search fields
Content/ Procedure	The Delphi method is a systematic interactive fore-casting method based on independent inputs of selected experts. Approach: 1) Questioning of internal and external experts concer-ning future trends in an subject area. 2) Analysis and evaluation of all expert opinions. 3) Summary of opinions. 4) Repeated questioning of all experts after the results of the preceding questioning were round made. 5) Accomplishment of further iterationsuntil all experts found a consent.

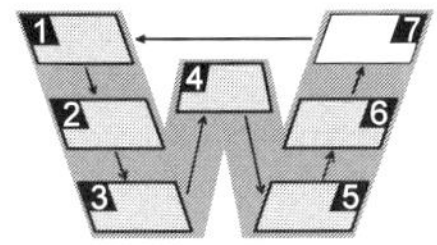

Pros	Cons
• No mutual influence of the respondents • Unlimited number of the team expert group • Use of distributed expert knowledge	• High expenditure of time due to the questioning of external experts and multiple compensation of opinions.

Literature/ Reference	• Haberfellner et al. 1999 • Questionnaire technique, survey
Notes:	

Design Best-Practice Guideline

Goal/ Result	Making successful concepts and drafts available
Informa-tion input	• Concepts/drafts recognized as successful
Content/ Procedure	The Best Practice Development Guide serves the development of knowledge management. As a first step after the conclusion of a project outstanding concepts are collected and recorded on a form. Then the information is entered into a data base, which supplies standard recommendations for the concentration on novelties for upcoming projects.

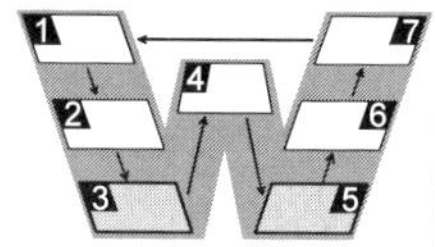

Pros	Cons
• Good overview and easy information access due to data base form	• Expenditure of the data base care and data collection

Literature/ Reference	• Mascitelli 2004
Notes:	

Design Review

Goal/ Result

Early recognition of weak points
⇒ Choice of concept

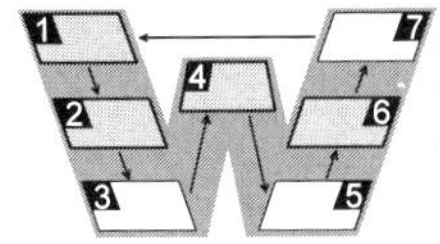

Information input

- Concepts

Content/ Procedure

For the completion of concept phases and measures all obtained results are examined on the fulfillment of all requirements and on potential errors by using checklist by interdivisional cooperated project-distant teams. Based on general questionnaires, which are updated in accordance to the experience level, the checklists are provided by the team members. The results are documented. For located errors the necessary changes and/or measures are arranged.

Pros	Cons
• Use of experiences with existing products • Avoiding errors before the implementation • Higher possibility of discovery via project external participants	• Only given requirements and defined questions are examined

Literature/ Reference

- Pfeifer 1996, Reinhardt 1996
- Quality survey

Notes:

Economic Efficiency Analysis

Goal/ Result	Estimation of the profitability of projects 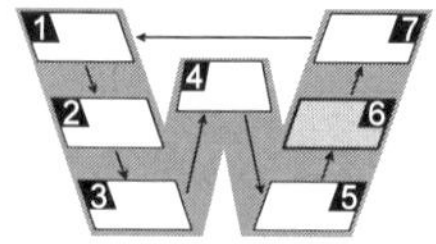
Information input	• Financial requirement • Utility of the project
Content/ Procedure	The economic efficiency analysis serves to examine profitability. First a complete financial plan is drawn up, which gives a project-status curve as an intermediate result. Finally, a cost-benefit analysis is carried out.

Pros	Cons
• Quick information on profitability of the project	• Lack of possibilities for adaptation to changed development conditions

Literature/ Reference	• EFA Assessment Tool
Notes:	

Efficiency Analysis

Goal/ Result

Identification of the best of multiple alternatives (in order to estimate project costs)
⇒ Efficiency of an alternative, Ranking of alternatives

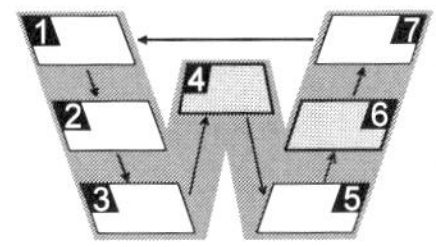

Information input

- Effort
- Complexity
- Resource need

Content/ Procedure

During this procedure a system of those factors which influence the effort crucially has to be provided. These factors have to be evaluated objectively or subjectively. This evaluation is based on adequate value parameters in order to yield the expected total expenditure via predetermined mathematical correlation. During a procedure a qualitative mathematical evaluation is necessary. The described problems should be considered and according to this dispositions must be made (for example divided appraisal of project items). Only in this case an error recognition and adaption from level to level is possible.

Pros	Cons
• Quick information about approximate project costs	• Localization of errors is barely possible • Barely possibilities of adaption to changed terms of development

Literature/ Reference

- Litke 1993
- Short calculation with similarity relations

Notes:

Experience Curve Analysis *Boston Consulting Group*

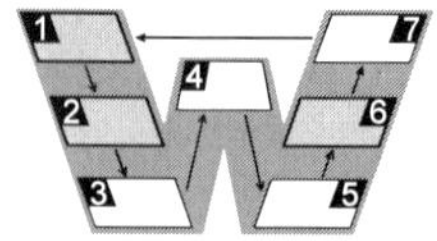

Goal/ Result	Calculation of price trends and cost trends ⇒ Height of cost reduction potential, intensity of market growth, profit potential
Informa-tion input	• Manufacturing costs • Experiences
Content/ Procedure	The experience curve effect states that the real average cost of a product is falling by a constant and predictable percentage (potentially 20-30%) each time when cumulative volume doubles. Thereby it is assumed that all cost reduction potentials are used consequently (learning effects, enterprise size and batch size declining effects). Thus predictions can be made about future cost trends and price trends as well as about profit potentials. According to this the experience analysis is very important for market shares and market growth and also for price policy design.

Pros	Cons
• The prognosis is long-term oriented • Forecasts of the price-political range of competitors and their market growth	• Complex measuring of experiences

Literature/ Reference	• Meffert 1998
Notes:	

Prognosis | **Analysis** | Creativity | Problem solving | Evaluation

Field of Force Analysis

K. Lewin

Goal/ Result

Determination of supporting and obstructing influencing factors on a problem solution
⇒ Positive and negative influence factors

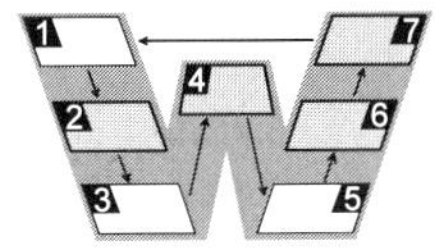

Information input

- Trends, potentials
- Tasks, problems

Content/ Procedure

Force field analysis is a useful technique for looking at all forces for and against a decision. If necessary forces and influences can be weighted depending on strength and they can be weighted into a row.

Pros	Cons
• Structured overview of influencing factors • Comparison of the problem solution power of different approaches to a problem	• Very unclear by complex problems

Literature/ Reference

- Brassard, Ritter 1994, Higgins, Wiese 1996

Notes:

Failure Mode and Effects Analysis and Eco-FMEA

Goal/ Result

Systematic collection of possible errors and estimation of the associated risks

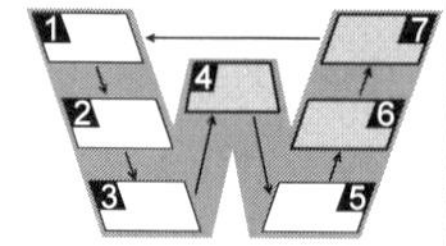

Informa-tion input

- Product data
- Data on error frequency, consequences and discovery in the past

Content/ Procedure

The FMEA serves the collection of possible errors and estimation of their risks. Here, first the probability of occurrence A of an error, its significance B and the probability of discovery E are evaluated using point values of 1 (very low) to 10 (very high). Subsequently, the risk priority number (RPN) is calculated according to the formula: RPN= A x B x E. Finally, the RPN is examined regarding affiliation to the following categories:

RPN < 50 : Error uncritical
RPN > 200: Error critical
50 < RPN < 200: situational consideration and decision.

The "Eco-FMEA" also includes ecological criteria.

Pros	Cons
• Error prevention less expensive than error correction • Preventive quality assurance	• High expenditure of time

Literature

- DIN EN 60812: Analysetechniken für die Funktionsfähigkeit von Systemen (FMEA), 2006

Notes:

Prognosis
Analysis
Creativity
Problem solving
Evaluation

Force-Fit Method

Goal/ Result

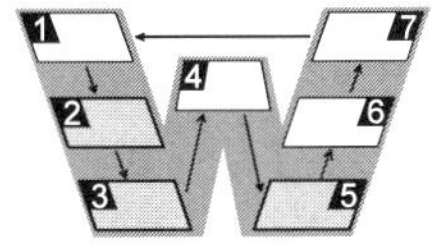

Generation of new solution ideas by bringing together two different terms via creative though processes

⇒ Ideas and suggestions for a solution

Information input

- Innovation tasks

Content/ Procedure

The method application takes place as a game with which the team is divided in two arrays. One array gives to the other one one stimulus-word which is far as possible from the problem. The array must compile a solution in a specific time window. For each approach different prints are given. The developed solutions are examined afterwards to its problem solution potential.

Pros	Cons
• Development of unusual solutions and solution combinations	• High expenditure of time • Verification and examination of results and their problem solution potential is necessary • Team has to get involved in game situation

Literature/ Reference

- Brauchlin 1995, Schlicksupp 1992
- Stimulus-word analysis

Notes:

Function analysis

Goal/ Result

Consideration of many function fulfillment possibilities with the selection of different product functions

⇒ Ideas and suggestions for a solution for product functions

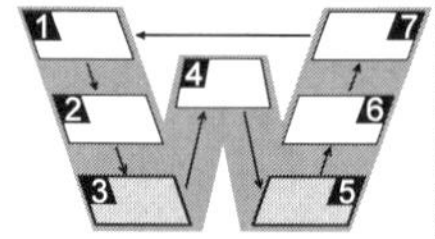

Information input

- Innovation tasks

Content/ Procedure

The function analysis can be used individually as well as in teams. It is comparable with the application of the morphologic matrix. In the matrix columns the product functions are specified, and not the possible product attributes. For each function alternative principles are offered in a list. From this list the best operational principle must be selected.

Pros	Cons
• Identification of multiple solutions • Structured procedure	• Restriction of free brainstorming and creativity as a result of technical emphasis • Difficulties in the integration of the different technical solutions for unique functions

Literature/ Reference

- Bugdahl 1990, Brauchlin 1995, Schlicksupp 1992
- Morphological matrix

Notes:

Function Modeling and/or Function Structure

Goal/ Result	Structuring of complex systems into solution-neutral sub-functions 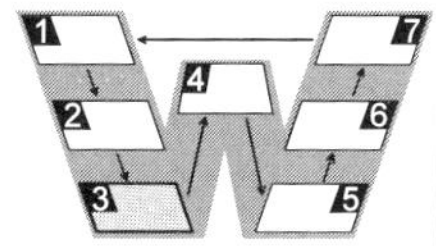
Information input	• Overall function
Content/ Procedure	Function modeling serves the structuring of complex systems. Here the functions are first divided into sub-functions and connections are visualized. Subsequently, the sub-functions are further sub-divided until the desired detail level is reached. Above all, attention is to be paid to a solution-neutral formulation.

Pros	Cons
• Creates new solution channels • Creates analogies with other disciplines	• Not particularly integrative

Literature	• Pahl. Beitz 1993 Ehrlenspiel 2003
Notes:	

Prognosis

Analysis

Creativity

Problem solving

Evaluation

Guidance Questions

Goal/ Result

Stimulation and focus of the participants' creativity

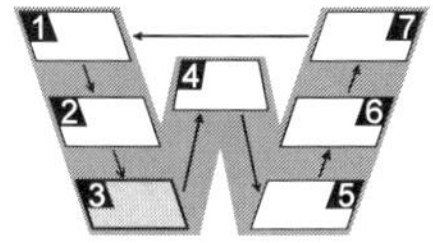

Informa-tion input

- Setting of tasks

Content/ Procedure

Guidance questions serve the stimulation and focus of creativity. As a first step, project-specific guidance questions are prepared by the moderator. The individual guidance questions are written down clearly or projected by a video projector and described. After that the participants answer the guidance questions on Metaplan cards, which are then all read out, explained and filed. Finally the answers to different guidance questions are combined and evaluated.

Pros	Cons
• Can be accomplished intuitively and fast	• Subjectivity on the part of the moderator can be very high

Literature/ Reference

- EFA Assessment Tool

Notes:

Ideality

Goal/ Result

Support of the creative processes by purposeful idea generation

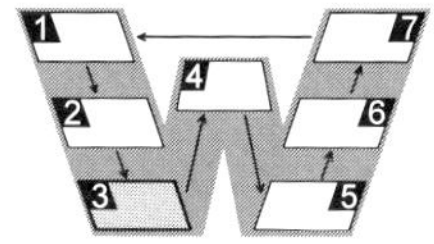

Information input

- Characteristics of an ideal product and/or task

Content/ Procedure

Ideality serves the support of creative processes. First the characteristics of an ideal product are described, followed by a search for potential solutions which lead towards the ideal product.

Pros	Cons
• Very simple to carry out • Structuring aid during the creative process	• Does not always deliver very useful results

Literature/ Reference

- EFA Assessment Tool

Notes:

Innovation Checklist

Goal/ Result

Overcoming an innovation problem by its precise description

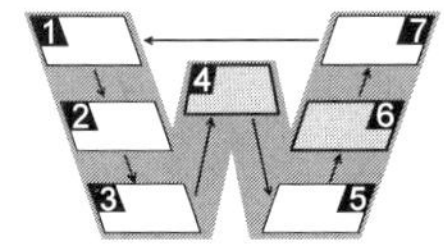

Information input

- Information about the system to be improved and its environment

Content/ Procedure

The innovation checklist serves as an aid to overcoming innovation problems. Here, six core aspects are to be answered by discussion and Metaplan technique. On the one hand is the collection of information about the system to be improved and its environment, followed by the analysis of the system environment for the identification and utilization of available resources. Then information is collected on the problem situation including the history of its origins.

Pros	Cons
• Systematic approach • Pointing out potential for innovative solution approaches	• Partly high expenditure during data collection

Literature/ Reference

- EFA Assessment Tool

Notes:

InnovationRoadMap

Fraunhofer IPT

Goal/ Result

Temporal structuring of the planning results in the innovation process
⇒ Planning table innovation activities

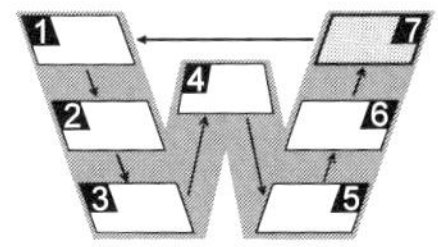

Information input

- Temporal classification of future projections
- Innovation tasks, concepts, detailed solutions

Content/ Procedure

The InnovationRoadMap is a structured method for innovation process results. It is stretched by the axes market, technology and time and their sub ranges market and technology. The technology section is divided into two sections. The product concepts are integrated in the upper range, and the detail solutions in the lower range. Before that the classification of the future projections and innovation tasks as well as the produced product concepts are chronologically structured according to their negotiability.

Pros	Cons
• Integration of strategic and operational levels • Visual representation of planning activities • Long-term controller	• Representation of several levels is difficult • A lot of preliminary work is necessary

Literature/ Reference

- Brandenburg 2001
- Roadmapping

Notes:

Ishikawa-Diagram

K. Ishikawa

Goal/ Result

The Ishikawa diagram shows all the main and secondary aspects which influence a problem
⇒ Cause coherences and effect coherences

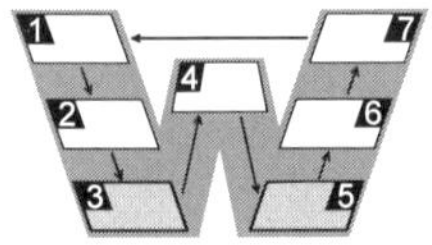

Informa-tion input

- Problems
- Solution concepts

Content/ Procedure

All potential and well-known causes which lead to a specific problem are divided top-down into main and secondary causes. Representation of causes is made by a fishbone-diagram. The determined effect (i.e. the problem) is shown down right. Individual causes are arranged vertically staggered along the main effect line. Thus it is possible to show dependency of influencing variables on the target variable and to identify positive and negative influencing variables.

Pros	Cons
• Clear visual representation of problem structure • Structured consideration of all causes	• Very complex with complex structures

Literature/ Reference

- Brassard, Ritter 1994, Higgins, Wiese 1996
- Cause-matrix, correlation diagram

Notes:

Kano Diagram

Goal/ Result

Illustration of customer behavior

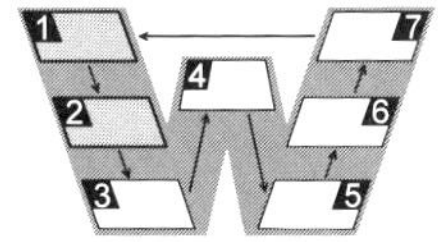

Information input

- Empirical values regarding the customer requirements
- Market observations

Content/ Procedure

The Kano diagram serves the illustration of customer behavior. Initially, three attributes classes are formed (enthusiasm, achievement, threshold attributes). The achieved customer satisfaction is then set in relation to the competitors and the results are analyzed.

Pros	Cons
• Representation of the temporal development of the customer needs	• No accurate values, temporal development only schematically representable

Literature/ Reference

- Gausemeier,Ebbesmeyer,Kallmeyer 2001

Notes:

K.O.-Criteria Catalogue

Goal/ Result	Early sorting out of unsuitable concepts on the basis of K.O. - Criteria
Informa-tion input	• Valuation criteria
Content/ Procedure	The K.O.-criteria catalogue serves the early sorting out of unsuitable concepts. First the criteria are selected which can reach critical values if they exceed or fall below certain parameters. The K.O.-criteria are defined by setting a lower or upper limit of the value of a specific characteristic. Finally each new product idea goes through the K.O.-criteria and is examined regarding its fulfillment thereof.
Literature/ Reference	
Notes:	

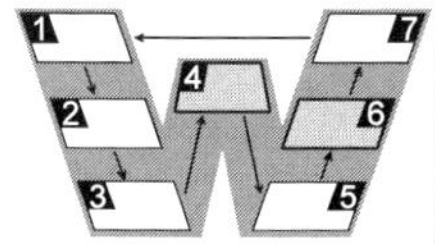

Pros	Cons
• Completely unsuitable ideas are quickly rejected at little expenditure	• Criteria usually only rough estimation • Typically not very detailed

Lead User Concept

Goal/ Result	Customer oriented product development and process development ⇒ Ideas and proposed solutions from experts 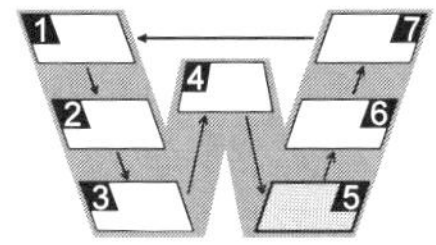
Information input	• Trends • Problems • Solution concepts
Content/ Procedure	With the lead user concept highly qualified experts are as early as possible involved in the development process. First of all lead user have to be identified, i.e. experts for trends, problem definition or application of products. These experts are involved in the development process, e.g. via workshops. A following acceptance examination of ideas is meaningful to average customers.
Literature/ Reference	• Herstatt 1998, von Hippel 1988
Notes:	

Pros	Cons
• Early involvement of experts • Constant customer contact during development	• Problems with identification of experts • Risk of too strong orientation on experts

Lean Development

Goal/ Result	Designing production development processes in such a way that the results relevant for the customer can be achieved faster and more efficiently 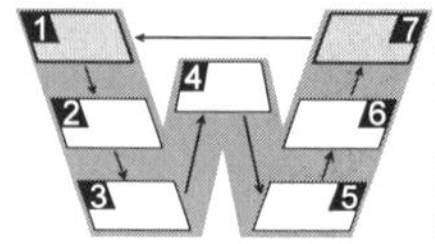
Information input	• Inefficient production development processes • Value analysis
Content/ Procedure	Lean development serves faster and more efficient process development. It is done in two main steps. First the results from the production development process are set in relation to customer value. As the second step, wasting is reduced by re-engineering and re-modeling processes, organization, administration and internal communication. This is done following the five principles of Lean Production: Value, Value Stream, Flow, Pull instead of Push, Perfection.

Pros	Cons
• Creation of a Continuous Improvement Process	• Possible occurrence of implementation barriers

Literature/ Reference	• Graebsch, Lindemann, Weiß 2007
Notes:	

Lean QFD

Goal/ Result

QFD version with quicker applicability

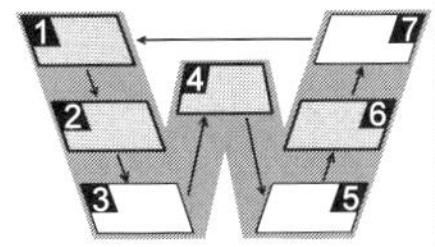

Informa-tion input

- Dependent on the phase in which one uses the Lean QFD

Content/ Procedure

The Lean QFD serves the supply of a QFD version with quicker applicability. Here one proceeds "QFD intuitively".

Pros	Cons
• Faster access to advantages of the QFD version	• Subjective

Literature/ Reference

- Mascitelli 2004

Notes:

Matrix of Priorities

S. Marjano

Goal/Result

Decision making
⇒ Weighting of solutions

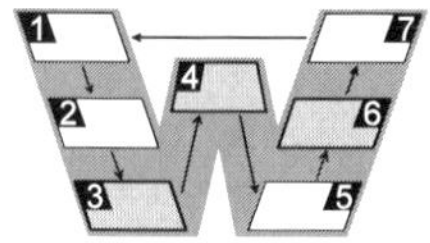

Information input

- Alternative solutions

Content/Procedure

Several solutions alternatives are evaluated in a matrix via weighted criteria. Alternatives are listed one below the other and each of which refer to two different lines. The first line contains exclusively the number of the referring alternative. The second line contains the numbers of alternative solutions that are yet to be compared. The regarded alternative solution has to be compared with each other alternative in the matrix via a pairwise comparison. A final total point evaluation is performed that is can be used as bsis for decision-making.

Pros	Cons
• Clear basis for decision-making • Balanced decisions	• More than 10 alternatives create confusing complexity

Literature/Reference

- Higgins 1996, Brassard 1994
- Pairwise comparison

Notes:

Metaplan Technique

Goal/ Result	Collection and structuring of the most diverse criteria 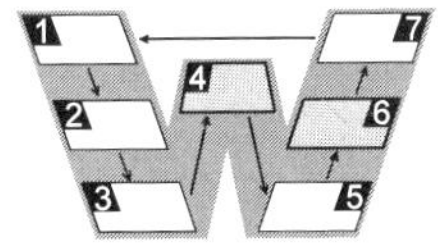
Information input	• Approaches and ideas of the staff involved
Content/ Procedure	The Metaplan technique serves to arrange and structure the most diverse criteria. First the central question is formulated and then answered by the appropriate co-workers. Next the collected cards are explained and structured in common agreement and visible to everyone.

Pros	Cons
• Quick results • Integration of each person	• Difficulties in describing certain problems with simple key words

Literature/ Reference	• EFA Assessment Tool
Notes:	

Method 635	*B. Rohrbach*
Goal/ Result	Finding and further development of ideas ⇒ Suggestions for ideas and solutions 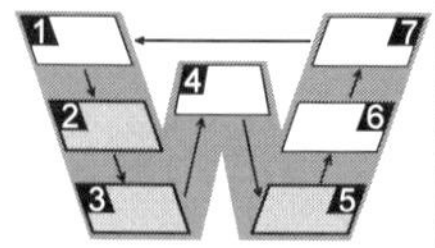
Informa-tion input	• Innovation tasks
Content/ Procedure	Based on a defined problem 6 people list 3 ideas each in a period of 5 minutes. Afterwards the sheets are passed on to the neighbor and the proceeding is repeated. After 6 iterations the team members have ideally generated 108 proposals. Subsequently all suggestions are analyzed and evaluated.

Pros	Cons
• Cooperation of a large number of participants (formation of several groups) • Postal execution of the method possible • Result documentation without additional expenditure • Increasing solution quality because of further development	• Negative effect of time pressure on creativity • Misunderstandings can occur because of short idea formulations • During the procedure inspiring effects can be missing because of missing direct communication

Literature/ Reference	• Eversheim, Schuh 1996, Schlicksupp 1992, Haberfellner 1999 • Brainwriting, Discussion 66, Synectics
Notes:	

Method for Concept Selection

Goal/ Result	Choice of a concept
Information input	• Ideas for solutions to set tasks
Content/ Procedure	The method for concept selection serves the selection of a concept. Here first schematic sketches of the concepts are made and valuation criteria determined. Subsequently, an evaluation matrix is drawn up and those concepts are selected which are to be further pursued (more + than -).

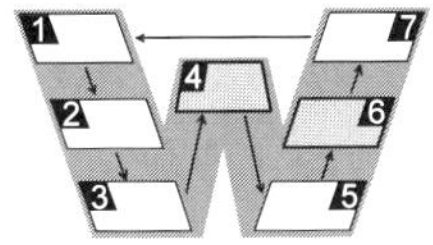

Pros	Cons
• Easily accomplishable	• Very subjective and heuristic

Literature/ Reference	• Mascitelli 2004
Notes:	

Mind Mapping	T. Buzan
Goal/ Result	Cartographic representation of mental contents and resulting chain of thoughts ⇒ Suggestions for ideas and solutions 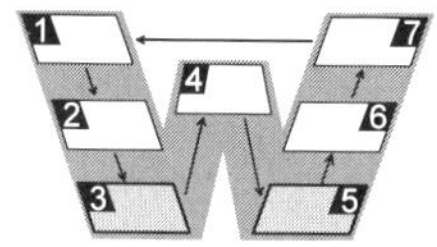
Information input	• Innovation tasks
Content/ Procedure	The main key word for a problem is arranged in the middle of a sheet. Based on the main key word associations and ideas that are generated in the group are spread over the whole sheet in form of limbs, branches and sprigs. Every new idea is written on a line (branch) starting from the referring key word. The visual representation shall promote generation of new ideas because of being adjusted to the structures of linkages in the human brain.

Pros	Cons
• Stimulation of creativity by use of psychological principles of thinking • Simultaneous documentation because of display format	• No systematic acquisition of connections because of the association linkage

Literature/ Reference	• Brauchlin 1995, Higgins, Wiese 1996 • Lotus blossom technique
Notes:	

Module Optimization Checklist

Goal/ Result

Consideration of the central aspects within the modularization

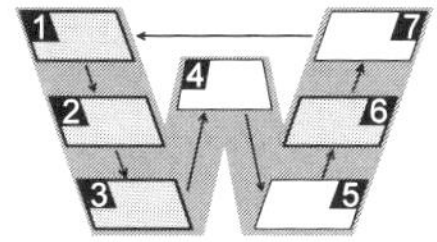

Informa- tion input

- Decision in favor of modularization of a product

Content/ Procedure

The module optimization check list serves the consideration of central aspects within the modularization. Here first a discussion (with following evaluation) regarding the individual statements of the check list takes place. Afterwards the valuations are summarized and need for revision derived from this summary.

Pros	Cons
• Shows where modularization is useful and where not	• Due to discussion & valuation very subjective and very heuristic

Literature/ Reference

- Mascitelli 2004

Notes:

Morphological Matrix — *F. Zwicky*

Goal/ Result

Development of new ideas
⇒ Solution concepts and product concepts, structuring of ideas

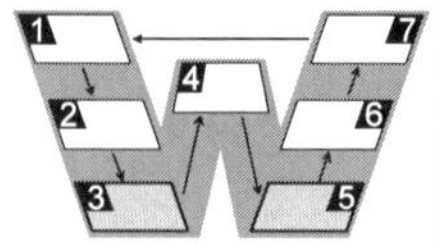

Information input

- Solution ideas
- Detailed solutions

Content/ Procedure

The method can be used individually as well as in a team. In the first step all functions of a product are listed. Afterwards principle solutions are developed for every single function. The functions are filled in on the vertical axis in a two-dimensional matrix. According to the referring functions the principle solutions are filled in on the horizontal axis. The combination of those partial solutions leads to a new total solution.

Pros	Cons
• Development of complex solution structures • High reusability of developed matrixes	• Difficulties in decision-making because of huge number of potential solutions • Decision support is not existent • Not all combinations are feasible

Literature/ Reference

- Haberfellner et al. 1999, Higgins, Wiese 1996, Schlicksupp 1992
- Morphological box, Attribute Listing

Notes:

Prognosis
Analysis
Creativity
Problem solving
Evaluation

Must/Should/Could Prioritization

Goal/ Result	Prioritization of product functions already in the development phase
Information input	• Product functions
Content/ Procedure	The Must/should/could prioritization corresponds to the Kano diagram, with the difference that a classification of the priority levels is possible.

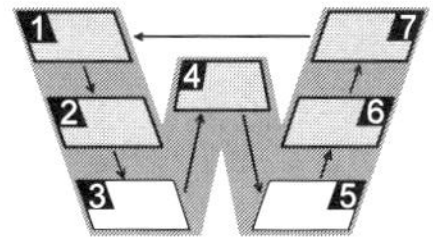

Pros	Cons
• Indication of customer-orientated prioritization for the development process	• General classification in the categories is almost impossible

Literature/ Reference	• Mascitelli 2004
Notes:	

Neighbor Field Integration

Goal/ Result	Development of solutions under consideration of environmental or marginal conditions ⇒ Future projections, innovation tasks, solution ideas, and suggestions for solutions 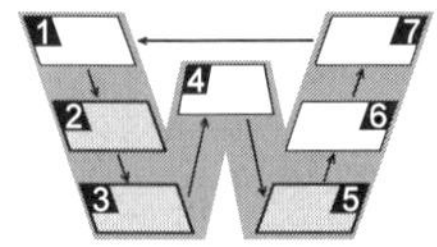
Informa-tion input	• Trends • Innovation tasks • Enterprise potential
Content/ Procedure	First of all neighbor fields of the wanted solution are defined, i.e. interdependency areas. Afterwards 5 to 15 elements are identified by association referring to neighbor fields. Based on these elements the solution design is derived.

Pros	Cons
• Successful method if problems occur because of interdependencies with problem field environment • Integrated consideration of the problem • Integration of environmental and marginal conditions	• Risk of wrong assumptions because of drawing conclusions from neighbor fields on the main problem

Literature/ Reference	• Schlicksupp 1992
Notes:	

Patent Analysis

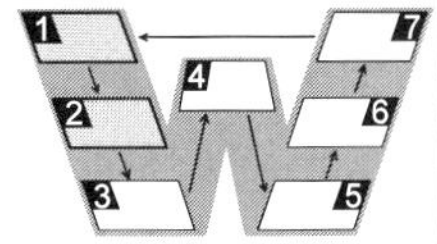

Goal/ Result	Knowledge about new products and technologies, in particular regarding lines of development and changes in the competitive environment
Information input	• Defined object of investigation • Filed patents from patent data bases
Content/ Procedure	The patent analysis serves the utilization of the information function of patents. The object of investigation is first distinguished on the basis of certain criteria such as technology fields, markets or special competitors and then subject field research is carried out by the use of patent data bases, e.g. WPI/L and INPADOC (STN). After the examination and analysis, strategically relevant data bases are extracted and evaluated.

Pros	Cons
• Well suitable for early recognition & competition analysis	• Many companies no longer file patents as protection against imitations

Literature/ Reference	• Schramm, Ludwig, Töpfer 1997
Notes:	

Pairwise Comparison

Goal/ Result	Determination of a criteria ranking ⇒ Prioritization of criteria
Information input	• Ideas • Concepts
Content/ Procedure	Using a pairwise comparison a larger number of criteria is arranged according to their importance. Through the direct comparison of two criteria at a time any criteria is compared to each other and evaluated as more, equal or less important. Summation of all weighting results in a ranking of all criteria. A list of all criteria that should be weighted or evaluated must be available to perform the comparison. Results of the pairwise comparison can be standardized in different ways, so that the weighting of criteria can be used for further methods directly, e.g. target costing.
Literature/ Reference	• Eversheim, Schuh 1996 • Priority matrix
Notes:	

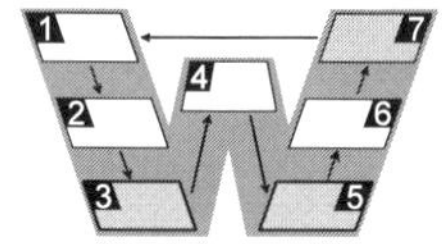

Pros	Cons
• Creation of a ranking is simple	• A use of this method makes sense if only a few criteria are compared. Not appropriate to complex comparisons

Platform Plan

Goal/ Result	Developing platforms from promising components 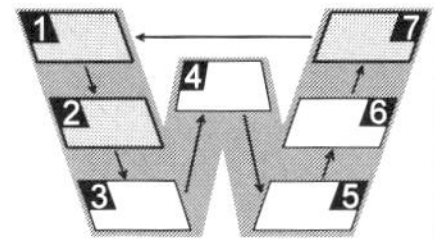
Information input	• Product structure
Content/ Procedure	The platform plan serves the development from promising components to platforms. Here first the platform potential of all subsystems and components is evaluated. Subsequently, the priorities are set according to potential, simple negotiability and attainable cost advantage, and represented in the platform matrix. Finally a conversion project plan is drawn up.

Pros	Cons
• All advantages of the modularization (cost savings, etc.)	• Platforms only useful in small numbers , since with too many platforms the advantage of the modularization disappears

Literature/ Reference	• Mascitelli 2004
Notes:	

Plan for Information Procurement

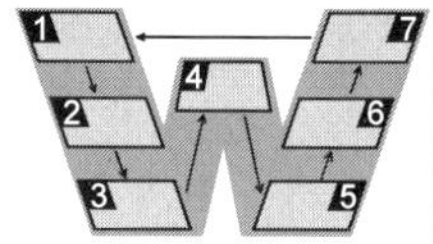

Goal/ Result	Overview of the information that should be procured ⇒ Information overview
Informa- tion input	• Project experiences
Content/ Procedure	The plan for information procurement is an accessory for the systematic collection of all relevant information within a complex project. In this plan all necessary information, their sources, methods for information procurement, competences etc. are listed. Thereby an overview of all activities is given that have to be performed procuring information.
Literature/ Reference	• Haberfellner et al. 1999 • Checklist
Notes:	

Pros	Cons
• Structured representation of necessary information	• Risk of incompleteness with new problems

Polarity Profile

Goal/ Result	Representation of several alternatives concerning their criteria fulfillment ⇒ Choice of solutions based on criteria fulfillment
Informa- tion input	• Trends • Potentials • Solution ideas
Content/ Procedure	For each solution alternative certain characteristics and criteria are estimated according to a grading key and the result are entered on a scale. The scale products of all criteria are connected to each other and create an evaluation profile. Evaluations profiles of solution alternatives can easily be compared to each other in a visual way.
Literature/ Reference	• Haberfellner et al. 1999 • Polar profiles (Parallel scales are not used but polar coordinates.)
Notes:	

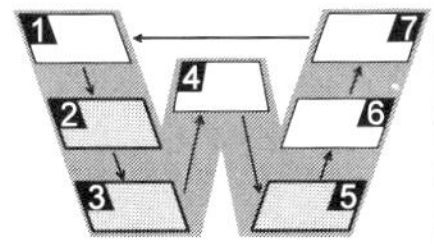

Pros	Cons
• Good graphical comparability of different methods of resolution	• High expenditure of time with many methods of resolution

Prognosis
Analysis
Creativity
Problem solving
Evaluation

Portfolio Analysis

Goal/ Result

Two-dimensional representation of evaluation results

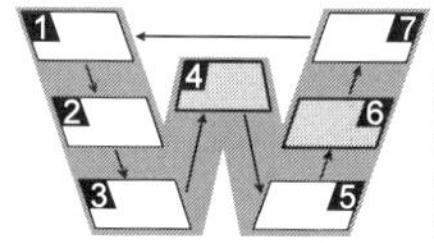

Information input

- Evaluation results

Content/ Procedure

The portfolio analysis serves the visualization of evaluation results. First the criteria are formulated in such a way that a strong fulfillment corresponds to a good result (e.g. “low risk"). Next the evaluation is carried out and two relevant factors are selected. Subsequently, the portfolio is developed.

Pros	Cons
• Good visualization aid • Strategic decision-making aid	• Concrete definition sometimes difficult

Literature/ Reference

- EFA Assessment Tool

Notes:

Problem Solution Tree

Goal/ Result

Graphical representation of complex correlations and facts
⇒ Identification of correlations, problem structure

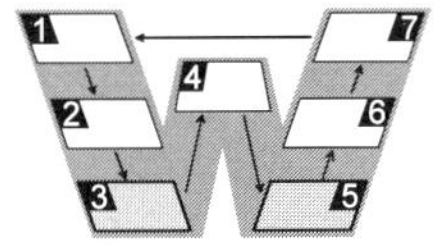

Information input

- Innovation tasks
- Problems

Content/ Procedure

The method can be used both, individually and in a team. Using a problem solution tree theoretical alternatives can be visualized in different abstraction levels. Procedure is divided into 5 steps:

1) Decompensation of problem structure
2) Composition of problem tree
3) Definition of relevant problem paths
4) Construction of a visual overview of the structure
5) Documentation

Pros	Cons
• Visual overview of the problem structure • Structured documentation of the problem structure	• Very extensive with complex products

Literature/ Reference

- Schlicksupp 1992, Brassard, Ritter 1994, Haberfellner et al. 1999
- Functional analysis

Notes:

Product Potential Evaluation

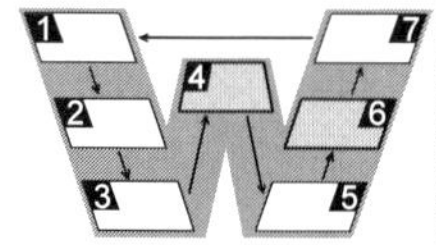

Goal/ Result	Ranking of the project ideas according to their chances and risks
Informa-tion input	• Standard NPV calculation
Content/ Procedure	The product-potential evaluation serves the classification of project ideas according to their chances/risk ratio. Here first the NPV is calculated with consideration of economic risks only. Consideration of special product risks (market risk etc.) is applied by means of proportional deductions from the NPV. Unexpected personnel bottlenecks are considered by the division of the adjusted NPV by the number of necessary working hours. Finally, the projects are ranked.
Literature/ Reference	• Mascitelli 2004
Notes:	

Pros	Cons
• Multidimensional view • Fast prioritization possible	• Subjective prioritization

Quality Function Deployment (QFD)

Y. Akao

Goal/ Result

Transformation of customer requirements into technical criteria
⇒ Feasibility of solutions

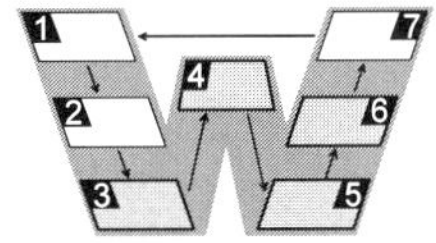

Information input

- Customer requirements
- Problems

Content/ Procedure

The QFD is used to transfer customer requirements into technical criteria through four stages (product, parts, processes, parameters). The weighted requirements are contrasted to the criteria and the strength of dependency is determined in a matrix (House of Quality). The significance of criteria is calculated by the weighting and the strength of dependency of the requirements. Furthermore the correlations of criteria to each other are recorded in the so-called roof of the matrix. Additionally technical and market positions in relation to competitors are documented in the matrix. In the following stage criteria of the first matrix are matched to requirements of the second matrix etc.

Pros	Cons
• Focus on relevant product criteria • Continuous applicability in all stages • Gradual particularization	• Subjective, partly divergent evaluations • High efforts needed • Knowledge of requirements needed

Literature/ Reference

- Pfeifer 1996, Akao 1992

Notes:

Quick-Look Value Engineering Event

Goal/ Result	Reduction of the production development down to value-creating functions (80-20 rule) and systematic cost reduction 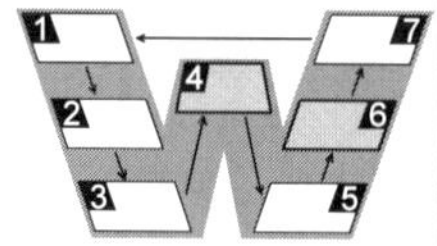
Informa-tion input	• Product idea and first task setting
Content/ Procedure	The quick look value engineering event serves the systematic cost reduction. Here as a first step customer problems are identified and utility functions defined. Subsequently, the alternatives are arranged as primary and secondary and prioritized according to relative costs, whereby the 20-cost-levers method is used. Finally, an action list is drawn up with promising alternatives, which are judged in more detail.
Literature/ Reference	• Mascitelli 2004
Notes:	

Pros	Cons
• Focus on creation of value	• Very heuristic procedure

R&D Portfolio

Goal/ Result	Visualization of the substantial decision criteria for R&D projects 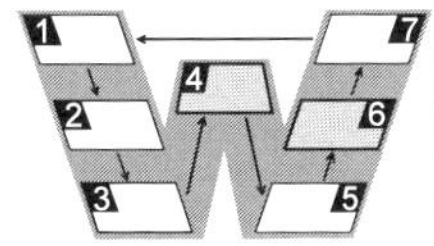
Information input	• Information for evaluating concepts
Content/ Procedure	The R&D portfolio serves as a visualization of the substantial decision criteria for R&D projects. Here first the substantial criteria are put in suitable portfolios, then the ideas and/or concepts are arranged in the portfolios.

Pros	Cons
• Intuitive • Good visualization	• Classification very subjective • Low comparability of the criteria

Literature/ Reference	• Gausemeier, Ebbesmeyer, Kallmeyer 2001
Notes:	

Requirement List

Goal/ Result	Orderly compilation of all qualitative and quantitative requirements in the developer's language 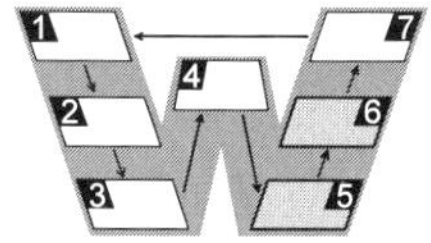
Information input	• Task
Content/ Procedure	The requirement list serves the orderly compilation of all requirements. Here first all requirements are collected (e.g. using check lists) and arranged according to specific criteria, e.g. personal record phases, request type etc. Subsequently, the requirements are formally documented in the requirement list, checked and finally confirmed.

Pros	Cons
Promotes smooth operations • Standardized list is easily transferable	• Must be kept up to date at all times

Literature/ Reference	• Ehrlenspiel 2002
Notes:	

Risk Analysis

Goal/ Result	Evaluation of different alternatives regarding certain aspects of risk. 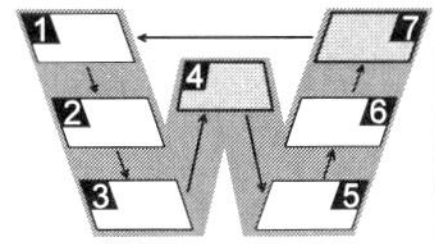
Informa-tion input	• Aspects of risk • Characteristics of the alternatives (sufficient and comparable information quality)
Content/ Procedure	The risk analysis serves the identification and evaluation of risks. As a first step the relevant aspects of risk are drawn up and the individual variants regarding the aspects of risk are evaluated, and the evaluations then weighted. Depiction is e.g. in the form of a spider net diagram.
	Pros: • Early recognition and prevention of possible problems **Cons**: • Subjective estimations always remain the basis of the analysis
Literature/ Reference	• EFA Assessment Tool
Notes:	

Scenario Management

J. Gausemeier

Goal/ Result	Creation of solid mission statements, goals and strategies ⇒ Future projections
Informa- tion input	• Environmental information • Enterprise potentials
Content/ Procedure	The scenario management supports entrepreneurial decisions and the creation of alternative future scenarios. The preparation is divided into five phases: **Scenario preparation** with project description and analysis of organization scopes **Scenario area analysis** with creation of influence areas and factors as well as determining key factors **Scenario prognostic** with preparation of key factors and generation of future projections **Scenario formation** with project bundling, creation of draft scenarios, future area mapping, scenario description **Scenario transfer** with impact analysis, possibility planning and solid planning

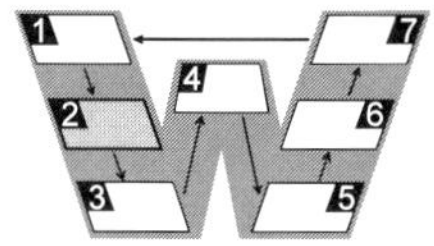

Pros	Cons
• Consideration of several trend possibilities	• Hardly clear action recommendations • Missing link to implantation planning

Literature/ Reference	• Kahn, Wiener 1967, Gausemeier, Fink, Schlake 1996
Notes:	

Prognosis
Analysis
Creativity
Problem solving
Evaluation

Selective List

Goal/ Result	Examination of basic feasibility of solutions ⇒ Feasible and infeasible solutions 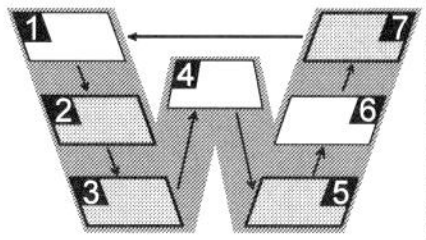
Information input	• Solution ideas • Solution concepts
Content/ Procedure	The acquired solution alternatives are verified by a checklist of fundamental issues: 1. Does the solution fulfill requirements and intentions? 2. Is compatibility with flanking solutions given? 3. Is the solution theoretically realizable? 4. Is complexity acceptable? 5. Is immediate security given? 6. Is the solution feasible according to the schedule? 7. Is sufficient know-how present? If the first two questions are both answered with "no", solution is rejected. The more questions are answered positively the more easily a solution is feasible.

Pros	Cons
• Systematical and structured evaluation of solution alternatives • Quick evaluation because of standardized procedure	• Some questions are too general and don't cover the problem sufficient

Literature/ Reference	• Ehrlenspiel 1995 • Checklist
Notes:	

Prognosis

Analysis

Creativity

Problem solving

Evaluation

SIS-Method

Batelle Institut, Frankfurt

Goal/ Result	Consolidation of single solutions to a total solution ⇒ Future projections, potentials, innovation tasks, solution concepts, product concepts
Informa-tion input	• Single solution ideas and elements
Content/ Procedure	The SIS-method (Systematic Integration of Solution elements) is a group technique. Each group member comes up with a potential solution for a defined problem. Two group members declaim their solutions which are compiled to a total solution by the whole team. Each solution proposal is processed likewise until a total solution has emerged for the given problem after a given time.

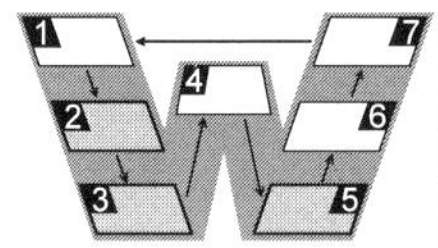

Pros	Cons
• Interdisciplinary knowledge integrated in the solution process • Integration of several solutions to a total solution	• High time comsumption due to sequential integration of solution approaches • Not a systematic development and structuring of solution approaches

Literature/ Reference	• Schlicksupp 1992, Higgins, Wiese 1996
Notes:	

Prognosis
Analysis
Creativity
Problem solving
Evaluation

Strategy Definition of the Innovative Strategy

Goal/ Result

Documented, strategic program with long-term orientation in the area of innovation

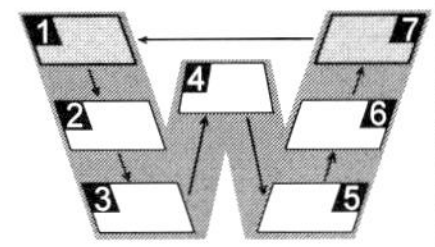

Information input

- Core competencies, chances and risks (core competency and SWOT analysis)

Content/ Procedure

The strategy definition of the innovative strategy serves the long-term orientation in the area of innovation. First the long-term innovation goals are planned with the help of portfolio presentations. Then implementation plans are drawn up for these goals and finally the strategic program is examined for its consistency.

Pros	Cons
• Long-term orientation of the innovation area • Conformity with the UN strategy	• Input for implementation is missing at detail level • Derivation of concrete measures difficult

Literature/ Reference

- EFA Assessment Tool

Notes:

Success Factor Portfolio

Goal/ Result

Show success factors as well as strengths and weaknesses in comparison with competitors

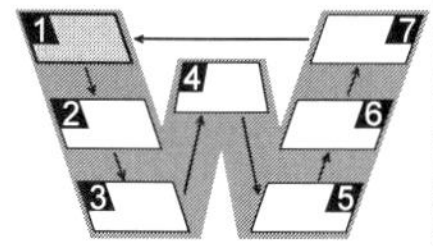

Information input

- Main business areas
- Total list of success factors

Content/ Procedure

The success factor portfolio serves the visualization of strengths and weaknesses in comparison with competitors. Here first success factors from the main business fields are determined and a questionnaire produced, which is sent to an external distributor. Subsequently, the internal/external results are compared in the portfolio and the portfolios analyzed for the desired information.

Pros	Cons
• Connection of operational and strategic knowledge • Results are also suitable for further analyses	• Relatively high expenditure during data gathering and analysis

Literature/ Reference

- Gausemeier, Ebbesmeyer, Kallmeyer 2001

Notes:

SWOT Analysis

Goal/ Result	Analysis of the relationships of strengths and weaknesses with opportunities and threats ⇒ Reference to strategic success factors
Informa-tion input	• Core competency analysis, if necessary combination of both methods
Content/ Procedure	The SWOT analysis is used to identify strengths and weaknesses within a company and to compare these with the main competitors. Then external opportunities and threats of the market are analyzed and the significance of these factors for the enterprise evaluated. Based on the results, chances and risks can be highlighted. Visualization of results is usually done with the aid of a portfolio.

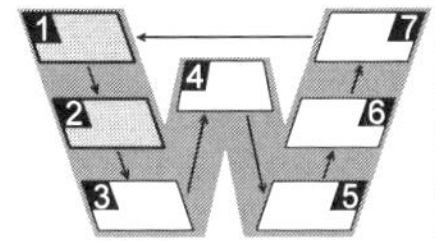

Pros	Cons
• Simple applicability	• High subjectivity when allocating to the SWOT categories

Literature/ Reference	• EFA Assessment Tool
Notes:	

Synectics

W.J.J. Gordon

Goal/ Result	Intensification of activities for solution search ⇒ Future projections, innovation potentials and solution proposals
Informa-tion	• Trends • Innovation tasks • Search area • Strategy
Content/ Procedure	The process consists of 4 phases: **Preparation:** First of all solution ideas are gathered spontaneously. For the creation of problem-extraneous structures and their combination alienations are conducted. **Incubation**: Analogies to the technique are acquired through personal, symbolic and visionary alienations. **Illumination:** Analogies are checked in respect to their adequacy. **Verification:** Finally solution concepts are developed.

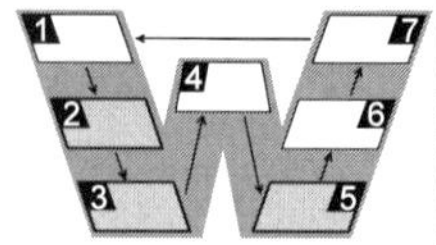

Pros	Cons
• Intensive development of solution concepts • Creation of solutions which are not easily recognized and self-suggestive	• A lot of practice needed • High level of explanation needed

Literature/ Reference	• Higgins, Wiese 1996, Brauchlin 1995, Haberfellner et al. 1999 • Brainstorming, Method 635, KJ-Method (working with post-its for idea generation)
Notes:	

Target Costing

Goal/ Result

Determination of manufacturing costs for system components
⇒ first cost estimation, selected ideas

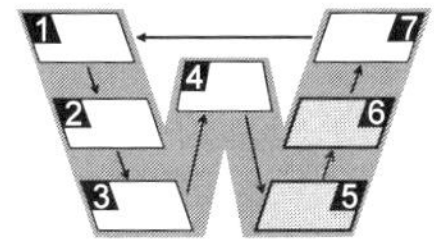

Informa-tion input

- Comparable products, market knowledge
- Target price, concepts, detailed solutions

Content/ Procedure

The basic principle is that manufacturing costs of a product are not determined by the production of a commodity but by the market and the defined profit. The first step of target costing is to determine the possible market price with market research methods or in comparison to competitive products (e.g. comparison oriented benchmarking). The target costs of the whole system can be calculated based on this information and the target costs of single components can be determined in the next step.

Pros	Cons
• Great coverage of product costs and attributes with functional requirements due to customer orientation	• Risk of exaggerated outsourcing of important system components to reduce costs

Literature/ Reference

- Eversheim, Schuh 1996, Horvath 1994, Ehrlenspiel 1995

Notes:

Technology Calendar

Goal/ Result	Recognition and full utilization of technological potential 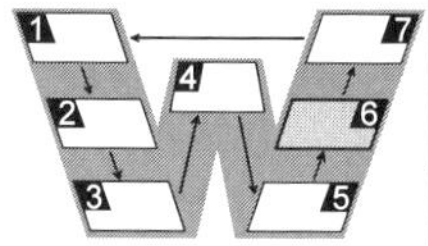
Information input	• Mastered and applied technology
Content/ Procedure	The technology calendar serves the collection and utilization of technological potential. First the requirements are identified in a product analysis, they are converted by abstraction to basic functions. Technology research follows to reveal a broad spectrum of technological solutions, which are assigned to the appropriate product. This serves above all for linking the constant manufacturing concepts and for formulating alternative technology-chains. The goal is to make production alternatives more comparable. Subsequently, the project plan/the technology calendar is developed, so that one achieves a program, which, within a short or longer period, initiates activities for the development of new and existing technologies.
Literature/ Reference	• EFA Assessment Tool
Notes:	

Pros	Cons
• Possibility of considering potential technologies in current development projects.	• Considerable expenditure necessary to gain information

Technology Tree

Goal/ Result

Collection of technological competencies

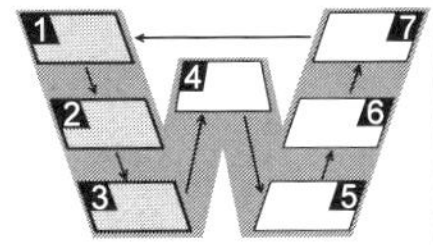

Informa- tion input

- Mastered technologies

Content/ Procedure

The technology tree serves the collection of the company-internal technological competencies. First the principle is described on the basis of the technology-tree scheme. Then the company-specific competencies within the individual areas are collected and analyzed regarding their special strengths and current application fields. The relationships are illustrated by arrows.

Pros	Cons
• Comprehensive analysis of the own, present competencies • Competency gaps are identified	• Confusing if too big

Literature/ Reference

- EFA Assessment Tool

Notes:

TILMAG-Method

H. Schlicksupp

Goal/ Result

Determination of new solution ideas with a multi-stage association process
⇒ Ideas and solution proposals

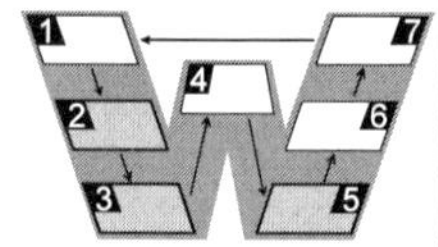

Information input

- Innovation tasks

Content/ Procedure

The TILMAG-Method (Transformation of ideal solution elements through the matrices of formation of associations and commonalities, german: **T**ransformation **i**dealer **L**ösungselemente durch **M**atrizen der **A**ssoziations- und **G**emeinsamkeitsbildung) is divided into several steps:

1. Determination of criteria of an ideal solution
2. Search for associations to these solution criteria
3. Derivation of possible solutions out of these associations
4. Search for commonalities between associations and possible solutions
5. Joining of commonalities to total solutions

Pros	Cons
• Target oriented approach to the ideal solution	• Only applicable when ideal solution is visible

Literature/ Reference

- Schlicksupp 1992

Notes:

Trend Extrapolation

Goal/ Result

Creation of forecasts of future trends

⇒ Future trends

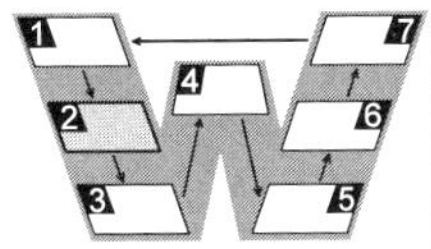

Information input

- Historical Data
- Monitored Values

Content/ Procedure

Trend extrapolation is based on observations of the past. The assumption is that those are also valid for the future. Different function types can be taken for this purpose.

The linear trend assumes that the linear development of the past is continued in the future.

The exponential trend assumes that the relative growth is constant.

A logistical trend is also possible. The logistical curve assumes a slow growth in the beginning which increases progressively until the turning point of the curve, growing digressively afterwards and decreases after the maximum.

Pros	Cons
• Comprehensible, systematic derivation	• Very elaborate

Literature/ Reference

- Koppelmann 1993, Meffert 1998

Notes:

TRIZ Contradiction Matrix

G. Altschuller

Goal/ Result	Overcoming of contradictory product attributes ⇒ Ideas and solution proposals
Information input	• Technical contradiction
Content/ Procedure	A technical contradiction exists when the simultaneous realization of at least two attribute parameters - both to be optimized – with the known technical means doesn't allow a satisfactory compromise. Both attributes leading to this contradiction are going to be abstracted in such a way that the so-called technical standard parameters are applicable. Starting from the matrix solution principles can be found. Those can be transferred to the real contradiction situation and can be worked out in more details.

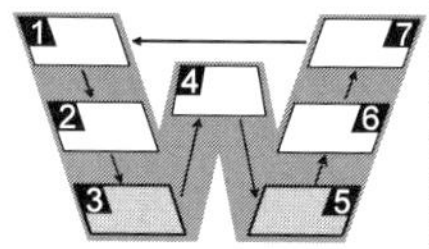

Pros	Cons
• Structured and simple determination of solution ideas • New suggestions for solutions	• Detailing of solution principles requires experience • Knowledge about the method is required

Literature/ Reference	• Altschuller 1984, Terninko et al. 1998 • WOIS, Bionic
Notes:	

Prognosis
Analysis
Creativity
Problem solving
Evaluation

Value Analysis
L.D. Miles

Goal/ Result

Cost optimization of object functions with simultaneous increase of value of the object
⇒ Function and cost structure of the object

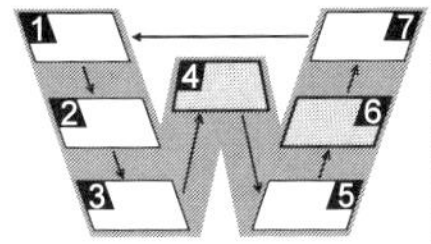

Information input

- Rough concept of the object

Content/ Procedure

The value analysis is used to reduce costs of existing products (value increase) and to avoid unnecessary costs of products to be developed (value design). The following approach is recommended (DIN 69910):

1. Prepare projects (structure tasks)
2. Analyze object situation (analyze tasks)
3. Determine nominal condition (formulate tasks)
4. Develop solution ideas (solution search)
5. Determine solutions (evaluate solutions)

The execution of these tasks is done in interdisciplinary teams.

Pros	Cons
• Function-oriented approach • Systematic research of deeper cost causes • Holistic view of the object	• Very time consuming • Intensive experience needed

Literature/ Reference

- Stippel 1999, Ehrlenspiel 1995
- Conjoint Analysis

Notes:

Weighted Point Evaluation

Goal/ Result	Ranking of the alternatives according to weighted criteria 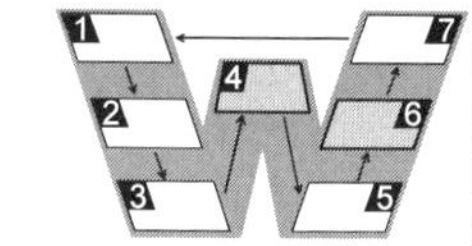
Information input	• Alternatives to be evaluated • Characteristics of the alternatives • Evaluation criteria
Content/ Procedure	The weighted point evaluation serves to generate a ranking for the alternatives to be evaluated. Here first the criteria are specified and weighted. From the point values for each criterion the total point values of each alternative are then determined.
	Pros • Ranking of the alternatives according to special, personal preferences **Cons** • Higher expenditure than with simpler methods
Literature/ Reference	• Ehrlenspiel 2007
Notes:	

Prognosis · Analysis · Creativity · Problem solving · **Evaluation**

Appendix B
Selected tools of the TRIZ-Methodology

Chapter 4.4 describes the TRIZ-methodology and its contradiction analysis. The analysis and resolution of technical contradictions need the following materials. The contradiction is described abstractly with the help of 39 technical parameters. Innovation principles can be chosen with the contradiction matrix. These principles determine a promising solution direction (Altschuller 1984; Terniko et al. 1998).

The 39 technical parameters of the TRIZ-Methodology

1. Weight of a moving object

The measurable force, caused by gravity, which a moving body exerts on a support which keeps it from falling. A moving object changes its position inherently or due to external forces.

2. Weight of a stationary object

The measurable force, caused by gravity, which a stationary body exerts on a support which keeps it from falling. A moving object doesn't change its position neither inherently nor due to external forces.

3. Length of a moving object

The linear measured value of length, height or width of a body in the direction of movement. The movement can caused internal or external forces.

4. Length of a stationary object

The linear measured value of length, height or width of a body in the direction of no movement.

5. Area of a moving object

The laminar measured value of a plane or particular plane of an object, which changes its position in space due to internal or external forces.

6. Area of a stationary object

The laminar measured value of a plane or particular plane of an object, which cannot change its position in space neither due to internal nor due to external forces.

7. Volume of a moving object

The cubical measured value of an object, which changes its position in space due to internal or external forces.

8. Volume of a stationary object

The cubical measured value of a plane or particular plane of an object, which cannot change its position in space neither due to internal nor due to external forces.

9. Speed

Speed with which an action or process is advanced with time.

10. Force

The ability to induce physical changes to an object or in a system. The change can be total or partial, permanent or temporary.

11. Pressure or tension

The intensity of forces affecting an object or system, measured as compression or tension per area.

12. Form

The external aspect or contour of an object or system. The form can change totally or partially, permanently or temporarily due to external forces affecting the object or system.

13. Stability of an object

The durability of a whole object or system against external effects.

14. Strength

The ability of an object or system to endure forces or strain within defined limits without breaking.

15. Durability of a moving object

The time span in which a moving object in space is able to fulfil its function.

16. Durability of a stationary object

The time span in which a stationary object in space is able to fulfil its function.

17. Temperature

The loss or gain of heat as a possible cause for changes on an object. System or product during the demanded operation sequence.

18. Brightness

Light energy per illuminated area, quality and characteristic of the light; level of illumination.

19. Energy consumption of a moving object

The energy demand of an object or system which moves in space due to internal or external forces.

20. Energy consumption of a stationary object

The energy demand of an object or system which doesn't move in space despite of internal or external forces.

21. Power

The ratio of effort and time for the action concerned. Serves for the characterization of needed but unwanted changes in power of the system or object.

22. Energy waste

Inability of a system or object to induce force, especially when work or production is not in progress.

23. Material waste

Decrease or loss of material, especially when work or production is not in progress.

24. Information loss

Decrease or loss of information or data.

25. Time waste

Increasing time consumption for the fulfilment of a specified function.

26. Material amount

The needed number of elements or amount of an element for the creation of an object or system.

27. Reliability

The ability to fulfil the specified function adequately over a certain time or number of cycles.

28. Measuring accuracy

The level of congruence between measured and true value of the attribute to be measured.

29. Production accuracy

The level of congruence with specifications.

30. External negative influences on an object

The external factors affecting an object influencing quality and efficiency.

31. Negative side effects of the object

Effects created internally which compromise the quality and efficiency of an object or system.

32. Manufacturing friendliness

Comfort and simplicity with which a product can be manufactured.

33. User friendliness

Comfort and simplicity with which an object or system can be operated or used.

34. Repair friendliness

Comfort and simplicity with which an object or system can be put back into operating conditions after damage or wear.

35. Adaptability

The ability to adapt to changing external conditions.

36. Complexity of the structure

Number and diversification of single parts including their links. Furthermore the difficulty of controlling a system as a user.

37. Complexity in control and supervision

Number and diversification of elements for controlling and supervision of the system but also the effort needed for a measurement with an acceptable accuracy.

38. Automation level

Ability to function without human interaction.

39. Productivity

The ratio between concluded actions and the time needed.

The Contradiction Matrix (Temiko et al. 1998)

	Optimization factor \ Problem factor	1 Weight of a moved object	2 Weight of a stationary object	3 Length of a moved object	4 Length of a stationary object	5 Expanse of a moved object	6 Expanse of a stationary object	7 Volume of a moved object	8 Volume of a stationary object	9 Speed	10 Force
1	Weight of a moved object	+	-	15, 8, 29,34	-	29, 17, 38, 34	-	29, 2, 40, 28	-	2, 8, 15, 38	8, 10, 18, 37
2	Weight of a stationary object	-	+	-	10, 1, 29, 35	-	35, 30, 13, 2	-	5, 35, 14, 2	-	8, 10, 19, 35
3	Length of a moved object	8, 15, 29, 34	-	+	-	15, 17, 4	-	7, 17, 4, 35	-	13, 4, 8	17, 10, 4
4	Length of a stationary object		35, 28, 40, 29	-	+	-	17, 7, 10, 40	-	35, 8, 2,14	-	28, 10
5	Expanse of a moved object	2, 17, 29, 4	-	14, 15, 18, 4	-	+	-	7, 14, 17, 4		29, 30, 4, 34	19, 30, 35, 2
6	Expanse of a stationary object	-	30, 2, 14, 18	-	26, 7, 9, 39	-	+	-		-	1, 18, 35, 36
7	Volume of a moved object	2, 26, 29, 40	-	1, 7, 4, 35	-	1, 7, 4, 17	-	+	-	29, 4, 38, 34	15, 35, 36, 37
8	Volume of a stationary object	-	35, 10, 19, 14	19, 14	35, 8, 2, 14	-		-	+	-	2, 18, 37
9	Speed	2, 28, 13, 38	-	13, 14, 8	-	29, 30, 34	-	7, 29, 34	-	+	13, 28, 15, 19
10	Force	8, 1, 37, 18	18, 13, 1, 28	17, 19, 9, 36	28, 10	19, 10, 15	1, 18, 36, 37	15, 9, 12, 37	2, 36, 18, 37	13, 28, 15, 12	+
11	Pressure or tension	10, 36, 37, 40	13, 29, 10, 18	35, 10, 36	35, 1, 14, 16	10, 15, 36, 28	10, 15, 36, 37	6, 35, 10	35, 24	6, 35, 36	36, 35, 21
12	Shape	8, 10, 29, 40	15, 10, 26, 3	29, 34, 5, 4	13, 14, 10, 7	5, 34, 4, 10		14, 4, 15, 22	7, 2, 35	35, 15, 34, 18	35, 10, 37, 40
13	Stability of an object	21, 35, 2, 39	26, 39, 1, 40	13, 15, 1, 28	37	2, 11, 13	39	28, 10, 19, 39	34, 28, 35, 40	33, 15, 28, 18	10, 35, 21, 16
14	Resistance	1, 8, 40, 15	40, 26, 27, 1	1, 15, 8, 35	15, 14, 28, 26	3, 34, 40, 29	9, 40, 28	10, 15, 14, 7	9, 14, 17, 15	8, 13, 26, 14	10, 18, 3, 14
15	Durability of a moved object	19, 5, 34, 31	-	2, 19, 9	-	3, 17, 19	-	10, 2, 19, 30	-	3, 35, 5	19, 2, 16
16	Durability of a stationary object	-	6, 27, 19, 16	-	1, 40, 35	-		-	35, 34, 38	-	
17	Temperature	36,22, 6, 38	22, 35, 32	15, 19, 9	15, 19, 9	3, 35, 39, 18	35, 38	34, 39, 40, 18	35, 6, 4	2, 28, 36, 30	35, 10, 3, 21
18	Brightness	19, 1, 32	2, 35, 32	19, 32, 16		19, 32, 26		2, 13, 10		10, 13, 19	26, 19, 6
19	Energy consumption of a moved object	12,18,28 ,31	-	12, 28	-	15, 19, 25	-	35, 13, 18	-	8, 35, 35	16, 26, 21, 2

Contradiction Matrix part 1

		11	12	13	14	15	16	17	18	19	20
	Problem factor / Optimization factor	Pressure or tension	Shape	Stability of an object	Resistance	Durability of a moved object	Durability of a stationary object	Temperature	Brightness	Energy consumption of a moved object	Energy consumption of a stationary object
1	Weight of a moved object	10, 36, 37, 40	10, 14, 35, 40	1, 35, 19, 39	28, 27, 18, 40	5, 34, 31, 35	-	6, 29, 4, 38	19, 1, 32	35, 12, 34, 31	-
2	Weight of a stationary object	13, 29, 10, 18	13, 10, 29, 14	26, 39, 1, 40	28, 2, 10, 27	-	2, 27, 19, 6	28, 19, 32, 22	19, 32, 35	-	18, 19, 28, 1
3	Length of a moved object	1, 8, 35	1, 8, 10, 29	1, 8, 15, 34	8, 35, 29, 34	19	-	10, 15, 19	32	8, 35, 24	-
4	Length of a stationary object	1, 14, 35	13, 14, 15, 7	39, 37, 35	15, 14, 28, 26	-	1, 10, 35	3, 35, 38, 18	3, 25	-	
5	Expanse of a moved object	10, 15, 36, 28	5, 34, 29, 4	11, 2, 13, 39	3, 15, 40, 14	6, 3	-	2, 15, 16	15, 32, 19, 13	19, 32	-
6	Expanse of a stationary object	10, 15, 36, 37		2, 38	40	-	2, 10, 19, 30	35, 39, 38		-	
7	Volume of a moved object	6, 35, 36, 37	1, 15, 29, 4	28, 10, 1, 39	9, 14, 15, 7	6, 35, 4	-	34, 39, 10, 18	2, 13, 10	35	-
8	Volume of a stationary object	24, 35	7, 2, 35	34, 28, 35, 40	9, 14, 17, 15	-	35, 34, 38	35, 6, 4		-	
9	Speed	6, 18, 38, 40	35, 15, 18, 34	28, 33, 1, 18	8, 3, 26, 14	3, 19, 35, 5	-	28, 30, 36, 2	10, 13, 19	8, 15, 35, 38	-
10	Force	18, 21, 11	10, 35, 40, 34	35, 10, 21	35, 10, 14, 27	19, 2		35, 10, 21	-	19, 17, 10	1, 16, 36, 37
11	Pressure or tension	+	35, 4, 15, 10	35, 33, 2, 40	9, 18, 3, 40	19, 3, 27		35, 39, 19, 2	-	14, 24, 10, 37	
12	Shape	34, 15, 10, 14	+	33, 1, 18, 4	30, 14, 10, 40	14, 26, 9, 25		22, 14, 19, 32	13, 15, 32	2, 6, 34, 14	
13	Stability of an object	2, 35, 40	22, 1, 18, 4	+	17, 9, 15	13, 27, 10, 35	39, 3, 35, 23	35, 1, 32	32, 3, 27, 16	13, 19	27, 4, 29, 18
14	Resistance	10, 3, 18, 40	10, 30, 35, 40	13, 17, 35	+	27, 3, 26		30, 10, 40	35, 19	19, 35, 10	35
15	Durability of a moved object	19, 3, 27	14, 26, 28, 25	13, 3, 35	27, 3, 10	+	-	19, 35, 39	2, 19, 4, 35	28, 6, 35, 18	
16	Durability of a stationary object			39, 3, 35, 23		-	+	19, 18, 36, 40		-	
17	Temperature	35, 39, 19, 2	14, 22, 19, 32	1, 35, 32	10, 30, 22, 40	19, 13, 39	19, 18, 36, 40	+	32, 30, 21, 16	19, 15, 3, 17	
18	Brightness		32, 30	32, 3, 27	35, 19	2, 19, 6		32, 35, 19	+	32, 1, 19	32, 35, 1, 15
19	Energy consumption of a moved object	23, 14, 25	12, 2, 29	19, 13, 17, 24	5, 19, 9, 35	28, 35, 6, 18	-	19, 24, 3, 14	2, 15, 19	+	-

Contradiction Matrix part 2

		21	22	23	24	25	26	27	28	29	30
	Problem factor / Optimization factor	Power	Waste of energy	Waste of material	Loss of information	Waste of time	Material quantity	Reliability	Measuring accuracy	Production accuracy	Negative external influences on the object
1	Weight of a moved object	12, 36, 18, 31	6, 2, 34, 19	5, 35, 3, 31	10, 24, 35	10, 35, 20, 28	3, 26, 18, 31	1, 3, 11, 27	28, 27, 35, 26	28, 35, 26, 18	22, 21, 18, 27
2	Weight of a stationary object	15, 19, 18, 15	18, 19, 28, 15	5, 8, 13, 30	10, 15, 35	10, 20, 35, 26	19, 6, 18, 26	10, 28, 8, 3	18, 26, 28	10, 1, 35, 17	2, 19, 22, 37
3	Length of a moved object	1, 35	7, 2, 35, 39	4, 29, 23, 10	1, 24	15, 2, 29	29, 35	10, 14, 29, 40	28, 32, 4	10, 28, 29, 37	1, 15, 17, 24
4	Length of a stationary object	12, 8	6, 28	10, 28, 24, 35	24, 26,	30, 29, 14		15, 29, 28	32, 28, 3	2, 32, 10	1, 18
5	Expanse of a moved object	19, 10, 32, 18	15, 17, 30, 26	10, 35, 2, 39	30, 26	26, 4	29, 30, 6, 13	29, 9	26, 28, 32, 3	2, 32	22, 33, 28, 1
6	Expanse of a stationary object	17, 32	17, 7, 30	10, 14, 18, 39	30, 16	10, 35, 4, 18	2, 18, 40, 4	32, 35, 40, 4	26, 28, 32, 3	2, 29, 18, 36	27, 2, 39, 35
7	Volume of a moved object	35, 6, 13, 18	7, 15, 13, 16	36, 39, 34, 10	2, 22	2, 6, 34, 10	29, 30, 7	14, 1, 40, 11	26, 26, 28	25, 28, 2, 16	22, 21, 27, 35
8	Volume of a stationary object	30, 6		10, 39, 35, 34		35, 16, 32 18	35, 3	2, 35, 16		35, 10, 25	34, 39, 19, 27
9	Speed	19, 35, 38, 2	14, 20, 19, 35	10, 13, 28, 38	13, 26		10, 19, 29, 38	11, 35, 27, 28	28, 32, 1, 24	10, 28, 32, 25	1, 28, 35, 23
10	Force	19, 35, 18, 37	14, 15	8, 35, 40, 5		10, 37, 36	14, 29, 18, 36	3, 35, 13, 21	35, 10, 23, 24	28, 29, 37, 36	1, 35, 40, 18
11	Pressure or tension	10, 35, 14	2, 36, 25	10, 36, 3, 37		37, 36, 4	10, 14, 36	10, 13, 19, 35	6, 28, 25	3, 35	22, 2, 37
12	Shape	4, 6, 2	14	35, 29, 3, 5		14, 10, 34, 17	36, 22	10, 40, 16	28, 32, 1	32, 30, 40	22, 1, 2, 35
13	Stability of an object	32, 35, 27, 31	14, 2, 39, 6	2, 14, 30, 40		35, 27	15, 32, 35		13	18	35, 24, 30, 18
14	Resistance	10, 26, 35, 28	35	35, 28, 31, 40		29, 3, 28, 10	29, 10, 27	11, 3	3, 27, 16	3, 27	18, 35, 37, 1
15	Durability of a moved object	19, 10, 35, 38		28, 27, 3, 18	10	20, 10, 28, 18	3, 35, 10, 40	11, 2, 13	3	3, 27, 16, 40	22, 15, 33, 28
16	Durability of a stationary object	16		27, 16, 18, 38	10	28, 20, 10, 16	3, 35, 31	34, 27, 6, 40	10, 26, 24		17, 1, 40, 33
17	Temperature	2, 14, 17, 25	21, 17, 35, 38	21, 36, 29, 31		35, 28, 21, 18	3, 17, 30, 39	19, 35, 3, 10	32, 19, 24	24	22, 33, 35, 2
18	Brightness	32	13, 16, 1, 6	13, 1	1, 6	19, 1, 26, 17	1, 19		11, 15, 32	3, 32	15, 19
19	Energy consumption of a moved object	6, 19, 37, 18	12, 22, 15, 24	35, 24, 18, 5		35, 38, 19, 18	34, 23, 16, 18	19, 21, 11, 27	3, 1, 32		1, 35, 6, 27

Contradiction Matrix part 3

		31	32	33	34	35	36	37	38	39
	Problem factor / Optimization factor	Negative side effects of the object	Ease of production	Ease of operation	Ease of repairing	Adaptability	Device complexity	Complexity of detecting and measuring	Automation level	Productivity
1	Weight of a moved object	22, 35, 31, 39	27, 28, 1, 36	35, 3, 2, 24	2, 27, 28, 11	29, 5, 15, 8	26, 30, 36, 34	28, 29, 26, 32	26, 35 18, 19	35, 3, 24, 37
2	Weight of a stationary object	35, 22, 1, 39	28, 1, 9	6, 13, 1, 32	2, 27, 28, 11	19, 15, 29	1, 10, 26, 39	25, 28, 17, 15	2, 26, 35	1, 28, 15, 35
3	Length of a moved object	17, 15	1, 29, 17	15, 29, 35, 4	1, 28, 10	14, 15, 1, 16	1, 19, 26, 24	35, 1, 26, 24	17, 24, 26, 16	14, 4, 28, 29
4	Length of a stationary object		15, 17, 27	2, 25	3	1, 35	1, 26	26		30, 14, 7, 26
5	Expanse of a moved object	17, 2, 18, 39	13, 1, 26, 24	15, 17, 13, 16	15, 13, 10, 1	15, 30	14, 1, 13	2, 36, 26, 18	14, 30, 28, 23	10, 26, 34, 2
6	Expanse of a stationary object	22, 1, 40	40, 16	16, 4	16	15, 16	1, 18, 36	2, 35, 30, 18	23	10, 15, 17, 7
7	Volume of a moved object	17, 2, 40, 1	29, 1, 40	15, 13, 30, 12	10	15, 29	26, 1	29, 26, 4	35, 34, 16, 24	10, 6, 2, 34
8	Volume of a stationary object	30, 18, 35, 4	35		1		1, 31	2, 17, 26		35, 37, 10, 2
9	Speed	2, 24, 35, 21	35, 13, 8, 1	32, 28, 13, 12	34, 2, 28, 27	15, 10, 26	10, 28, 4, 34	3, 34, 27, 16	10, 18	
10	Force	13, 3, 36, 24	15, 37, 18, 1	1, 28, 3, 25	15, 1, 11	15, 17, 18, 20	26, 35, 10, 18	36, 37, 10, 19	2, 35	3, 28, 35, 37
11	Pressure or tension	2, 33, 27, 18	1, 35, 16	11	2	35	19, 1, 35	2, 36, 37	35, 24	10, 14, 35, 37
12	Shape	35, 1	1, 32, 17, 28	32, 15, 26	2, 13, 1	1, 15, 29	16, 29, 1, 28	15, 13, 39	15, 1, 32	17, 26, 34, 10
13	Stability of an object	35, 40, 27, 39	35, 19	32, 35, 30	2, 35, 10, 16	35, 30, 34, 2	2, 35, 22, 26	35, 22, 39, 23	1, 8, 35	23, 35, 40, 3
14	Resistance	15, 35, 22, 2	11, 3, 10, 32	32, 40, 25, 2	27, 11, 3	15, 3, 32	2, 13, 25, 28	27, 3, 15, 40	15	29, 35, 10, 14
15	Durability of a moved object	21, 39, 16, 22	27, 1, 4	12, 27	29, 10, 27	1, 35, 13	10, 4, 29, 15	19, 29, 39, 35	6, 10	35, 17, 14, 19
16	Durability of a stationary object	22	35, 10	1	1	2		25, 34, 6, 35	1	20, 10, 16, 38
17	Temperature	22, 35, 2, 24	26, 27	26, 27	4, 10, 16	2, 18, 27	2, 17, 16	3, 27, 35, 31	26, 2, 19, 16	15, 28, 35
18	Brightness	35, 19, 32, 39	19, 35, 28, 26	28, 26, 19	15, 17, 13, 16	15, 1, 1, 19	6, 32, 13	32, 15	2, 26, 10	2, 25, 16
19	Energy consumption of a moved object	2, 35, 6	28, 26, 30	19, 35	1, 15, 17, 28	15, 17, 13, 16	2, 29, 27, 28	35, 38	32, 2	12, 28, 35

Contradiction Matrix part 4

	Problem factor / Optimization factor	1 Weight of a moved object	2 Weight of a stationary object	3 Length of a moved object	4 Length of a stationary object	5 Expanse of a moved object	6 Expanse of a stationary object	7 Volume of a moved object	8 Volume of a stationary object	9 Speed	10 Force
20	Energy consumption of a stationary object	-	19, 9, 6, 27	-		-		-		-	36, 37
21	Power	8, 36, 38, 31	19, 26, 17, 27	1, 10, 35, 37		19, 38	17, 32, 13, 38	35, 6, 38	30, 6, 25	15, 35, 2	26, 2, 36, 35
22	Waste of energy	15, 6, 19, 28	19, 6, 18, 9	7, 2, 6, 13	6, 38, 7	15, 26, 17, 30	17, 7, 30, 18	7, 18, 23	7	16, 35, 38	36, 38
23	Waste of material	35, 6, 23, 40	35, 6, 22, 32	14, 29, 10, 39	10, 28,24	35, 2, 10, 31	10, 18, 39, 31	1, 29, 30, 36	3, 39, 18, 31	10, 13, 28, 38	14, 15, 18, 40
24	Loss of information	10, 24, 35	10, 35, 5	1, 26	26	30, 26	30, 16		2, 22	26, 32	
25	Waste of time	10, 20, 37, 35	10, 20, 26, 5	15, 2, 29	30, 24, 14, 5	26, 4, 5, 16	10, 35, 17, 4	2, 5, 34, 10	35, 16, 32, 18		10, 37, 36,5
26	Material quantity	35, 6, 18, 31	27, 26, 18, 35	29, 14, 35, 18		15, 14, 29	2, 18, 40, 4	15, 20, 29		35, 29, 34, 28	35, 14, 3
27	Reliability	3, 8, 10, 40	3, 10, 8, 28	15, 9, 14, 4	15, 29, 28, 11	17, 10, 14, 16	32, 35, 40, 4	3, 10, 14, 24	2, 35, 24	21, 35, 11, 28	8, 28, 10, 3
28	Measuring accuracy	32, 35, 26, 28	28, 35, 25, 26	28, 26, 5, 16	32, 28, 3, 16	26, 28, 32, 3	26, 28, 32, 3	32, 13, 6		28, 13, 32, 24	32, 2
29	Production accuracy	28, 32, 13, 18	28, 35, 27, 9	10, 28, 29, 37	2, 32, 10	28, 33, 29, 32	2, 29, 18, 36	32, 23, 2	25, 10, 35	10, 28, 32	28, 19, 34, 36
30	Negative external influences on the object	22, 21, 27, 39	2, 22, 13, 24	17, 1, 39, 4	1, 18	22, 1, 33, 28	27, 2, 39, 35	22, 23, 37, 35	34, 39, 19, 27	21, 22, 35, 28	13, 35, 39, 18
31	Negative side effects of the object	19, 22, 15, 39	35, 22, 1, 39	17, 15, 16, 22		17, 2, 18, 39	22, 1, 40	17, 2, 40	30, 18, 35, 4	35, 28, 3, 23	35, 28, 1, 40
32	Ease of production	28, 29, 15, 16	1, 27, 36, 13	1, 29, 13, 17	15, 17, 27	13, 1, 26, 12	16, 40	13, 29, 1, 40	35	35, 13, 8, 1	35, 12
33	Ease of operation	25, 2, 13, 15	6, 13, 1, 25	1, 17, 13, 12		1, 17, 13, 16	18, 16, 15, 39	1, 16, 35, 15	4, 18, 39, 31	18, 13, 34	28, 13 35
34	Ease of repairing	2, 27 35, 11	2, 27, 35, 11	1, 28, 10, 25	3, 18, 31	15, 13, 32	16, 25	25, 2, 35, 11	1	34, 9	1, 11, 10
35	Adaptability	1, 6, 15, 8	19, 15, 29, 16	35, 1, 29, 2	1, 35, 16	35, 30, 29, 7	15, 16	15, 35, 29		35, 10, 14	15, 17, 20
36	Device complexity	26, 30, 34, 36	2, 26, 35, 39	1, 19, 26, 24	26	14, 1, 13, 16	6, 36	34, 26, 6	1, 16	34, 10, 28	26, 16
37	Complexity of detecting and measuring	27, 26, 28, 13	6, 13, 28, 1	16, 17, 26, 24	26	2, 13, 18, 17	2, 39, 30, 16	29, 1, 4, 16	2, 18, 26, 31	3, 4, 16, 35	30, 28, 40, 19
38	Automation level	28, 26, 18, 35	28, 26, 35, 10	14, 13, 17, 28	23	17, 14, 13		35, 13, 16		28, 10	2, 35
39	Productivity	35, 26, 24, 37	28, 27, 15, 3	18, 4, 28, 38	30, 7, 14, 26	10, 26, 34, 31	10, 35, 17, 7	2, 6, 34, 10	35, 37, 10, 2		28, 15, 10, 36

Contradiction Matrix part 5

		11	12	13	14	15	16	17	18	19	20
	Problem factor / Optimization factor	Pressure or tension	Shape	Stability of an object	Resistance	Durability of a moved object	Durability of a stationary object	Temperature	Brightness	Energy consumption of a moved object	Energy consumption of a stationary object
20	Energy consumption of a stationary object			27, 4, 29, 18	35				19, 2, 35, 32	-	+
21	Power	22, 10, 35	29, 14, 2, 40	35, 32, 15, 31	26, 10, 28	19, 35, 10, 38	16	2, 14, 17, 25	16, 6, 19	16, 6, 19, 37	
22	Waste of energy			14, 2, 39, 6	26			19, 38, 7	1, 13, 32, 15		
23	Waste of material	3, 36, 37, 10	29, 35, 3, 5	2, 14, 30, 40	35, 28, 31, 40	28, 27, 3, 18	27, 16, 18, 38	21, 36, 39, 31	1, 6, 13	35, 18, 24, 5	28, 27, 12, 31
24	Loss of information					10	10		19		
25	Waste of time	37, 36,4	4, 10, 34, 17	35, 3, 22, 5	29, 3, 28, 18	20, 10, 28, 18	28, 20, 10, 16	35, 29, 21, 18	1, 19, 26, 17	35, 38, 19, 18	1
26	Material quantity	10, 36, 14, 3	35, 14	15, 2, 17, 40	14, 35, 34, 10	3, 35, 10, 40	3, 35, 31	3, 17, 39		34, 29, 16, 18	3, 35, 31
27	Reliability	10, 24, 35, 19	35, 1, 16, 11		11, 28	2, 35, 3, 25	34, 27, 6, 40	3, 35, 10	11, 32, 13	21, 11, 27, 19	36, 23
28	Measuring accuracy	6, 28, 32	6, 28, 32	32, 35, 13	28, 6, 32	28, 6, 32	10, 26, 24	6, 19, 28, 24	6, 1, 32	3, 6, 32	
29	Production accuracy	3, 35	32, 30, 40	30, 18	3, 27	3, 27, 40		19, 26	3, 32	32, 2	
30	Negative external influences on the object	22, 2, 37	22, 1, 3, 35	35, 24, 30, 18	18, 35, 37, 1	22, 15, 33, 28	17, 1, 40, 33	22, 33, 35, 2	1, 19, 32, 13	1, 24, 6, 27	10, 2, 22, 37
31	Negative side effects of the object	2, 33, 27, 18	35, 1	35, 40, 27, 39	15, 35, 22, 2	15, 22, 33, 31	21, 39, 16, 22	22, 35, 2, 24	19, 24, 39, 32	2, 35, 6	19, 22, 18
32	Ease of production	35, 19, 1, 37	1, 28, 13, 27	11, 13, 1	1, 3, 10, 32	27, 1, 4	35, 16	27, 26, 18	28, 24, 27, 1	28, 26, 27, 1	1, 4
33	Ease of operation	2, 32, 12	15, 34, 29, 28	32, 35, 30	32, 40, 3, 28	29, 3, 8, 25	1, 16, 25	26, 27, 13	13, 17, 1, 24	1, 13, 24	
34	Ease of repairing	13	1, 13, 2, 4	2, 35	11, 1, 2, 9	11, 29, 28, 27	1	4, 10	15, 1, 13	15, 1, 28, 16	
35	Adaptability	35, 16	15, 37, 1, 8	35, 30, 14	35, 3, 32, 6	13, 1, 35	2, 16	27, 2, 3, 35	6, 22, 26, 1	19, 35, 29, 13	
36	Device complexity	19, 1, 35	29, 13, 28, 15	2, 22, 17, 19	2, 13, 28	10, 4, 28, 15		2, 17, 13	24, 17, 13	27, 2, 29, 28	
37	Complexity of detecting and measuring	35, 36, 37, 32	27, 13, 1, 39	11, 22, 39, 30	27, 3, 15, 28	19, 29, 39, 25	25, 34, 6, 35	3, 27, 35, 16	2, 24, 26	35, 38	19, 35, 16
38	Automation level	13, 35	15, 32, 1, 13	18, 1	25, 13	6, 9		26, 2, 19	8, 32, 19	2, 32, 13	
39	Productivity	10, 37, 14	14, 10, 34, 40	35, 3, 22, 39	29, 28, 10, 18	35, 10, 2, 18	20, 10, 16, 38	35, 21, 28, 10	26, 17, 19, 1	35, 10, 38, 19	1

Contradiction Matrix part 6

	Optimization factor \ Problem factor	21 Power	22 Waste of energy	23 Waste of material	24 Loss of information	25 Waste of time	26 Material quantity	27 Reliability	28 Measuring accuracy	29 Production accuracy	30 Negative external influences on the object
20	Energy consumption of a stationary object			28, 27, 18, 31			3, 35, 31	10, 36, 23			10, 2, 22, 37
21	Power	+	10, 35, 38	28, 27, 18, 38	10, 19	35, 20, 10, 6	4, 34, 19	19, 24, 26, 31	32, 15, 2	32, 2	19, 22, 31, 2
22	Waste of energy	3, 38	+	35, 27, 2, 37	19, 10	10, 18, 32, 7	7, 18, 25	11, 10, 35	32		21, 22, 35, 2
23	Waste of material	28, 27, 18, 38	35, 27, 2, 31	+		15, 18, 35, 10	6, 3, 10, 24	10, 29, 39, 35	16, 34, 31, 28	35, 10, 24, 31	33, 22, 30, 40
24	Loss of information	10, 19	19, 10		+	24, 26, 28, 32	24, 28, 35	10, 28, 23			22, 10, 1
25	Waste of time	35, 20, 10, 6	10, 5, 18, 32	35, 18, 10, 39	24, 26, 28, 32	+	35, 38, 18, 16	10, 30, 4	24, 34, 28, 32	24, 26, 28, 18	35, 18, 34
26	Material quantity	35	7, 18, 25	6, 3, 10, 24	24, 28, 35	35, 38, 18, 16	+	18, 3, 28, 40	13, 2, 28	33, 30	35, 33, 29, 31
27	Reliability	21, 11, 26, 31	10, 11, 35	10, 35, 29, 39	10, 28	10, 30, 4	21, 28, 40, 3	+	32, 3, 11, 23	11, 32, 1	27, 35, 2, 40
28	Measuring accuracy	3, 6, 32	26, 32, 27	10, 16, 31, 28		24, 34, 28, 32	2, 6, 32	5, 11, 1, 23	+		28, 24, 22, 26
29	Production accuracy	32, 2	13, 32, 2	35, 31, 10, 24		32, 26, 28, 18	32, 30	11, 32, 1		+	26, 28, 10, 36
30	Negative external influences on the object	19, 22, 31, 2	21, 22, 35, 2	33, 22, 19, 40	22, 10, 2	35, 18, 34	35, 33, 29, 31	27, 24, 2, 40	28, 33, 23, 26	26, 28, 10, 18	+
31	Negative side effects of the object	2, 35, 18	21, 35, 2, 22	10, 1, 34	10, 21, 29	1, 22	3, 24, 39, 1	24, 2, 40, 39	3, 33, 26	4, 17, 34, 26	
32	Ease of production	27, 1, 12, 24	19, 35	15, 34, 33	32, 24, 18, 16	35, 28, 34, 4	35, 23, 1, 24		1, 35, 12, 18		24, 2
33	Ease of operation	35, 34, 2, 10	2, 19, 13	28, 32, 2, 24	4, 10, 27, 22	4, 28, 10, 34	12, 35	17, 27, 8, 40	25, 13, 2, 34	1, 32, 35, 23	2, 25, 28, 39
34	Ease of repairing	15, 10, 32, 2	15, 1, 32, 19	2, 35, 34, 27		32, 1, 10, 25	2, 28, 10, 25	11, 10, 1, 16	10, 2, 13	25, 10	35, 10, 2, 16
35	Adaptability	19, 1, 29	18, 15, 1	15, 10, 2, 13		35, 28	3, 35, 15	35, 13, 8, 24	35, 5, 1, 10		35, 11, 32, 31
36	Device complexity	20, 19, 30, 34	10, 35, 13, 2	35, 10, 28, 29		6, 29	13, 3, 27, 10	13, 35, 1	2, 28, 10, 34	28, 24, 32	22, 19, 29, 40
37	Complexity of detecting and measuring	18, 1, 16, 10	35, 3, 15, 19	1, 18, 10, 24	35, 33, 27, 22	18, 28, 32, 9	3, 27, 29, 18	27, 40, 28, 8	26, 24, 32, 28		22, 19, 29, 28
38	Automation level	28, 2, 27	23, 28	35, 10, 18, 5	35, 33	24, 28, 35, 30	35, 13	11, 27, 32	28, 26, 10, 34	28, 26, 18, 23	2, 33
39	Productivity	35, 20, 10	28, 10, 29, 35	28, 10, 35, 23	13, 15, 23		35, 38	1, 35, 10, 38	1, 10, 34, 28	18, 10, 32, 1	22, 35, 13, 24

Contradiction Matrix part 7

		31	32	33	34	35	36	37	38	39
	Problem factor / Optimization factor	Negative side effects of the object	Ease of production	Ease of operation	Ease of repairing	Adaptability	Device complexity	Complexity of detecting and measuring	Automation level	Productivity
20	Energy consumption of a stationary object	19, 22, 18	1, 4					19, 35, 16, 25		1, 6
21	Power	2, 35, 18	26, 10, 34	26, 35, 10	35, 2, 10, 34	19, 17, 34	20, 19, 30, 34	19, 35, 16	28, 2, 17	28, 35, 34
22	Waste of energy	21, 35, 2, 22		35, 32, 1	2, 19		7, 23	35, 3, 15, 23	2	28, 10, 29, 35
23	Waste of material	10, 1, 34, 29	15, 34, 33	32, 28, 2, 24	2, 35, 34, 27	15, 10, 2	35, 10, 28, 24	35, 18, 10, 13	35, 10, 18	28, 35, 10, 23
24	Loss of information	10, 21, 22	32	27, 22				35, 33	35	13, 23, 15
25	Waste of time	35, 22, 18, 39	35, 28, 34, 4	4, 28, 10, 34	32, 1, 10	35, 28	6, 29	18, 28, 32, 10	24, 28, 35, 30	
26	Material quantity	3, 35, 40, 39	29, 1, 35, 27	35, 29, 25, 10	2, 32, 10, 25	15, 3, 29	3, 13, 27, 10	3, 27, 29, 18	8, 35	13, 29, 3, 27
27	Reliability	35, 2, 40, 26		27, 17, 40	1, 11	13, 35, 8, 24	13, 35, 1	27, 40, 28	11, 13, 27	1, 35, 29, 38
28	Measuring accuracy	3, 33, 39, 10	6, 35, 25, 18	1, 13, 17, 34	1, 32, 13, 11	13, 35, 2	27, 35, 10, 34	26, 24, 32, 28	28, 2, 10, 34	10, 34, 28, 32
29	Production accuracy	4, 17, 34, 26		1, 32, 35, 23	25, 10		26, 2, 18		26, 28, 18, 23	10, 18, 32, 39
30	Negative external influences on the object		24, 35, 2	2, 25, 28, 39	35, 10, 2	35, 11, 22, 31	22, 19, 29, 40	22, 19, 29, 40	33, 3, 34	22, 35, 13, 24
31	Negative side effects of the object	+					19, 1, 31	2, 21, 27, 1	2	22, 35, 18, 39
32	Ease of production		+	2, 5, 13, 16	35, 1, 11, 9	2, 13, 15	27, 26, 1	6, 28, 11, 1	8, 28, 1	35, 1, 10, 28
33	Ease of operation		2, 5, 12	+	12, 26, 1, 32	15, 34, 1, 16	32, 26, 12, 17		1, 34, 12, 3	15, 1, 28
34	Ease of repairing		1, 35, 11, 10	1, 12, 26, 15	+	7, 1, 4, 16	35, 1, 13, 11		34, 35, 7, 13	1, 32, 10
35	Adaptability		1, 13, 31	15, 34, 1, 16	1, 16, 7, 4	+	15, 29, 37, 28	1	27, 34, 35	35, 28, 6, 37
36	Device complexity	19, 1	27, 26, 1, 13	27, 9, 26, 24	1, 13	29, 15, 28, 37	+	15, 10, 37, 28	15, 1, 24	12, 17, 28
37	Complexity of detecting and measuring	2, 21	5, 28, 11, 29	2, 5	12, 26	1, 15	15, 10, 37, 28	+	34, 21	35, 18
38	Automation level	2	1, 26, 13	1, 12, 34, 3	1, 35, 13	27, 4, 1, 35	15, 24, 10	34, 27, 25	+	5, 12, 35, 26
39	Productivity	35, 22, 18, 39	35, 28, 2, 24	1, 28, 7, 10	1, 32, 10, 25	1, 35, 28, 37	12, 17, 28, 24	35, 18, 27, 2	5, 12, 35, 26	+

Contradiction Matrix part 8

The 40 Inventive Principles (according to the TRIZ methodology)

1. Segmentation

a. Decompose an object into independent parts.
b. Design the object decomposable.
c. Increase the degree of segmentation

Examples:
A. Decomposable furniture, modular computers, foldable leveling staff.
B. Hosepipes can be coupled to vary their range.

2. Extraction

a. Removal or extraction of the disturbing parts of the object.
b. Extracting the only necessary part respectively the significant attributes.

Example:
Using taped birdcalls to improve airport security.

3. Local Quality

a. Transition from a homogeneous structure of an object or its environment to a heterogeneous struture.
b. The different parts of the system have to fulfil different functions.
c. Employing every component of the system under its individually best conditions.

Examples:
A. To abate dust in subsurface mining a conical curtain of water ist sprayed around the tools. The smaller the drops are, the better is the absorbtion of the dust. But these smaller water droplets tend to build mist and thus complicate the working process. The solution following the Inventive Principle No. 3 says, the cone of smallest droplets has to be surrounded by a coat of bigger droplets.
B. Join pencil and etcher in one unit.

4. Asymmetry

a. Replace symmetrical forms by asymmetrical forms.
b. If asymmetry already exits, increase its degree.

Examples:
A. The outer side of a tyre is strengthened to better withstand the impacts of a contact with the kerbstone.
B. Wet sand poured through a funnel often forms an arch over the funnels opening with the result of a reduction of unsteady flow.

5. Combinig/ Merging

a. Arrange homogeneous or cooperating objects in the same area.
b. Synchronise homogeneous or cooperating objects. Combine the objects regarding time.

Example:
A rotatopn (dry) excavator has steam nozzles to unfreeze and moisten the undergroud surface.

6. Universality

The system fulfils several variable functions, which supersedes (redundantise) other systems or objects.

Examples:
A. A bed settee can be turned from a sofa by day to a bed by night.
B. The seat of a minivan can be positioned for seating, sleeping and transporting.

7. Nesting /Integration

a. An object is placed in another object, which is in turn placed inside a third object.
b. An object fits into or through the cavity of another object.

Examples:
A. Matryoshka.
B. Retractable antenna.
C. Stackable chairs.
D. Retractable pencil with integrated lead supply.

8. Anti-weight

a. The objects´ weight can be compensated by merging it to another stable object.
b. The objects´ weight can be compensated by aerodynamic or hydrodynamic forces.

Examples:
A. Hydrofoil boats.
B. Racing cars have a rear wing to increase traction.

9. Prior Counteraction

a. Before taking any action a counteraction needed has to be performed.
b. If an object works under stress, oposing stresses have to be created in advance.

Examples:
A. Pre-stressed concrete pillars for bridges.
B. Strengthened props: to increase stability these props consist of several joined pipes that have been distorted in a specific angle in advance.

10. Prior Action

a. Perform the required action in advance, either fully or partially.
b. Pre-arrange objects in a way tha allows their activity from the right place and without any time delay.

Examples:
A. Crafts cutter, whose blade is notches, so that edgeless parts can be broken off easily.
B. Bottled glue is poorly neat and evenly applicable, but applying the glue to a strip (gluestrip) makes it a lot easier.

11. Beforehand Cushioning

Compensate fort he poor reliability of a system by taking measures in advance.

Example:
To prevent shoplifting, magnetic encoded labels are tegged to the goods. The labels have to be demagnetised at the cash desk before the customer can leave the shop.

12. Equipotentiality

Change the conditions in a way, that the object can work with a constant potential of energy (e.g. so, that it doesn´t have to be either raised or lowered).

Example:
Changing cars´ motor oil takes always place above a pit. This way expensive lifting gear is unnecessary.

13. Inversion („The other way round")

a. Instead of implementing the predetermined action, solve the problem with the exact contrary action.
b. Make a fixed object movable or fix a moving object.
c. Turn the system upside down, reverse it.

Example:
Abrasive cleaning by vibrating the parts instead of vibrating the abrasive (e.g. sandblasting)

14. Spheroidality

a. Instead of linear parts and flat surfaces use curved ones and instead of cubic structures use spherical ones.
b. Use rollers, balls, spirals.
c. Substitute linear movements with rotating movements, utilise centrifugal force.

Example:
A PC-mouse uses a ball construction for the conversion of linear, biaxial movements to a vector.

15. Dynamism

a. Design a system or its environment in a way that it can be automatically adjusted to achieve an optimal performance.
b. Fragment a system into elements, which allows an optimal arrangement among each other
c. Make an immobile object mobile, adjustable or replaceable.

Examples:

A. The moveable interface between flashbulb and flashgun.

B. A transport ship consists of a cylindrical hulk. To reduce the draught at full cargo, the ship is manufactured with two interconnected half cylinders that are opened if required.

16. Partial or excessive action

If it is difficult to achieve 100% of a required function, realize a bit more or less to simplify the problem.

Example:

A. The lacquering of a cylinder is done by immersing in paint. Unfortunately, paint is wasted during this procedure. Excessive paint can easily be removed by rotation

B. To ensure a constant slipstream of powder out of a supply container, the inside part of the outlet is a upstanding funnel that is overfilled.

17. Shift to another dimension

a. Sidestep troubles of object movements along a line by using two-dimensional movements (in a plane). Analogically, use three dimensions instead of two.
b. Place objects in more then one plane.
c. Place the object in a tilted position.
d. Utilise projections in the surroundings or on the back of the object.

Examples:
A greenhouse with concave reflectors on the north side is able to utilise day light by reflection more efficient in that part of the building.

18. Mechanical vibration

a. Put an object in vibrations.
b. Is the object already oscillating, step up the frequency.
c. Utilise resonant frequency.
d. Substitute mechanical vibrations with piezo-vibrations.
e. Use supersonic in conjunction with electro-magnetic fields.

Examples:

A. A cast is removed with oscillating knifes instead of a common handsaw.

B. Casting mass is exposed vibration to facilitate its spreading and homogeneity.

19. Periodic action

a. Transition from continuous to periodic action.

b. Change periodical actions by varying the frequency.

c. Use breaks between separated impulses to insert other actions.

Examples:

A. Screws suffering from rust are easier to detach with force impulses instead of continuously high forces on a screw drive.

B. Warning lights are perceived better when the light is pulsed.

20. Continuity of useful action

a. Execute an action without interruption to ensure that all components work with constant load.

b. Eliminate idle time and interruptions.

Example:

A borer that is able to cut in both directions, forward and reverse.

21. Skipping

Perform hazardous or dangerous actions at a very high speed.

Example:

A cutting tool for flexible plastic parts cuts at a very high speed to prevent deformations of the work piece.

22. Converting harm into benefit

a. Utilise harmful factors to gain a positive effect.

b. Remove a harmful factor by combining it with another harmful factor.

c. Intensify a harmful factor to such a degree until it becomes irrelevant.

Example:

A. Sand and gravel freeze together when exposed to low temperature. Shock freezing the combination reverses the effect.

B. Heating metal pieces on the surface is done be high frequent alternating current. This effect is used for surface treatment.

23. Feedback

a. Introduce feedback.

b. If feedback already exists, modify it.

Example:

A. Water pressure at the outlet of a well is maintained by pressure measurements and with it a controlled connection of a pump.

B. For the production of defined ice-water-alloys, ice and water must be quantified separately. It is smarter first to scale the less easy to meter ice and then to use this reading for the control of a water dispenser.

C. Tools for the elimination of noise record it, move the phase and emit it again to obtain a superposition.

24. Mediator

a. Utilise a intermediate object to transfer or execute an action.

b. Temporarily connect the object with another, easy to remove one.

Example:

To avoid energy loss during the electrolysis of mouldings, electrolytes with a higher melting point are used.

25. Self service

a. The system shall service itself and carry out maintenance operations.

b. Utilise wasted resources and energy.

Example:

A. To dispense abrasive material on rolls and to protect them from attack, they are fabricated of the identical material as the abrasive.

B. The welding-rod in electrical welding is pushed through by a special feeding device. A creative simplification is a wire feed that is controlled by a magnetic coil and the welding current.

26. Copying

a. Use simple, low-cost copies instead of an expansive, fragile and difficult to operate object.

b. Replace a system or object with its optical copy. Here, the scale (increase, decrease) can be changed.

c. Do optical copies already exist, shift to infrared or ultraviolet copies.

Example:
The height of tall objects (buildings) can be determined by measuring the shadow.

27. Disposability

Substitute a cost-intensive system with a collection of inexpensive ones, forgoing certain features.

Example:

A. Disposable diapers.

B. One-way mousetraps consist of a plastic pipe provided with lure. The mouse runs through the tight and funnel-shaped opening without a chance to return.

28. Replacement of a mechanical system.

a. Replace a mechanical system with optical, acoustical or olfactory ones.

b. Use electrical, magnetic or electro-magnetic fields.

c. Replace fields: static fields by moveable fields, constant fields by periodic fields, unstructured fields by structured fields.

d. Utilise fields in conjunction with ferromagnetic parts.

Example:
To improve the adhesive force of a metallic treatment on a thermoplastic, the coating is done in an electro-magnetic field, whereby the metal is pressed on the work piece with a higher force.

29. Pneumatic and hydraulic construction

Substitute solid and heavy parts of a system with gaseous or liquid ones. Use water or air to inflate.

Examples:

A. To increase the airflow in an industry chimney, a spiral, porous pipe in the inside is used. The pipe carries air that streams out of the pores and thereby creates an air cushion. This improves the airflow.
B. Postages of fragile objects are packed with air bubble foil or foamed material.

30. Flexible shells and thin films

a. Substitute traditional constructions with those made of flexible shells or thin films.
b. Isolate objects from its environment by using a thin film or a membrane.

Examples:
To prevent plants from water loss, the leaves are sprayed with polyethylene. The advantage of polyethylene is that it is permeable to air but not to water.

31. Porous material

a. Design an object with porous material or elements.
b. Is a porous object already existing, fill the pores with a useful substance.

Examples:
To avoid laborious plumping of cooling agents into machines, porous parts with the cooling agents already in the pores are used. The cooling agent then vaporise in the machine during the operation.

32. Changing colour

a. Change the colour of an object or its environment.
b. Change the degree of transparency of an object or its environment.
c. Use suitable colour additives for observing difficult to see objects or processes.
d. Additional to colour additives, use phosphor, luminescent tracers or similar.

Examples:
A. Transparent plasters allow observing a wound without removing it.
B. Workers in a steelwork are protected from the heat by a water curtain. Unfortunately, water is only absorbing infrared rays (heat) but not the extreme brightness. The intensity can be reduced by adding colour dyestuff without having impact on the transparency.

33. Homogeneity

Make objects interacting with a given object of the same or similar material.

Example:
To dispense abrasive material on rolls and to protect them from attack, they are fabricated of the identical material as the abrasive.

34. Rejecting and regenerating parts

a. After completion of its function, reject or modify certain parts of an object.
b. Restore exhausted system parts immediately, even during operation

Examples:
A. Bullet casings are ejected after the shot.
B. Booster-rockets are removed from the main rocket after fulfilling their task.

35. Transformation of the physical or chemical properties.

Change on objects aggregate state: solid, liquid, gaseous but also quasi liquid or other properties like concentration, density, elasticity, temperature.

Example:
The transport screw of a transport system for brittle and fragile materials is made of elastic material. Thus, the flank lead of the screw can be increase, which changes the transport speed at a constant rpm.

36. Phase transitions

Utilise effects during phase transitions of a substance: change of volume, emission and absorption of heat.

Example:
To stretch ribbed tubes uniformly, they are first filled with water and then frozen.

37. Thermal expansion

a. Utilise thermal expansion or contraction of materials.
b. Use materials with different expansion coefficients.

Example:
To automatically open and close the roof of a greenhouse, the windows are made of bi-metallic trusses. Temperature changes mean bending of the tosses, which leads to opening or closing of the windows.

38. Strong oxidants

a. Replace common air with oxygen-enriched air.
b. Replace oxygen-enriched air with pure oxygen.
c. Expose air or oxygen to ionising radiation.
d. Use ozone.

Example:
A torch produces more light when provided with oxygen

39. Inert environment

a. Replace normal environment with inert ones.
b. Operate a process in a vacuum.

Example:
To avoid auto ignition of wool in a storehouse, the wool is treated with inert gas.

40. Composite materials

Replace homogenous materials with composite ones.

Example:
Highly stressed wings of military planes are fabricated of plastics with carbon fibre to obtain a composite material of high stability and low weight.

Appendix C
Product-idea data-sheet

The following pages show an example of product-idea data-sheets containing product-idea-related and planning-relevant information. The document shall inspire the development of enterprise specific product-idea data-sheets. The presented data-sheet contains data of the planning process of Mr. Stone, who executes the product innovation planning with guidance of the InnovationRoadMap-Methodology in the fictitious case study of the company "Center Positioning Systems Ltd").

Center Positioning Systems Ltd.	Data-Sheet Product-Idea	lfd. Nr.: 0 0 1

Organisational

Editor:	Mr. Stone	Department:	New Business Development	Date:	28.08.2002
Lemma:	Intelligent Endoscope	Business Unit:	Positioning Systems		
Contact Technique:	Mr. Plumb	Contact Market:	Mrs. Begovic		

Status:	Idea	Draft Design	R&D-Project	Serial Dev.	rejected*	deferred*
Datum:	16.01.2002	23.04.2002				
*Reason:						

General

Idea Description:

»Intelligent Endoscope – IntEnd«: A micro-positioning-system is inserted into the head of an endoscope, Where it moves an optical system. The optical system consists of a camera and an illumination unit. The optical system needs to be positioned step less to avoid acceleration or bucking, which would imply changes of the endoscope position. Furthermore, the used parts shall meet the hygienic requirements of medical standards. The positioning system shall be designed as an stand-alone system and compatible to standard endoscopes..

Idea Description Technique

What is the present solution ?

- the competitor solution ?

- the solution of similar problems in other branches ?

What is the technical improvement ?

How can the problem be solved ?

The activation must be applied via radio to design the endoscope neck as small as possible for minimal-invasive treatment. The power supply is done by a flexible wire inside the endoscope neck. Once in position, no exterior movement shall occur, but by the optical system. Thus, turnings of 320° In horizontal level and 120° in vertikal level must be realised.

Center Positioning Systems Ltd.	Data-Sheet Product-Idea	lfd. Nr.: 0 0 2

Idea Description Market

Market / Branch ?

The IntEnd can be launched in medical engineering, especially diagnostic

Which customer requirements shall be fulfilled ?

The IntEnd shall support minimal invasive treatments to achieve slightest injuries on patients caused by treatment tools. Furthermore, the patients feels less pain by movements of the endoscope because of The one-time positioning. The examination then is done by the flexible camera head in the inside.

How are customer requirements fulfilled at present ?

By what is the customer benefit increased ?

Minimum of pain by quick and bucking-free movements.

Which market segment shall be supplied ?

Endoscope Diagnostic

potential customers:	Competitor:
Mediklinik Schwarzwald, Hals & Beinbruch GmbH, Uniklinik Gipsfuß	Fokus Positioniersysteme GmbH, Centerpoint Positioning Systems

Market Evaluation:	x	increasing		stagnating		decreasing

Market Characteristics:

Medical Engineering is seen as an increasing future market. The demographically obvious increase of elder people as well as the awareness of health are the informative basis for the need of sensitive medical treatments for the future. This implies the need of adequate technical tools. Especially minimal invasive are suited to extremely shorten retention time of patients in hospitals by causing slight Injuries. Thus, the medical care costs can also be reduced.

Remarks

Idea Evaluation

Outcome

Indirect enterprise benefit: 4, direct enterprise benefit: 4.5, scale: 1 (low) to 5 (high)
Technology potential: 5, Promising potential: 5
Remark: A high indirect enterprise potential was identified: there is a very high synergy potential with existing R&D projects and serial products. A high benefit for corporal image also delivered a high value. When determining the direct enterprise benefit, the profitability potential was evaluated as extremely high. Although the sub criteria “expenses for realisation” was evaluated as bad, the idea has a very high technology potential. The promising potential was evaluated as very high due to the expected customer benefit.

Center Positioning Systems Ltd.	Data-Sheet Product-Idea	lfd. Nr.: 0 0 3

Draft Design Technique

Clarification of feasibility? x yes; no

If not, what tests are planned?

Prototype existing? x yes; now

If not, what is the expense?

Which existing parts are used?

No existing parts can be used

Open questions (Which problems are not solved yet)?

The maximum turning degree in the y-z-level is not realised completely (100° instead of 120°)
Material supplier (Titan) not clarified yet

Existing Documentations:

Construction documentations, results of material analysis, endoscope supplier catalogues, test data sheets, sales forecast, results of patent research

Development Expenses:				Participating Business Units / Task:
Invest [€]:	~ 160.000			New Business Development (realisation planning)
Stuff [M.St.]:	1570	[€]	144.000	
Sum [€]:	304.000			
Duration:	12		Month	

Patent Situation:

Patent research completed, no concurring concept found

Center Positioning Systems Ltd.	Data-Sheet Product-Idea	lfd. Nr.: 0 0 4

Draft Study Market

Market Data:

Sales Vol: ca. 500 [units/p.a.]; Price: n/a [€/unit]

Turn-over: n/a [€/p.a.]

Aiming market share: 1. year [%]: 5 ; 2. year [%]: 8 ; 3. year [%]: 20

monetary benefit (Rationalisation / Growth in Sales):

Expenses for Market Analysis: Internal Market Analysis ?: yes no

Investments [€]: Stuff [M.hr.]: [€]

Sum [€]: Duration: [Weeks/Month]

Remark

Evaluation Draft Study

Result

Appendix D
EDEN™ as software platform to support the Aachen Innovation Model

This is a brief summary of a software tool that can be used to improve the usability of the Aachen Innovation Model (AIM). EDEN™ is a collaborative, web-based platform in which the AIM methodology has been implemented to provide a dynamic working environment that utilises the structures, methods and tools described within the AIM methodology.

Introduction to EDEN™

EDEN™ is a company-wide Innovation Management software platform, used to manage innovation projects utilising roadmaps. It facilitates the capturing and sharing of knowledge within and between project teams and across different projects, ensuring concurrent collaboration throughout the planning and execution phases of an innovation project. As the software is web-based, it supports geographically dispersed teams and provides the flexibility necessary to execute and manage today's multi-national, multi-organisation projects.

Roadmaps, highlighting the various steps of an innovation project, serve as the guiding mechanism throughout execution. Various tools including definitions and descriptions, checklists, inputs and outputs, design and control objectives and additional information (such as guidelines, templates, methods, etc.) facilitate the execution of these project steps, while document management facilities ensure co-ordinated and efficient creation of project outputs. The software also provides a flexible working area used to store project execution and working documentation, allowing the user sufficient freedom to create his/her own working environment. In doing so, the software captures the experience gained during the project for reutilisation in future projects. The application of meta-data to all information and a powerful search engine (dtSearch®) allows for the rapid identification and location of relevant information within project context.

The contribution and benefits in using EDEN™ within the innovation project environment is categorised into the following main areas:

- **Document Management** – A project roadmap structure is provided in which users can capture, categorise, store and share project documentation and related information. Functionalities, including file management (folders and files), file security and version control, provide the controlled environment necessary to perform project tasks and generate outputs concurrently. It also enhances future re-use of project information and experience.

- **Knowledge Management** – The roadmap structure also ensures that all project information is captured, stored and maintained in context in the relevant location for future utilisation, while metadata and searching functionalities ensures that this information is easily retrievable.
- **Project Management** – The roadmap structure serves as guidance for the execution of project tasks with additional information, including guidelines, tools, templates and methods, supporting the project team in completing these tasks in a structured manner. Functionality in the form of checklists, inputs and outputs, and design and control objectives assist in managing project progress and the fulfilment of objectives. The web-based, collaborative and controlled environment allows large and geographically dispersed project teams to rapidly generate project outputs concurrently.
- **Programme Management** – Navigation between and viewing of multiple project roadmaps, and the ability to link project steps and documentation, assists the programme manager in his/her task of coordinating several projects concurrently. Project progress can be monitored using various functionalities including checklists, an activity summary and the project outputs. A report generation tool provides the EDEN™ user with an outline of the roadmaps' contents, rapidly providing a summary of the programme status quo.

The EDEN™ Graphical User Interface

A screenshot of the EDEN™ graphical user interface is depicted in Fig. D.1 and a brief summary of the various items and functionalities are provided below. The intention is to briefly introduce the reader to the EDEN™ working environment.

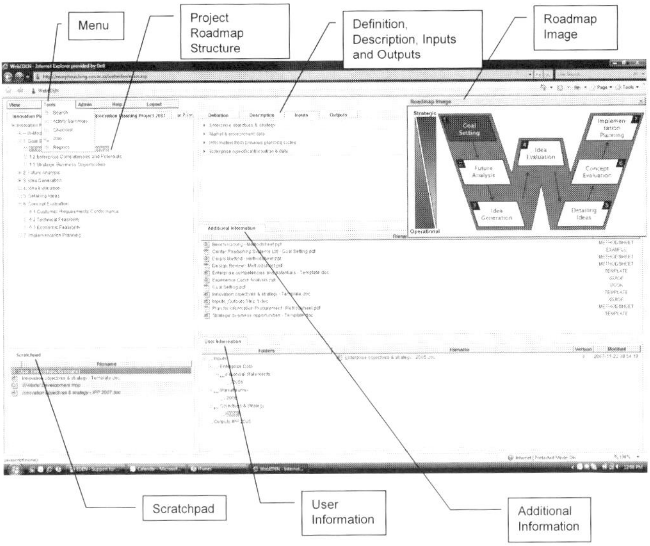

Fig. D.1 EDEN™ Graphical User Interface

- **Project Roadmap Structure** – the guidance mechanism defining the tasks/steps that need to be completed (at minimum considered) for the project to be successfully completed.
- **Definition, Description, Inputs and Outputs** – define and describe to the user what is required from them in a particular task/step, what documentation they can use to assist them and what the final output must be.
- **Additional Information** – provides the user with various material, including guidelines, templates, methods, etc., that can facilitate and expedite the process of completing a particular task/step.
- **User Information** – the dynamic user-defined working area in which both working documentation and completed outputs are stored and shared with other project team members.
- **Scratchpad** – serves as the portal between the web-based environment of EDEN™, in which content is held on a central server, and the local storage of the user, such as a desktop computer, internet cafe computer, PDA, etc. It is also used to show documentation that is being worked on by users, and which may not be changed by other users.
- **Roadmap Image** – a graphical representation of the tasks/steps that is used to depict the relationship- and facilitate the navigation between project tasks/steps

- **Menu** – contains the various functionalities that are used to improve the users' ability to execute and manage the project. The "View" tab provides various EDEN™ viewing options including the Roadmap Image, the Workspace (shown in the screenshot) and the Roadmap Navigator (used to manage the roadmaps). The "Admin" tab provides the functionalities to administrate the EDEN™ environment pertaining to user accounts, security setting, etc. (these only open to an administration account) and the user with the ability to alter his/her own account. The "Help" and "Logout" tabs are exactly that – EDEN™ help and system logoff. In the figure above, the "Tools" menu is shown, containing the following functionalities:
    - Search – to find any project roadmap content in Additional Information, User Information and/or the Scratchpad using either full text or keyword search.
    - Activity Summary – summarise active and recently updated project documentation and outputs, and allowing the user to monitor pre-defined project documentation.
    - Checklist – manage and depict project progress.
    - Wiki – provides the opportunity to link to an organisational wiki site.
    - Reports – provides the user with a summarised report of the roadmap's contents.

How EDEN™ is supporting the Aachen Innovation Model

How EDEN™ supports the Aachen Innovation Model (AIM) is described in a typical innovation project scenario. The Schott-Glas case study, described in this book is used to create the context for the application of EDEN™ within this scenario. This scenario is used to demonstrate how EDEN™ may improve the efficiency of the model by guiding new users in their understanding and application of AIM. It describes briefly how a new AIM user is able to rapidly understand the activities associated with an innovation planning exercise and contextualise their role therein.

The Schott-Glas Background

Schott-Glas (as described in the case study) was the world leader in the manufacture of special glass products in the mid-nineties, including such products as laboratory glass, television glass, pharmaceutical glass, pipe glass and optical glass in their range. However, increasing cost pressures and a diversifying product range were creating gaps for competitors. They therefore decided to re-evaluate their market positioning.

The team leader tasked with managing this project has asked one of the team members to determine how their current innovation objectives and strategy addresses the current market conditions, i.e., to re-evaluate their innovation strategy based on a recently conducted market survey.

Two basic assumptions are made:

- The team member has very little exposure to AIM, but does have a basic understanding of the EDEN™ environment and its functionality.
- AIM and EDEN™ have been integrated and utilised in earlier projects (EDEN™ becomes more beneficial the more it is used – a principle of all Knowledge Management systems).

The process within EDEN™

Elaborating on this "new user" scenario, the events that could typically take place are now described, along with how EDEN™ is used to overcome potential obstacles during the process.

The team member, John, receives an email from the team leader, Sue, with instructions to re-evaluate their innovation strategy based on a recently conducted market survey within the most recent Innovation Planning Project 2007 (IPP 2007).

- Included is a *EDEN™ Link* to the *Innovation objectives and strategy* document of IPP 2007
- He "clicks" on the link and is directed to the relevant document in EDEN™ and within the relevant *EDEN™ Project Step*

John then browses through roadmaps of current and previous innovation planning projects to familiarise himself with the environment.

- He looks at the *EDEN™ Checklists* to establish the progress of the projects and to better understand the context.
- He browses through the *Innovation objectives and strategy* documents of these projects to understand the current strategy and objectives – opens and compares both 2006 & 2007 documents.

John then uses EDEN™ Search to find the relevant new and old market survey information and data.

- He searches for "2006 market survey" and "2007 market survey", finding the relevant information within the projects' context, i.e., within the relevant EDEN™ Project Step.
- He studies the EDEN™ Metadata of the documents and navigates to their locations using the EDEN™ Link.
- He also compares the 2006 & 2007 surveys simultaneously looking for changes in the market situation.

He also searches for other inputs to the objectives and strategy document and evaluates what has changed and how this impacts the objectives.

- He reads the EDEN™ Definition and Description of the relevant input and output steps for the document.
- He browses through the EDEN™ Additional Information (templates and methodsheets) of the inputs and outputs.

John now creates a new version of the Innovation objectives and strategy document using the template, the old document for structural guidance and the 2007 market survey for content.

- He places the new document in the relevant EDEN™ Project Step and updates the metadata to keep the document's context and facilitate later searches.
- He sends an EDEN™ Link of the document in EDEN™ to Sue for review.

Sue navigates to the document using EDEN™ Link, opens document, reviews and makes relevant changes, with EDEN™ tracking all version changes, ensuring proper control thereof.

- She signs off the document.
- She updates EDEN™ Checklist, maintaining project progress.

Conclusion

Using EDEN™ to support the AIM methodology allowed John to quickly familiarise himself with the project and the methodology, quickly find the relevant information in context, and perform the assigned task more effectively having all the necessary information and guidelines available to him within a single environment. Further, he was able to share that document with Sue, wherever she may have been, while control of the document's version is continuously maintained as they go through the review and refine process.

Accessing EDEN™

For more information about EDEN™ and getting online access to a demonstration version of the software visit: www.eden.indutech.co.za. This web site offers a downloadable, static web-browser version, suitable to accompany the text of the book, which includes handy template and example documents. In addition, access is provided to the dynamic EDEN™ version, including document management features, enabled for two of the roadmap steps of the AIM methodology.

Alternatively, contact Niek du Preez at niek@indutech.co.za.

Index

Editor and Authors

Dipl.-Ing. Elke Baessler

Born in 1972, Mrs. Elke Baessler studied psychology and mechanical engineering at RWTH Aachen University, as well as biochemistry and molecular biology at Boston University and Harvard Extension School. Mrs. Baessler holds a Bachelor Degree in Chemistry as well as a Masters Degree in Mechanical Engineering. Between 2002 and 2006 Mrs. Baessler was a research assistant at the Chair of Production Engineering of the Laboratory for Machine Tools and Production Engineering at RWTH Aachen University. She focused her research on R&D and Innovation Management. Since 2007 she is in charge of the corporate innovation management at HILTI AG in Schaan, Liechtenstein.

Dr.-Ing. Thomas Bauernhansl

Born in 1969, obtained his Diploma in Mechanical Engineering at the RWTH Aachen in 1998. He received a Ph. D. at the WZL of the RWTH in 2002 where he worked as a research assistant and chief engineer until 2003. In 2003 he joined Freudenberg and worked until 2007 as general manager for the Freudenberg Anlagen- und Werkzeugtechnik GmbH. Since 2007 he is Director of the Production Center of the Freudenberg Dichtungs- und Schwingungstechnik GmbH & Co. KG.

Dr.-Ing. Dipl.-Wirt. Ing. Uwe H. Böhlke

Born in 1964, studied mechanical engineering and economics at RWTH Aachen University. After finishing his studies he workd for six years in research and science management for Fraunhofer Institute for Production Technology IPT, Aachen. During this period he received a Ph. D. in engineering and finally was assigned chief engineer and as head of group "Technology- and Innovation Management" member of the Institute Management. Following after his scientific career he worked for eleven years with Schott AG, Mainz. Starting as Vice President "Corporate Engineering and Special Machines". In his professional career he became responsible for Global Intellectual Properties" and finally was appointed Executive Vice President of the "Business Unit Home Tech". Since 2007 Dr. Böhlke is member of the Management Committee Lonza AG, Basel (CH) and acts as Chief Operative Officer "Exclusive Synthesis".

Dr.-Ing. Frank Brandenburg

Born in 1968, finished his studies in mechanical engineering at RWTH Aachen University in 1995. He then worked as a research assistant and project manager at the Fraunhofer Institute for Production Technology IPT, Aachen, in the field of technology and innovation management. In parallel he finished his dissertation in 2001. Since then he is working for the Hilti Corporation, Liechtenstein, in various responsibilities. Actually he is the general manager for Hilti´s plastics production & competence center in Strass, Germany.

Dr.-Ing. Thomas Breuer

Born in 1973, has studied mechanical engineering at RWTH Aachen University, Germany. From 2000 to 2005 he was scientific staff member and project leader in the department "Planning and Organization" and later in the department "Technology Management" at the Fraunhofer Institute for Production Technology IPT in Aachen, Germany, with main focus on technology and innovation management as well as Procurement. Since 2006, Thomas Breuer has been working at Hilti Corporation in Schaan, Liechtenstein. His current position is Head of Supply for Mechanical Components.

Prof. em. Dr.-Ing. Dr. h.c. mult. Dipl.-Wirt. Ing. Walter Eversheim

Born in 1937, has studied mechanical engineering and economics at RWTH Aachen University. After finishing his Ph. D. as Dr.-Ing. he was appointed chief engineer and reporting to Prof. Dr. Herwart Opitz in the Laboratory for Machine Tools and Production Engineering (WZL) at RWTH Aachen University. From 1969 until 1973 he gained professional expertise as head of department in two global acting European enterprises. In the year 1973 he was appointed full professor and director of the Laboratory for Machine Tools and Production Engineering, RWTH Aachen University. In 1980 he became head of department and director of Fraunhofer Institute for Production Technology IPT, Aachen. In 1989 Prof. Eversheim was additionally appointed guest professor and director of the Institute for Technology Management (ITEM) at University of St. Gallen, Switzerland, and in 1992 he took over the role of topmanagements director of the Institute for Operations Management (FIR), Aachen. The University of Trondheim (NTH), Norway, dedicated the award of Dr. honoris causa to Prof. Eversheim in the year 1992.

In the same year he was appointed Honorary Professor from Tianjin University, China. In the year 2000 the University of St. Gallen (HSG), Switzerland, honoured Prof. Eversheim as Dr. oec. h.c. for his merits in economics and management. Same year he became Honorary Professor of Huazhong University, Wuhan (China). In the year 1997 Prof. Eversheim has been elected "Spokesman of the Board of Directors for the International Charlemagne Prize" at Aachen.

Dr.-Ing. Dipl.-Wirt. Ing. Dipl. Mediatorin Anne Gerhards

Born in 1968, studied mechanical engineering, RWTH Aachen University, and business administration, Distance University of Hagen, and traines as Dipl. Mediator, Institute for Applied Mediation in Liechtenstein. From 1997 to 2001 she was Scientific Assistant and Project Engineer at the Fraunhofer Institute for Production Technology IPT, Aachen, Department Organization and Planning. After her dissertation, between 2002 and 2005 she worked as Manager Marketing Project at the Hilti AG in Liechtenstein, Business Area Electric Tools. Today she has her own business for consultancy and conflict management.

Dr.-Ing. Markus Grawatsch

Born in 1975, studied mechanical engineering at RWTH Aachen University, Germany and at University of New South Wales (UNSW), Australia. Between 2001 and 2006 he worked as research assistant and as consultant at Fraunhofer IPT with focus on Innovation and Technology Management. After his Ph. D. with the topic "TRIZ-based Technology Intelligence" he continued his career at Procter & Gamble/ Braun GmbH as a Project Manager for Household Devices.

Dipl.-Ing. Michael Hilgers

Born in 1970, has studied mechanical engineering and human medicine at RWTH Aachen University, Germany. From 2002 to 2007 he was scientific staff member and project leader in the department "Planning and Organization" and later in the department "Technology Management" at the Fraunhofer Institute for Production Technology IPT in Aachen, Germany, with main focus on technology and innovation management. Since 2007, Mr. Hilgers has been working at SEW-EURODRIVE GmbH & Co. KG, Bruchsal, Germany. His current position is head of the "WIEPROconsulting Business Process Management" department. In this function Mr. Hilgers is responsible for optimizing business processes in production and assembly plants of SEW-EURODRIVE worldwide.

Prof. Dr. Winfried J. Huppmann

Born in 1945, studied physics at the University of Vienna, Austria, and completed his studies there with the Ph. D. in 1969. He gathered professional experience in industry in Canada and Germany, as well as in basic research (Max-Planck-Institute for Metals Research). He obtained the venia docendi for powder metallurgy at the Technical University of Vienna (1979) and gave lectures there as well as at the University of Karlsruhe, Germany. In his main function he was a member of the Corporate Management Group of Hilti AG in the Principality of Liechtenstein, where he directed the research activities of that company for almost fifteen years. He was in charge of the innovation process management until his retirement in 2005. Today Prof. Huppmann acts as a consultant for innovation management and materials science. Among the offices of general interest he acted as president of the Deutsche Gesellschaft für Materialkunde DGM in 2003-2004 and holds the position as chairman of the Kuratorium of the Max-Planck-Institute for Metals Research in Stuttgart.

Dr.-Ing. Markus Knoche

Born in 1973, studied mechanical engineering at RWTH Aachen University and made his PhD at RWTH Aachen. His PhD thesis on Technology Management was supervised by Professor W. Eversheim and Professor G. Schuh. After completing his studies he started at Fraunhofer-Institute for Production Technology IPT as a consultant specialized for lean production and supply management. In 2005 he took an assignment at Hilti Corporation as a project manager responsible for a global marketing and supply project reporting to an executive vice president before being promoted to his current function.

He is currently employed as Head of Supply of Business Unit Installation Systems responsible for production, sourcing, and logistics at Hilti Corporation in Schaan, Liechtenstein. In this function he is also member of the Management Team of the BU Installation.

Dipl.-Ing. Rudolf-Henning Lohse

Born in 1963, is a mechanical engineer from the University Dortmund. He joined Dräger in 1992 and was since then responsible for several change management and reengineering projects. In his role as Head of Quality & Processes from 1999 to 2002 in Dräger Medical he headed the implementation of Process Management in a global acting company. In addition he acted as strategic Process Owner for the Business Process Innovation and implemented a global Innovation process based on a Product Platform. From 2003 to 2004 he headed the two integrations projects within Dräger Medical (2003: Joint Venture with Siemens AG and 2004: Acquisition of Air-Shields). After successful integration of Air-Shields he acted until end of 2007 as General Manager for the new acquired business. During this time he renewed 70 % of the whole organization and consolidated the business at a new site, while growing the business by 30 % and turning the business from loss to profit. Today Henning Lohse reports to the newly introduced COO for Drägerwerk AG and acts as project manager to drive Globalization within Dräger.

Dipl.-Ing. Christian Rosier

Born in 1976, studied mechanical engineering at RWTH Aachen and University of California in Berkeley. He began his professional career at the Fraunhofer Institute for Production Technology IPT in Aachen. After being awarded a doctorate for a thesis on centralized technology development in diversified enterprises, he changed to an international market leader in series machine construction. There, he was responsible for innovation, change request management, product documentation, and product variants configuration. In 2007 Christian joined Management Engineers International Consultants where he works mainly for companies in the automotive industry as well as machine and plant engineering. His primary focus is on strategic and operational aspects of R+D, production, and purchasing.

Dr.-Ing. Dipl.-Kfm. Sebastian Schöning

Born in 1975, studied mechanical engineering and economics at RWTH Aachen University, Germany, and Ecole Centrale Paris (ECP), France. He continued his scientific career at Fraunhofer Institute for Production Technology IPT, Aachen and was appointed head of group "Technology Forecast". As chief engineer he became responsible for the department of "Innovation Management" at the "Laboratory for Machine Tools and Production Engineering (WZL)" at RWTH Aachen University.

Since December 2007 he is assigned "Vice President Platform Management" at MAG Industrial Automation Systems, and responsible for coordination of Global Platform Design.

Dr.-Ing. Jens Schröder

Born in 1969, studied mechanical engineering at UGH Paderborn and RWTH Aachen University. He worked as a scientific assistant at the Laboratory for Machine Tools and Production Engineering in Aachen and subsequently as head of the Technology Management Division at Fraunhofer Institute for Production Technology IPT, Aachen.

Since 2006, he is Director of the Nuclear Casks Division at GNS Gesellschaft für Nuklear-Service, Essen.

Dr.-Ing. Daniel Spielberg

Born in 1971, studied mechanical engineering at the Universities of Bochum and RWTH Aachen University, Germany. From 1996 to 2001 he worked as project engineer particularly with regard to innovation processes for the Fraunhofer Institute for Production Technology IPT, Aachen, department of Planning and Organization. He finished his career in Aachen with a Ph. D. Since 2001, he is working for SUSPA Holding GmbH, an automotive supplier with headquarter in Altdorf near Nuremberg. Starting as an Assistant to the Board, from 2005 on he is heading SUSPA's business segment Powered Systems as General Manager.

Dr.-Ing. Carsten Voigtländer

Born in 1963, Dr. Voigtländer is CEO of Oerlikon Textile. Since January 2007 he has been running the entire textile division of the Saurer Group – now Oerlikon Textile. As CEO of Neumag Saurer, Neumünster, Germany, Dr. Voigtländer has been primarily responsible for the transformation of a classic machinery manufacturer to an engineering company with turn-key competence and consistent establishment of the nonwoven sector – an area of strategic growth in the textile chain. Dr. Voigtländer studied mechanical engineering and process engineering at Technical University Braunschweig, specialising in production engineering, and gained his doctorate in thermodynamics. He began his working career in 1994 in the development sector of Neumag, then leading the project management of the company and has been a member of the board of directors of Neumag since 1998 – since 2004 the sole managing director.

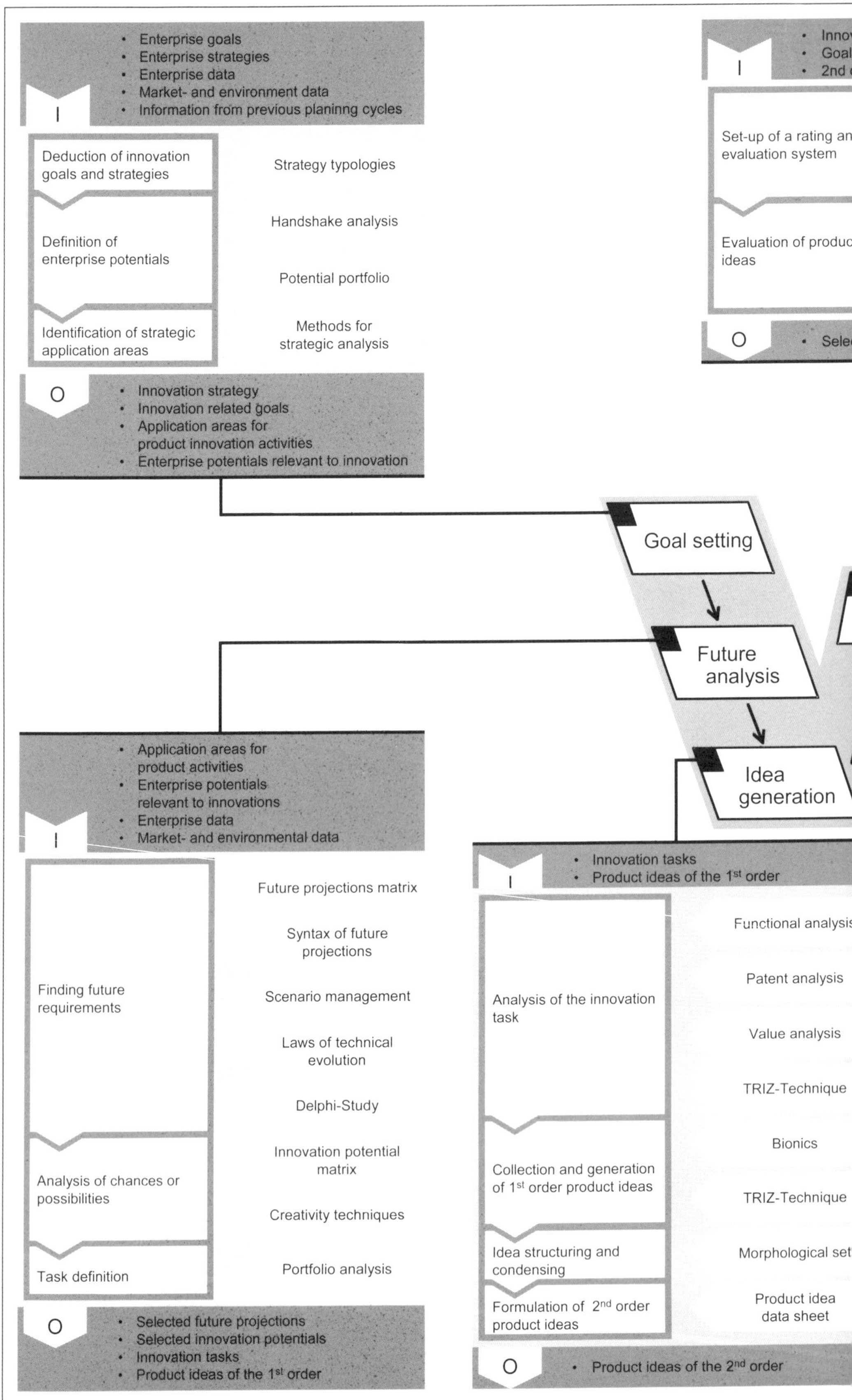
I
Enterprise goals
Enterprise strategies
Enterprise data
Market- and environment data
Information from previous planinng cycles
Deduction of innovation goals and strategies
Definition of enterprise potentials
Identification of strategic application areas
Strategy typologies
Handshake analysis
Potential portfolio
Methods for strategic analysis
O
Innovation strategy
Innovation related goals
Application areas for product innovation activities
Enterprise potentials relevant to innovation
Goal setting
Future analysis
Idea generation
I
Application areas for product activities
Enterprise potentials relevant to innovations
Enterprise data
Market- and environmental data
Finding future requirements
Analysis of chances or possibilities
Task definition
Future projections matrix
Syntax of future projections
Scenario management
Laws of technical evolution
Delphi-Study
Innovation potential matrix
Creativity techniques
Portfolio analysis
O
Selected future projections
Selected innovation potentials
Innovation tasks
Product ideas of the 1st order
I
Innovation tasks
Product ideas of the 1st order
Analysis of the innovation task
Collection and generation of 1st order product ideas
Idea structuring and condensing
Formulation of 2nd order product ideas
Functional analysis
Patent analysis
Value analysis
TRIZ-Technique
Bionics
TRIZ-Technique
Morphological set
Product idea data sheet
O
Product ideas of the 2nd order
I
Set-up of a rating and evaluation system
Evaluation of product ideas
O